Natural Materials of the Holy Land and the Visual Translation of Place, 500–1500

Natural Materials of the Holy Land and the Visual Translation of Place, 500–1500, focuses on the unique ways that natural materials carry the spirit of place. Since early Christianity, wood, earth, water and stone were taken from *loca sancta* to signify them elsewhere. Academic discourse has indiscriminately grouped material tokens from holy places and their containers with architectural and topographical emulations, two-dimensional images, and bodily relics. However, unlike textual or visual representations, natural materials do not describe or interpret the Holy Land; they are part of it. Tangible and timeless, they realize the meaning of their place of origin in new locations.

What makes earth, stones or bottled water transported from holy sites sacred? How do they become *pars pro toto*, signifying the whole from which they were taken? This book will examine natural media used for translating loca sancta, the processes of their sanctification and how, although inherently abstract, they become charged with meaning. It will address their metamorphosis, natural or induced; how they change the environment to which they are transported; their capacity to translate a static and distant site elsewhere; the effect of their relocation on users/viewers; and how their containers and staging are used to communicate their substance.

Renana Bartal (PhD, Hebrew University, 2010) is Senior Lecturer at the Department of Art History, Tel Aviv University. She co-edited *Between Jerusalem and Europe: Essays in Honour of Bianca Kühnel* (Leiden: Brill, 2015) and is the author of *Gender, Piety and Production in Fourteenth Century Apocalypse Manuscripts* (New York: Ashgate/Routledge, 2016).

Neta Bodner (PhD, Hebrew University, 2016) is a Post-PhD fellow in the ERC-funded project 'Beyond the Elite: Jewish Daily Life in Medieval Europe' at the Hebrew University of Jerusalem. Bodner teaches in the Art History Department of the Hebrew University and at the Open University in Raanana. She has published several articles about Pisan monuments and architectural translations of Jerusalem to Vienna and northern Italy.

Bianca Kühnel is Jack Cotton Professor Emerita of Fine Arts and Architecture at the Department of the History of Art, Hebrew University of Jerusalem. Her publications include *From the Earthly to the Heavenly Jerusalem: Representations of the Holy City in Christian Art of the First Millennium* (Rome-Freiburg/Br.-Vienna: Herder, 1987); *The Real and Ideal Jerusalem in Jewish, Christian, and Islamic Art*, Special Issue of Jewish Art 23/24, 1997/1998; and *Visual Constructs of Jerusalem*, ed. with G. Noga-Banai and H. Vorholt (Turnhout: Brepols Publishers, 2014).

Natural Materials of the Holy Land and the Visual Translation of Place, 500–1500

Edited by
Renana Bartal, Neta Bodner,
and Bianca Kühnel

Routledge
Taylor & Francis Group
LONDON AND NEW YORK

First published 2017
by Routledge
2 Park Square, Milton Park, Abingdon, Oxon OX14 4RN

and by Routledge
711 Third Avenue, New York, NY 10017

Routledge is an imprint of the Taylor & Francis Group, an informa business

British Library Cataloguing-in-Publication Data
A catalogue record for this book is available from the British Library

Library of Congress Cataloging-in-Publication Data
Names: Bartal, Renana, 1977– editor. | Bodner, Neta, editor. | Kühnel, Bianca, editor.
Title: Natural materials of the Holy Land and the visual translation of place, 500–1500 / Renana Bartal, Neta Bodner, Bianca Kühnel.
Description: New York : Routledge, 2017. | Includes bibliographical references and index.
Identifiers: LCCN 2016044330 | ISBN 9781472451774 (alk. paper)
Subjects: LCSH: Christian pilgrims and pilgrimages—Palestine—History—To 1500. | Souvenirs (Keepsakes)—Palestine. | Natural history—Religious aspects—Christianity. | Palestine—Description and travel.
Classification: LCC BX2320.5.P19 N38 2017 | DDC 263/.04256940902—dc23
LC record available at https://lccn.loc.gov/2016044330

ISBN: 978-1-4724-5177-4 (hbk)
ISBN: 978-1-315-21031-5 (ebk)

Typeset in Sabon
by Apex CoVantage, LLC

Contents

Color plates

Figures

Chapter 6

Chapter 8

Chapter 9

Chapter 10

Chapter 11

Chapter 13

Chapter 14

Contributors

Michele Bacci (PhD, Scuola Normale Superiore, Pisa, 1999) is Professor of Medieval Art History at Fribourg University, Switzerland. He studies the dynamics of art-historical contacts between the different Mediterranean traditions in the broader context of their cultural and religious relationship and the history of cult objects and holy sites from a phenomenological-comparative viewpoint. He is editor of the journal *Iconographica* and a member of several cultural institutions. Among his most important publications are the monographic works *Il pennello dell'Evangelista* (1998), *Pro remedio animae* (2000), *Investimenti per l'aldilà* (2003), *Lo spazio dell'anima* (2005), and *San Nicola. Il grande Taumaturgo* (2009). Professor Bacci's latest book, *The Many Faces of Christ: Portraying the Holy in the East and West, 300 to 1300*, was published in London by Reaktion Books (2014).

Renana Bartal (PhD, The Hebrew University of Jerusalem, 2010) is Senior Lecturer in the Department of Art History at Tel Aviv University and former member of the research group SPECTRUM: Visual Translations of Jerusalem to Europe. Her research explores devotional practices related to illuminated manuscripts, sacred space, and holy matter. She has published articles in *Gesta*, *Journal of Medieval History*, *Viator*, and the *Bodleian Library Record*. Dr. Bartal co-edited *Between Jerusalem and Europe: Essays in Honour of Bianca Kühnel* (Brill, 2015) and is the author of *Gender, Piety and Production in Fourteenth Century English Apocalypse Manuscripts* (Ashgate/Routledge, 2016).

Kobi Ben-Meir (PhD, The Hebrew University of Jerusalem, 2016) is a scholar of contemporary art, who studies how Christian visual traditions originating in medieval Europe reappear in the twenty-first century in new manifestations. He is currently a Minerva post-doc fellow at the Freie Universität Berlin. He is a curator of contemporary art at the Israel Museum, who assisted the artist Susan Hiller in curating her 2010 exhibition *Susan Hiller: Work in Progress*. Dr. Ben-Meir curated various exhibitions at the Israel Museum, in which contemporary artworks were juxtaposed to artifacts ranging from prehistory to the nineteenth century, juxtapositions that underscored common ground between the modern and the premodern.

Neta Bodner (PhD, The Hebrew University of Jerusalem, 2016) is a post-doctoral fellow in the ERC-funded project Beyond the Elite: Jewish Daily Life in Medieval Europe headed by Professor Elisheva Baumgarten at The Hebrew University of Jerusalem. Her PhD dissertation, which examines the influence of Jerusalem on Pisan architecture in the eleventh and twelfth centuries, was written under the auspices of Bianca Kühnel's Advanced ERC Research Grant SPECTRUM: Visual Translations of Jerusalem to Europe. Dr. Bodner teaches in the Department of Art History at The Hebrew University and at the Open University in Raanana. Her research focuses on medieval architecture in Western Europe, especially from the eleventh to thirteenth centuries. She has published articles about Pisan monuments that derive from her dissertation and about architectural translations of Jerusalem to Vienna and northern Italy: '*Why are there Two Medieval Copies of the Holy Sepulchre in Pisa? A Comparative Analysis of San Sepolcro and the Baptistery*', in: Pisa: The Mechanical City, Gerhard Wolf and Hanna Baader eds. (Pisa: Edizioni ETS, forthcoming 2017). '*Earth from Jerusalem in the Pisan Camposanto*', in: Between Jerusalem and Europe: Essays in Honour of Bianca Kühnel, Renana Bartal and Hana Vorholt eds. (Leiden: Brill, 2015), pp. 74–93. '*The Baptistery of Pisa and the Rotunda of the Holy Sepulchre in Jerusalem – a reconsideration*', in: Visual Constructs of Jerusalem, Bianca Kuehnel, Galit Noga-Banai, and Hanna Vorholt eds. (Turnhout: Brepols, 2014), pp. 95–105. '*The Kreuzweg of Vienna: Local History and Universal Salvation*' in: From the Industrial Revolution to World War II in East Central Europe, Europa Orientalis, vol.12 Marija Wakounig and Karlo Ruzicic-Kessler eds. (Vienna: Universität Wien, 2011), pp. 225–239. '*Transcending Geography: the Architectural Transportation of Sanctity from the Holy Land to the Home Land*', in: United in Visual Diversity. Images and Counter-Images of Europe, Benjamin Drechsel and Claus Leggewie eds. (Innsbruk: Studien Verlag, 2010), pp. 238–249.

Caroline Walker Bynum is Professor Emerita at the Institute for Advanced Study in Princeton, New Jersey, and University and Professor Emerita at Columbia University in New York City. She studies the religious history of the European Middle Ages. Her most recent book is *Christian Materiality: An Essay on Late Medieval Religion* (2011). Recent articles include a piece on academia in the 1960s, "Gender, Generations, and Faculty Conflict," *The Chronicle of Higher Education* (November 19, 2012); "Avoiding the Tyranny of Morphology, Or, Why Compare?," *History of Religions* 53 (2014), which compares objects in Hindu and Christian devotion; "'Crowned with Many Crowns:' Nuns and Their Statues in Late-Medieval Wienhausen," *The Catholic Historical Review* 101 (2015), a study of devotional objects in north German convents; and "Are Things Indifferent? How Objects Change Our Understanding of Religious History," *German History: The Journal of the German History Society* 34 (2016), which explores the recent "material turn" and its implications for understanding the transition between the Middle Ages and the Protestant Reformation in northern Europe.

Lucy Donkin is Lecturer in History and History of Art at the University of Bristol. Her research interests lie in medieval and early modern perceptions of place, with particular reference to Italy and the Mediterranean. Her published work has focused on sacred space, especially the creation, use, and decoration of holy ground. Intersecting most closely with the themes of *Natural Materials of the Holy Land* are a paper in *Matter of Faith* (British Museum, 2015) on earthen contact relics of the Archangel Michael, and an article in *Word & Image* (2016), "Following the Footsteps of Christ in Late Medieval Italy," which brings together research on Christological imprints in the Holy Land and the liturgical functions of church pavements. She has also co-edited, with Hanna Vorholt, a collection of essays on maps and plans of Jerusalem: *Imagining Jerusalem in the Medieval West* (OUP, 2012). Her chapter in this volume derives from a new project on the materiality and portability of place, which explores the spiritual and political connotations of moving earth to and from Rome.

Elina Gertsman is Professor of Medieval Art at Case Western Reserve University. She is the author of *The Dance of Death in the Middle Ages: Image, Text, Performance* (2010), which won the 2014 John Nicholas Brown Prize, and of *Worlds Within: Opening the Medieval Shrine Madonna* (2015), which was shortlisted for the Charles Rufus Morey Award; the editor of *Visualizing Medieval Performance: Perspectives, Histories, Contexts* (2008), *Crying in the Middle Ages: Tears of History* (2011, 2013), and *Animating Medieval Art*, a special issue of *Preternature* (2015); and co-editor of *Myth and Mystique: the Cleveland's Gothic Table Fountain* (2016) and *Thresholds of Medieval Visual Culture: Liminal Spaces* (2012). Professor Gertsman's articles have appeared in many peer-reviewed collections and journals such as *Gesta*, *Studies in Iconography*, and *Art History*. A recipient of several prestigious fellowships, including awards from the National Endowment for the Humanities, the Mellon Foundation, the Kress Foundation, and the American Council of Learned Societies, she currently serves on the board of directors of the International Center for Medieval Art.

Brad Hostetler is a Jane and Morgan Whitney Fellow in the Department of Medieval Art and the Cloisters at the Metropolitan Museum of Art. In 2016, he received his PhD in Art History from Florida State University with a specialization in Late Antique and Byzantine Art and Architecture. His research interests include word and image relationships, the cult of saints and relics, epigrams and epigraphy, imperial ceremonies, and patronage. His current book project, *Enshrining Sacred Matter: The Form, Function, and Meaning of Reliquaries in Byzantium, 843–1204*, is the first synthetic study of reliquaries produced in the Byzantine Empire after Iconoclasm and before the Fourth Crusade. Combining visual, textual, and epigraphic evidence, he presents a paradigm for understanding the use and meaning of Middle Byzantine reliquaries within their social, religious, and ceremonial contexts. Dr. Hostetler has been a recipient of grants from the Gladys Krieble Delmas Foundation, the A.G. Leventis Foundation, and the Kress Foundation, and from 2015 to 2016 was a Junior Fellow in Byzantine Studies at the Dumbarton Oaks Research Library and Collection.

Bianca Kühnel is Jack Cotton Professor Emerita of Fine Arts and Architecture, Department of the History of Art at The Hebrew University of Jerusalem. She is the founding director of the European Forum at The Hebrew University. Her current research interests include visual translations of Jerusalem, spiritual and material bridges between the Holy Land and Europe, and the medieval image of the world. Professor Kühnel is the recipient of prestigious awards and grants, including the Alexander von Humboldt and Max Planck Award; an ERC Advanced Research Grant for the study Projections of Jerusalem in Europe: A Monumental Network; and the Austrian Cross of Honour for Science and Art, First Class. Her publications include *From the Earthly to the Heavenly Jerusalem; Representations of the Holy City in Christian Art of the First Millennium* (Rome-Freiburg/Br.-Vienna: Herder Verlag, 1987) (Römische Quartalschrift für christliche Altertumskunde und Kirchengeschichte, 42. Supplementheft); *Crusader Art of the Twelfth Century: A Geographical, an Historical, or an Art-Historical Notion?*

(Berlin: Gebr. Mann Verlag, 1994); *The Real and Ideal Jerusalem in Jewish, Christian, and Islamic Art*, special double issue of the annual *Jewish Art* 23/24, 1997/1998 (editor and author of the introduction); *The End of Time in the Order of Things: Science and Eschatology in Early Medieval Art* (Regensburg: Schnell & Steiner, 2003); *Visual Constructs of Jerusalem*, ed. with G. Noga-Banai and H. Vorholt (Turnhout: Brepols Publishers, 2014).

Ora Limor is Professor Emerita of History at the Open University of Israel. She specializes in Medieval History, focusing on two topics: Christian-Jewish encounters in the Middle Ages; and pilgrimage, travel narratives, and sacred space in Christian and Jewish culture. She has published numerous articles in scientific periodicals in Israel and abroad, as well as several books, among them *Die Disputationen zu Ceuta (1179) und Mallorca (1286): Zwei antijudische Schriften aus dem mittelalterlichen Genua*, published in the Monumenta Germaniae Historica (1994). Professor Limor was awarded the Ben-Zvi Prize for her book *Holy Land Travels: Christian Pilgrims in Late Antiquity*. Her most recent book, *Pilgrimage: Jews, Christians, Moslems*, was published by the Open University of Israel. In the years 2005–2009, she served as Vice President for Academic Affairs of the Open University of Israel.

Manfred Luchterhandt studied History, Art History and Archeology at the Universities of Wurzburg, Padova, and Rome. In 1997, he earned his PhD with a dissertation on the Romanesque Cathedral of Parma. From 1994 to 2000, he worked as a scientific assistant at the Max Planck Institute for the History of Art, Rome (Bibliotheca Hertziana). In 1998, he received the Max Planck Award for "outstanding scientific PhD research," and a one-year travel grant in the Eastern Mediterranean. From 2001 to 2011, he served as Assistant Professor at the University of Münster. In 2007, he attained habilitation on the Early Medieval Court Culture of the Popes. Since 2011, he serves as Professor of Medieval Art History, especially Mediterranean Art History, at the University of Göttingen. Professor Luchterhandt's publications include works on Romanesque art, medieval Rome, Mediterranean court cultures, art and liturgy, and the performative use of images. He has curated and collaborated on several exhibitions.

Nadine Mai studied Art History, Italian, and Museum Management at the Universities of Leipzig, Pisa, and Hamburg. In 2016, she completed her PhD at the University of Hamburg with a dissertation entitled "Jerusalem – Transformationen," focusing on the Chapel of Jerusalem in Bruges. Between 2013 and 2015, Dr. Mai was affiliated with the international project, SPECTRUM: Visual Translations of Jerusalem. Currently, she is organizing the interdisciplinary research project 'Making (in)visible' – Cultural Perspectives on Covering and Revealing at the Isa Lohmann-Siems Foundation in Hamburg. She has published on the mediality and materiality of winged altarpieces as well as on the strategies of lighting within the chapel of Jerusalem in Schlie, Heike, and Koch, Elke, eds., *Orte der Imagination – Räume des Affekts. Die mediale Formierung des Sakralen* (München: Fink, 2016).

Asa Simon Mittman is Professor of Art History at California State University, Chico. His publications include *Maps and Monsters in Medieval England* (Routledge, 2006; paperback 2008), *Inconceivable Beasts: The Wonders of the East in the Beowulf*

Manuscript (co-written with Susan Kim, ACMRS, 2013), and, co-edited with Peter Dendle, *The Ashgate Research Companion to Monsters and the Monstrous* (2012; paperback 2013), as well as a number of articles on the subject of monstrosity and marginality in the Middle Ages. Professor Mittman is co-director, with Martin Foys, of Virtual Mappa. His forthcoming work will concern the Franks Casket and images of Jews on medieval world maps.

Lawrence Nees (PhD 1977, Harvard University) has taught in the Department of Art History at the University of Delaware since 1978, where he now serves as Professor and Chair. He is a specialist in early medieval art, especially ca. 500–1000 CE. Professor Nees has authored *From Justinian to Charlemagne. European Art, A.D. 565–787*; *The Gundohinus Gospels*; *A Tainted Mantle: Hercules and the Classical Tradition at the Carolingian Court*; *Early Medieval Art*, in the Oxford History of Art; *Perspectives on Early Islamic Art in Jerusalem*; and edited *Approaches to Early-Medieval Art*. He has written many articles on medieval topics and has been the fortunate recipient of many fellowships, for which he is very grateful. He is a Fellow of the Medieval Academy of America and the Society of Antiquaries, and a past President of the International Center of Medieval Art.

Yamit Rachman-Schrire received her PhD from The Hebrew University of Jerusalem in 2016. Her dissertation engages with natural stones and rocks from Christian holy places of Jerusalem, and their reception and imagination in late medieval Latin cultures, from material, visual, and textual perspectives. Her research interests encompass cover late medieval pilgrimage to the Holy Land, with particular focus on the intersection between art and nature. Currently, she holds a post-doctoral fellowship at the Yad Ben-Zvi Institute in Jerusalem. Her recent publications include "The Rock of Golgotha in Jerusalem and Western Imagination," in: *Räume der Passion Raumvisionen, Erinnerungsorte und Topographien des Leidens Christi in Mittelalter und Früher Neuzeit*, eds. Hans Aurenhammer und Daniela Bohde, Vestigia Bibliae – Jahrbuch des Deutschen Bibel-Archivs, 2015; "Sinai Stones on Mount Zion: Mary's Pilgrimage in Jerusalem," in: *Between Jerusalem and Europe*: Essays in Honour of Bianca Kühnel, eds. Renana Bartal and Hana Vorholt (Leiden: Brill, 2015).

Zaza Skhirtladze is Professor and Head of the Institute of the History and Theory of Art at Tbilisi State University, Georgia. He has written extensively on the art of medieval Georgia and its relations with Byzantium and the Christian East. Professor Skhirtladze also works on desert monasticism, particularly the monastic complex of the Gareja Desert, Georgia. He is the author of the *Fresco Inscriptions of Sabereebi* (1985), *Epigraphic Monuments of the Gareja Desert* (1999), *Historical Figures at Kolagiri Monastery in the Gareja Desert* (2000), *The Tomb of Saint David Garejeli* (2001), *Early Medieval Georgian Mural Painting: The Frescoes of Telovani Church* (2008), *The Murals of Otkhta Eklesia* (2009), *Illuminated Historical Documents in Georgian Museums and Archives* (2011), and *An Old Copy of the Life of Saint Shio Mgvimeli* (2014). He has published articles in *Revue des Etudes Géorgiennes et Caucasiennes*, *Cahiers Archéologiques*, *Jahrbuch der Österreichschen Byzantinistik*, *Oriens Christianus*, *Iconographica*, *Anatolian Studies*, *Eastern Christian Art*, *Le Museon*, and *Dumbarton Oaks Papers*. Currently, Professor Skhirtladze is heading the project *Historical Figures in Medieval Georgian Art* – a multi-volume corpus with all known images of the donors in medieval Georgian murals, reliefs, metalwork, icon-painting, miniature painting, embroidery, and numismatics.

Julia M.H. Smith took up the Chichele Professorship of Medieval History at Oxford in 2016, having previously held the Edwards Chair of Medieval History at the University of Glasgow. Her current project, *Christianity in Fragments: Relics From Constantine to the First Crusade*, examines the instrumental use of small material objects in late antique and early medieval Christian practice from a variety of perspectives, and offers a new analysis of the origin and growth of relic cults. Recent publications include an essay on relics as "portable Christianity," an edition of all the relic labels surviving at Saint-Maurice d'Agaune from the seventh to the seventeenth centuries, and a historical analysis of the relic labels preserved in the Sancta Sanctorum. Professor Smith has also published on politics, gender, hagiography, and saints' cults in late antiquity and the early Middle Ages, ca. 300–1100, including *The Cambridge History of Christianity III: Early Medieval Christianities, 600–1100* (with Thomas F X Noble, 2008); *Europe after Rome: A New Cultural History 500–1000* (2005); *Gender in the Early Medieval World: East and West 300–900* (with Leslie Brubaker, 2004); *Early Medieval Rome and the Christian West* (2000); and *Province and Empire: Brittany and the Carolingians* (1992).

Tanja Trška is a Senior Research and Teaching Assistant in the Department of Art History, Faculty of Humanities and Social Sciences, University of Zagreb. In 2014, she received her PhD in Art History from Scuola Normale Superiore di Pisa and the University of Zagreb. Dr. Trška has been awarded research and study grants from the Fondazione di Studi di Storia dell'Arte Roberto Longhi, Ministry of Science of the Republic of Croatia, Fondazione Giorgio Cini Onlus and Centro Internazionale di Studi di Architettura Andrea Palladio. She is currently collaborating on the international research project *Visualizing Nationhood: The Schiavoni/Illyrian Confraternities and Colleges in Italy and the Artistic Exchange with South East Europe (15th–18th c.)* supported by the Croatian Science Foundation. Her publications and research focus on Renaissance and Baroque art, and patronage and artistic exchanges between the eastern Adriatic coast and Italy.

Laura Veneskey is Assistant Professor of Art History at Wake Forest University. Her research explores the visual culture of the late antique and medieval Mediterranean, with special focus on issues of materiality, sacred geography, medieval image theory, and the cult of relics. She is currently preparing a book manuscript examining the material dimensions of early medieval image-making. Dr. Veneskey has received grants from the Mellon Foundation, the Max Planck Society, the Samuel H. Kress Foundation, the Graham Foundation, and the Warburg Institute.

Acknowledgments

This volume has been produced under the auspices of the research project SPECTRUM Visual Translations of Jerusalem (www.spectrum.huji.ac.il), based at the Hebrew University of Jerusalem and supported by the European Research Council under the European Union's Seventh Framework Programme (FP7/2007–2013) / ERC grant agreement no. 249466.

The interest in natural materials of the Holy Land as agents of dissemination of places emerged over several project seminars. First thanks thus go to the project's team members. During the five years of the project, these weekly meetings evolved from discussions of general topics connected to the transport of Jerusalem elsewhere, to presentations of particular case studies, and finally to the delineation of still-unexplored and promising subjects for future research.

We were extremely pleased with the wide international echo our call for papers has found and wish to thank the contributors to this volume for their valuable work. We are especially grateful to Caroline Walker Bynum for taking an interest in our project, for reading the book in manuscript, and for honoring us by writing the foreword.

We thank the reviewers of the separate articles, as well as the two anonymous reviewers of the publishing house for their constructive suggestions. They greatly improved the quality of the book. Many thanks go to our language editor, Evelyn Grossberg, for her careful editing.

Finally, the editors wish to thank each other for a fruitful and pleasant cooperation.

Renana Bartal
Neta Bodner
Bianca Kühnel
Jerusalem, July 2016

Foreword

If we cast our minds back twenty or twenty-five years, we encounter a scholarly world in which "relics" tended to mean human remains or body parts, and the study of relics tended to be a study of their containers or "reliquaries." In Christian studies, the post-Tridentine categorization of relics into first, second, and third class, depending on whether they were parts of a body or not, tended to be elaborated into a distinction between relics (body parts) and contact relics or brandea (objects which had touched bodies) – a medieval distinction, it is true, but not one that organized relic cult. Indeed, actual relics were little studied. In northern Europe, many had been discarded in the aftermath of the Reformation or Puritan iconoclasm or the French Revolution; even in Catholic areas their presence was often slightly embarrassing. Historians limited themselves to textual references to relics; museum curators tended to concentrate on the precious materials or iconography of their containers. Much attention was given to reliquaries that made body parts visible, so that the vast array of early medieval containers that did not hold bits of body was often ignored.[1] Several historians went so far as to argue that it is the reliquary that makes the relic a relic.[2]

If we cast our minds back in a similar fashion to the Anglophone writers of the early 1990s, we find a great enthusiasm for the study of nature, but "nature" tended to mean human nature, and human nature tended to mean "the body." Although the Christian doctrine of the Incarnation (that is, of the manifestation of the ultimately unknowable divine in human form) seemed to underline and authorize the increasing importance of bodies and depictions of bodies in Western art, little attention was given to the prominence of the doctrine of creation in both medieval Judaism and medieval Christianity. The tendency of theologians in both traditions to understand body as part of creation and "matter" as stuff extending in a continuum from spiritual beings to lowly worms and dust, all labile and capable of metamorphosis, was similarly ignored. When scholarly attention turned, ten years ago, to materiality instead of body, the study of religious objects still seemed to hesitate before full encounter with stuff. Indeed, art historians continued to stress seeing as the basic issue, so much so that "looking beyond" or "through" has been a popular book title recently.[3] How (and whether) art (image or "Bild") emancipated itself from liturgy or devotion ("Kult") has dominated much recent discussion.[4] Despite the so-called "material turn" or the even more recent "thing theory," there has been very little work on the way in which matter itself carries the power of the holy.[5]

It is against this historiographical background that the present volume makes such a striking claim. For here is a set of studies that do not project backward anachronistic categories or art historical concerns but rather turn to the significance and power of things themselves. Gone is the assumption that the interesting relics are bodies or that "the body" should be the locus of the historian's investigation of religious meaning. Gone is the tendency to privilege container over contents, or outside over inside, although the relationship of container to contents becomes more complex than ever before. In this volume, the questions are different from those of two decades ago. How, ask the authors, do materials, especially the materials of the Holy Land, convey and even bestow holiness on other places? How and to what extent do earth, stone, architectural fragments as *spolia*, even vials of water and oil or marks of blood and tears convey not only the memory of significant places but their actual presence and power as well? How did the Holy Land come to Europe in the Middle Ages either through stuff itself or through the replication of stuff? And how do replica and material relate?

Whether the process that is studied here is called "transportable topography," "substitutional," "metonymic not metaphoric," or "mnemonic" – and these are all terms the authors of these essays use – the objects they analyze with such precision and sophistication alter our understanding of the nature of relic cult and its place in medieval developments. Considering strategies of authentication ranging from labeling to physical contact and fragmentation, they challenge earlier understandings of what makes a relic a relic. They also contribute to broadening current discussion of materiality, for these things are performative. As Lucy Donkin points out, earth not only transports significance when it is moved from one place to another; earth also differentiates the bodies buried in it. Some decay and some do not. Earth itself acts. Although historians such as Peter Brown, Patrick Geary, Lester Little, and Thomas Head have studied the ways in which relics operate in ritual, the materials considered here have a wider range of roles and impacts than earlier historians fully realized.[6] These materials transport and mirror place; they offer proof of presence and authenticate miracles; they construct and preserve memory; they provide opportunities for political control and authority (for example, the issuing of indulgences). Moreover, the studies introduce concerns of scale that are becoming important in history, as they have been in anthropology. As Julia M. H. Smith points out below, many relics are very small. Hence, consideration of how they have been collected, bundled, and grouped has greater pertinence than we have realized before.

Exactly because of the innovative character of this collection, some important questions are raised, not answered. The editors stress that they have chosen inorganic not organic material (such as ivory). This distinction is surely appropriate when considering objects from the Holy Land. But further work may suggest that we can use the approach pioneered here to ask how important such a line really is for the Middle Ages, when stones and earth were theorized as what we would call organic because generative and the significance of earth, wood, and oil was owing partly to contact with blood and tears.

Furthermore, the issue of "representation," much discussed recently by art historians, is considered here in fascinating ways that go beyond current generalizations yet challenge us to go still further. Some of the authors describe the materials they treat as representational; other deny this. Yet the question of similitude is a fraught one and the objects considered here do much to elucidate it. In Nadine Mai's essay, for example, it is the similarity of the column at Bologna to the flagellation column that gives it power and not its provenance. Yet clearly many of the objects treated in these essays do not signal what they are or where they come from by how they appear. As Smith notes, some are relics because they are labeled as such. Thus, these essays problematize the issue of representation by reminding us that "representation" can mean several different things, including "stand in for." When we speak in ordinary discourse, for example, of an ambassador as representing his or her country, we do not mean that John Kerry or Hillary Clinton looks like the United States. The measure of Christ's height that pilgrims brought back from the Holy Sepulchre by bringing a piece of string did not look like Christ. But it made him present. A stone might represent the Holy Sepulchre or a bit of wood the True Cross but not so much by how they looked as by the stuff they were.

This complication of terminology relates to another. Historians have often tried to distinguish relic from icon or image in the development of Christianity by using C.S. Pierce's distinction between icon, index, and symbol.[7] In such classification, icon has power because it looks like; symbol has power because it refers to, by some sort of agreed upon code; index has power because of an actual, often physical relation to the object referred to. This has sometimes led to the assumption that images are icons and relics are indexes, that is, that icons image or represent or look like, whereas relics are. These essays show the inadequacies of such distinctions for the European Middle Ages. Historians have recently come to realize that such categories overlap both because we often find relic and image combined in medieval sites (as, for example, Zaza Skhirtladze shows for Mtskheta, an old capital of Georgia) and because iconic images such as the *arma Christi* (discussed in the essay by Nadine Mai) are often images of objects that are themselves also relics. These essays go further in deepening our understanding of the inappropriateness of such distinctions as icon and index by making it clear that we have paid far too little attention to the nature of materials themselves. What, for example, can earth do and not do? As Donkin points out, it may be able to bring Jerusalem to Rome but not, at second remove, bring Jerusalem from Rome to Ireland. Why, we must ask, is this so? Even if we can never fully express in words the power of such an object, the example shows us that the earth of Akeldama is not exactly icon or index or symbol, yet it has the power of all three. The earth, oil, water, marble, and wood considered in such detail in these essays point beyond themselves as a symbol does; medieval people knew how to read *pars pro toto* – to see oil, for example, as a paradoxical sign of both royal power and preparation for death. These materials are also the land they come from, pieces of (that is, indexes of) the Potter's field or the hill of Golgotha that brought those places to Italy or the Balkans or Eastern Europe. And they are also (for example, as arranged in the box of stones from the Sancta Sanctorum) images that bring before the eyes of medieval Europeans the specificity of far off places and events they cannot visit except in the visualization of spiritual pilgrimage.

The essays assembled here break new ground by significantly expanding the geographical area usually considered as Europe. By drawing on cases from Eastern Europe as well as Western, the European North as well as the Mediterranean, they add new evidence to considerations of borders, of evangelizing and confessionalization, and of how ideas and practices are spread by things. Methodologically sophisticated and extensively documented, they display the new material history at its best. Historians, art historians, curators, anthropologists, travelers, pilgrims, and all those interested in the present-day Near East will find in these articles a challenge to assumptions that have long shaped the history of Christianity. Each part here truly stands for a whole that reaches even beyond the volume itself and induces us to ask new questions.

Caroline Walker Bynum
January 2016

Notes

1 See Cynthia Hahn, "The Voices of the Saints: Speaking Reliquaries," in *Body-Part Reliquaries*, eds. Caroline Bynum and Paula Gerson, *Gesta* 36, no. 1 (1997): 20–31.

2 Jean-Claude Schmitt, "Les images et le sacré" in *La performance des images*, eds. Alain Dierkins, Gil Bartholeyns, and Thomas Golsenne [Problèmes d'histoire des religions, 19], (Brussels: Éditions de l'université de Bruxelles, 2010), 29–46. See also Martina Bagnoli, Holger A. Klein, C. Griffith Mann, and James Robinson, eds., *Treasures of Heaven: Saints, Relics, and Devotion in Medieval Europe* (Cleveland, Baltimore and London: Yale University Press, 2010) and Cynthia Hahn, *Strange Beauty: Issues in the Making and Meaning of Reliquaries, 400-circa 1204* (University Park, PA: Pennsylvania State University Press, 2012).

3 See, for example, Colum Hourihane, ed., *Looking Beyond: Visions, Dreams, and Insights in Medieval Art and History* (Princeton and University Park, PA: Pennsylvania State University Press, 2010). See also Herbert L. Kessler, *Spiritual Seeing: Picturing God's Invisibility in Medieval Art* (Philadelphia, PA: University of Pennsylvania Press, 2000).

4 The classic argument is Hans Belting, *Bild und Kult: Eine Geschichte des Bildes vor dem Zeitalter der Kunst* (Munich: Beck, 1990); in English as *Likeness and Presence: A History of the Image before the Era of Art*, trans. Edmund Jephcott (Chicago: The University of Chicago Press, 1994).

5 See, for example, Bill Brown, "Thing Theory," *Critical Inquiry* 28 (2001): 1–22; Lorraine Daston, ed., *Things That Talk: Object Lessons from Art and Science* (New York: Zone Books, 2004); Daniel Miller, ed., *Materiality* (Durham, NC: Duke University Press, 2005); David Morgan, ed., *Religion and Material Culture: The Matter of Belief* (London: Routledge, 2010); and Martha Rosler et al., "Notes from the Field: Materiality," *The Art Bulletin* 95 (2013): 11–37.

6 Peter Brown, *Cult of the Saints: Its Rise and Function in Latin Christendom* (Chicago: The University of Chicago Press, 1981); Patrick J. Geary, *Furta Sacra: Thefts of Relics in the Central Middle Ages*, 2nd ed. (Princeton: Princeton University Press, 1990); Lester Little, *Benedictine Maledictions: Liturgical Cursing in Romanesque France* (Ithaca, NY: Cornell University Press, 1993), and Thomas Head, "Relics," in *Dictionary of the Middle Ages*, ed. Joseph Strayer, vol. 10 (New York: Scribner, 1988), 296–299.

7 See Digital facsimile: http://www.cs.indiana.edu/~port/teach/103/sign.symbol.short.html (accessed December 13, 2015). I thus need to rethink what I say in: "Avoiding the Tyranny of Morphology, Or, Why Compare?" *History of Religions* 53 (May, 2014): 341–368, esp. 361 n. 59.

Natural materials, place, and representation

Renana Bartal, Neta Bodner, and Bianca Kühnel

In the Hinnom Valley in Jerusalem, to the south of Mount Zion, is an area known as Akeldama, which is associated with the potter's field mentioned in the Gospel of Matthew (27:7–8) and the Book of Acts (1:19). In Matthew's version, Judas returns the thirty pieces of silver received for the betrayal to the priests. As blood money, they say, it cannot be returned to the Temple treasury, and they use it instead to buy the potter's field for the burial of foreigners. The Book of Acts adds that Judas' bowels gushed out there, and so the place became known as *Acheldemach*, or "field of blood" in Aramaic.[1]

Despite its name and bloody history, Akeldama became a sacred site, drawing pilgrims as early as the fourth century.[2] In 1143, the site was given to the order of the Knights Hospitaller so that they might use the land to build a charnel house for the burial of pilgrims who died in Jerusalem, a role fitting the Gospel account.[3]

As the site became sanctified its indigenous soil became sacred as well. Many visitors ascribed miraculous powers to earth from Akeldama, especially that it decomposed bodies buried in it very quickly and without a foul smell.[4] As a result of these transformative powers, Akeldama's earth was coveted by individuals and communities abroad, so travelers hoarded it and transported it home.[5] According to a commemorative plaque in Rome's Church of Santa Croce in Gerusalemme, Empress Helena transported earth from Jerusalem to a chapel there, which became known as "Gerusalemme."[6] Helena's earth was perceived to be soaked with blood and thus could have carried an association with two sites in Jerusalem, both Calvary and Akeldama.

The earth of Akeldama was placed under the foundations of churches and local burial sites in Europe, which became infused with its characteristics and sanctified through the earthen relic.[7] The earth communicated its significance to the new locations by miraculous connections to its place of origin. Thus, for example, the anonymous English author of the *Ynformacio*, written between 1480 and 1526, describes the ability of the soil from Akeldama (replanted near Saint Peter's in Rome) to decompose bodies within fourteen days and to regurgitate any locals buried there.[8]

Belief that the earth at a holy site is enduringly imbued with the sanctity of its origin suggests that holiness itself could be multiplied and disseminated far and wide through its natural components. Enduring connections between natural materials and the sacred sites of the Holy Land from which they were taken underlie the subject of this book. The essays in this volume discuss the means of procuring natural materials such as earth, water, stones, and wood in the Holy Land;[9] track their transportation; and consider new associations acquired once they were removed and taken elsewhere. They describe evolving relationships between places and their constituent materials and consider the agency of specific substances in representing and disseminating loca sancta. They also look at the place of natural materials of the Holy Land among relics of other provenance and within the framework of architectural and artistic representation of the Holy Land in Europe.

Natural material and the Holy Land

Our definition of natural materials embraces raw matter indigenous to the land, including wood, earth and other dust or dirt from a site, stones, rocks, and water, fruit and other flora.[10] In addition to these are their derivatives – such as oil pressed from local olives, or stone columns. Also included are raw materials brought into contact with other locative natural materials such as water or oil poured over bedrock or large stones and then collected.[11] Fragments of materials modeled into man-made objects such as the wood of the cross or of Christ's manger, stones of the Sepulchre or of the unction, or the rod of Aaron are also included.

The essays in the present volume highlight the way in which all of these types of natural materials – whether found *in situ*, framed in a reliquary, sanctified through touch and blessing, chiseled or gnawed from the original, or wrapped in labeled bundles – assimilated meaning associated with their origins and embodied the sanctity of sacred places in new locations.

Holy Land natural materials and objects containing natural materials are treated here as relics of place or, to use a term coined by Bruno Reudenbach, loca sancta relics.[12] These are often neglected in more general discussions of relics or are seen as separate categories of "devotionalia" or "pilgrim souvenirs." By focusing on natural materials from the Holy Land, the present book supports existing attempts to counter the idea that the cult of relics was mainly about bodies and saints. The distinction drawn in scholarly literature between corporal and the other relics does not seem to be in evidence in collections and inventories.[13] As demonstrated by Julia M. H. Smith, in early medieval Rome, *reliquiae* was a collective notion for remains of Christian saints, churches, and holy sites. Early medieval collections of relics such as the ones in the Lateran Palace, Sens, and Chelles include Holy Land materials together with remains of the bodies of saints. It appears that both sorts of relics, corporeal and site-related, were collected, stored, labeled, and presented in similar ways.[14]

Indeed, as demonstrated by the various chapters herein, natural materials of the Holy Land share many attributes with corporeal relics. Like the latter, they are sacred, are nonrepresentational, and translate sanctity elsewhere. Like bodily relics, natural remnants have no intrinsic worth, but their documented origins, made known through labeling or appropriate containers, make them priceless. They are charged with unnatural potency, for example, the ability to behave in an unnatural way, to perform miracles, or otherwise to evoke enchantment. Similar to a body part such a relic can serve as *pars pro toto*, a part signifying the whole, representing its place of origin in the same way as a saint is represented by a fragment of bone or a wisp of hair. Thus, such a relic can evoke the memory of the event associated with its site of origin and convey related hopes of salvation.[15]

As fragments from the Holy Land, however, the prominent feature of the natural materials we discuss here is their ability to dislocate memory from place and carry it elsewhere. *Terra sancta* from Akeldama, water from the Jordan River, and stones and wood from the Mount of Calvary act as synecdoche of their respective places.

Considering the term "holy place," Mircea Eliade defined it as a cosmological center ritually enacted and manifested to the religious believer in an act he describes as hierophany, that is, distinguishing the sacred from the profane.[16]

Although Eliade's paradigm is still a cornerstone of the discourse on holy places at large, other scholars have considerably refined it as it applies to the Holy Land. Emphasis has since been laid on the specificities of the biblical sacred sites, on the one hand, and the fluidity rather than strict dichotomy between sacred and profane spaces, on the other.[17] Robert Markus calls attention to the lack of historical context in Eliade's claims, noting that in describing the nature of holy places in religions in general, he doesn't consider the specific development of the concept of holy places in Christianity: from sacred community to sacred sites and monuments.[18] Whereas early Christians referred to virtuous men and women and to Scripture as holy, the term *loca sancta* entered the Christian lexicon with Empress Helena's rediscovery of such sites and the Constantinian building campaigns around them.[19] The fourth century saw a notable increase in the number of pilgrims who flocked to see and to touch the places where Christ walked.[20] As traced by Maurice Halbwachs, the sites of biblical events met the early Christian community's need for a tangible way to identify with the past.[21]

Scholarly opinion is divided as to whether the Christian concept of holy sites originated in the Holy Land or elsewhere, especially Rome, under the influence of the cult of martyrs.[22] The loca sancta function differently from martyrs' graves. Sites dedicated to the memory of saints became potent through the presence of their bodies; in the Holy Land the place itself is venerated.[23] Pilgrims venerating a bodily or other relic elsewhere interact with an object, whereas Holy Land pilgrims interact with an item representing space and its associated biblical story. Indeed, the celebration of an event as a component of a *locus sanctus* is a significant addition to the definition: only in Jerusalem, notes Jonathan Z. Smith, "story, ritual, and place could be one."[24]

Recent discussions go on to portray sacred sites as dynamic spaces, continually being generated and shaped by action, movement, and use rather than limited to a mere geographical dimension.[25] The sites of the Holy Land are certainly more than static locations and more than backdrops for the stories attached to them. Their definitions encompass what is done and imagined as well as what is actually found in these places, including both their developed and their natural environments.[26] Indeed, it is the materiality of the holy sites that makes the practice of loca sancta relics possible, and it is the portability of these material remnants of the holy sites that allows for their sanctity to be carried off to distant places.[27]

The fluidity and dynamism of the loca sancta are witnessed by the habit of translating and representing them elsewhere. Throughout the Christian ecumene buildings and complexes representing the sacred sites of the Holy Land were built and used to evoke the originals.[28] Often, landscape elements within these monumental representations were used to mark the topography of the Holy Land: a local valley was named "Jehoshaphat" and a hill became the "Mount of Olives."[29] Loca sancta relics were often incorporated within such architectural and topographical transpositions and played key roles in providing authenticity, serving as tangible links between the sacred spaces defined by the architecture and the original sites they signified.[30] Yet, whereas recent years have seen a growing awareness of the role of architecture and monumental art in translating loca sancta elsewhere, much less attention has been devoted to the unmediated means of translating the Holy Land, that is, the carrying off of its natural materials.[31]

Materiality and representation

In pursuing the intrinsic association between matter and place, this volume contributes to the study of what has been termed "the material turn," that is, the growing scholarly interest in matter and materiality as subjects for interrogation. For students of religion, the interest in matter comes from the understanding that it is, to use Ann Taves' term, a fundamental "building block" of religion,[32] or a way of "making invisible faith visible," by focusing on practices and objects rather than on abstract beliefs.[33] Embracing material manifestations of religion as a new analytic concept has brought objects such as relics and images as well as the study of sensory and bodily experiences they engender to the fore of scholarly inquiry.[34] The concern with works of art as "things" that act upon beholders and generate responses was developed by Hans Belting, David Freedberg, Alfred Gell, and W. J. T. Mitchell.[35] Indeed, art historians have always manifested a particular interest in the material aspect of things, although this interest has certainly intensified in recent years.[36] Herbert Kessler, for example, calls attention to the "overt materiality of medieval art," posing new questions about the role of different materials in the making and meaning of medieval objects.[37]

Caroline Walker Bynum demonstrated that medieval objects call attention to their meaning through a combination of material and shape.[38] She shows that the later Middle Ages were characterized by an intense awareness of the power of materials manifest in works of art as well as in theological, devotional, and philosophical writings.[39] She points to the phenomenon of "living holy matter," meaning material that behaved in a way particular to living things, such as animated statues, bleeding hosts, and holy dust that itself mediated transformation.[40] This medieval diffusion (or even rejection) of the difference between something that is alive and inanimate material is explored by Jane Bennet, who searches for natural instances of "material vitalism."[41]

Historians of the Middle Ages have found a deep confluence between belief in the Incarnation and interest in materiality, as it was seen as a filter through which the natural world could be understood with religious eyes. Sara Ritchey argues that the Incarnation, through the presence of God in the physical world made all natural matter holy.[42] Ritchey's reading places emphasis on the sanctity of the natural world as something that embodies God's creation, enduring presence, and promise of salvation, whereas art historians have focused more on the use of matter in art and object and less on its natural form.

The essays herein address instances when matter stood on its own, with its own associations and meanings, rather than as a medium, but was made part of an artistic installation that enhanced its properties. This sets the present volume apart from other investigations of matter in art history and religious studies, and the confluence of matter and place distinguishes it from studies on the materiality of objects.

Of course, this does not mean that the natural materials were not framed or manipulated. Generic, even worthless, matter gained its identity as "holy matter" through narrative, visual, epigraphic, architectural, and ritual constructs. Elaborate reliquaries added interpretation and content to the natural (though significant) materials they held. Reliquaries could conceal simple material remnants, call attention to their presence, or elevate their worth by the use of precious materials that evinced their hidden meanings, bridging the gap between their earthly manifestations and heavenly significance.[43] The artistic forms of the reliquaries were so prominent that Cynthia Hahn suggests that the relics themselves hold only a small part in the reception and understanding of these objects.[44]

However, it was not in all cases that sacred matter was displayed in reliquaries. Earth from Akeldama in Pisa and in Rome was spread on the ground, fruits from the trees of the Holy Land were eaten, and handfuls of earth and tiny stones were kept in bundles of silk and vellum that were labeled with their origin and significance. As demonstrated by the essays in this volume, the meaning of material remnants was heavily dependent not only on their containers but to the same extent on such signifying acts as labeling, placement, presentation, illustration, architectural surroundings, or other means of making their significance known. Natural materials could be completely invisible, their worth signaled by verbal communication, written or spoken. They could be presented in a collection with other Holy Land relics and juxtaposed with various other materials, images, or inscriptions. What was done with the materials dictated their perception and the engagement of their beholders – whether they were placed within a trove, displayed as individual exemplars, or given as political gifts.

Structure of the book

Before a remnant of a natural material could become a sacred relic, it had to be isolated from the landscape and defined as such. The first section of the book, "Collecting and collections," approaches the issue of isolation, considering the act of collecting materials at loca sancta and their later grouping in elaborate collections. The included essays deal with the various stages that were crucial in the process of defining material remnants as sacred, such as having them blessed by clerical authorities at the local sites, bringing them into contact with other objects, or labeling them by narratives and inscriptions. The essays further highlight the institutional use of natural materials at the sacred sites, such as their organized distribution among pilgrims by clerics, their use as political currency between rulers, or their exchange between clerical institutions.

Ora Limor traces evidence of individual collections of natural fragments of the Holy Land as found in early pilgrim accounts. The materials collected by these pilgrims, such as local stones or branches, fruits and vegetables, and water and earth, were accessible, found in abundance, and not carefully guarded. Their value was entirely dependent on their identification with their source locality, which gave them power and significance. Pilgrims were not only the collectors but also the reporters and commentators regarding these objects. Egeria, the Piacenza Pilgrim, and Willibald told stories about the Holy Land, its marvels, virtues, and sacred traditions, and brought home objects that validated these stories, which could be accessed through the senses – touch, sight, hearing, taste, and smell. Both acts, collecting and writing, commemorated the visit and transmitted the essence of the holy places in didactic, sentimental, and sensual ways. That some of these objects, such as fruit collected in the Holy Land, would have eventually rotted raises the question of how and if they were preserved or if they were, rather, eaten. If the latter, one would wish to know whether it was with any ceremony or on a specific occasion.

At the opposite extreme were the fragments collected within silk or vellum wrappers, labeled, hoarded, sealed, and stored within an altar for posterity. The relics included in these collections were tiny, amorphous, and indeterminate objects, rarely visible. The two eleventh-century relic troves described by Julia M.H. Smith were accompanied by prominently located lists that recorded all the items in the collection. In contrast to narrative descriptions of the Holy Land in pilgrim itineraries, these relic lists do not constitute coherent accounts from a geographical point of view, nor are their significance and raison d'être self-evident. Smith explores the relic collections of Bishop Leofric of Exeter (1046/50–1072) and Bishop Gundechar of Eichstätt (1057–1075), which they used to promote the reputation and resources of their dioceses in the third quarter of the eleventh century. Gundechar cemented the Holy Land relics into the altars of his church, where they were hidden from sight but marked by inscriptions. By contrast, Leofric had the relics of Jerusalem carried around the church and city for all to see, and their significance was expounded aloud. The two collections raise questions about how readily Holy Land relics were available, if and how they were differentiated from the corporal relics held with them, and how their significance was communicated. Smith explores the ways in which the act of recording the name of every particle in a relic collection provided its components with durable identities and enduring significance, a task then completed by religious ritual and splendid settings.

Architectural settings built to house relics were not limited to single collections and often became accumulative "monumental reliquaries" in their own right. Manfred Luchterhandt explores this aspect of the additive and varying relic collections within the chapel of the Sancta Sanctorum in Rome. A wide range of substances had reached Rome from the Holy Land in the seventh, eighth, and ninth centuries. The papal palace possessed one of the largest documented assemblages of relics from Palestine during the Middle Ages, larger than the ampullae of Theodelinda, the relic collections of Charlemagne, or the imperial Pharos Chapel at Constantinople. Although the collection includes more than 50 relics from the Holy Land, previous scholarly discussion has focused on a single object[45]: the wooden reliquary box painted with scenes from the life of Christ, discussed in detail in this volume by Elina Gerstman and Asa Mittman. Luchterhandt explores this relic collection as a whole, reconstructing the content of that huge assemblage, which was eventually dispersed in the twentieth century. The collection serves as evidence for pilgrimage activity in the Holy Land after the Muslim Conquest and of surprising political exchanges in the form of diplomatic gifts of loca sancta relics by the Roman popes, attesting to the political connections between Rome and Jerusalem.

Relic collections are thus shown to have been manipulated on a broad scale, extending beyond the immediate topographical setting of Jerusalem. Laura Veneskey investigates the status of fragments of the True Cross as a mobile marker of Jerusalem's sacred topography. The mobility of the cross relics allowed for them to accrue multiple associations across different locations and to be deployed not only as potent spiritual objects, but also as signifiers of power and authority, both religious and political. Veneskey argues that equating the relics of the True Cross with the Holy City was complicated by the seemingly endless division and distribution of its splinters. Focusing on three sixth- and seventh-century accounts of the cross's translation, those of Avitus of Vienne, Queen Radegund of Poitiers, and Theodelinda, Queen of Lombard, she demonstrates that authenticity of a splinter could be constructed in a variety of ways, influenced by geographic context and historical conditions. These relics continued to communicate aspects of Jerusalem's topography and the passion narrative across time and space, but they also provided an axis around which new sacred narratives could be told.

The second section, "Agents of translation," focuses on the mechanisms by which natural material, when transported, signified a sacred site in another location. As the first two chapters of this section demonstrate, the story told by holy matter is multilayered, rooted in times predating the witnessed biblical event and extending well afterward. The translated material acquires a mythical dimension in its new location, entwined with local miracles and history and enriched by the legend of its own translation, developing a legacy of local pride in light of its Holy Land origins. Tanja Trška and Zaza Skhirtladze focus on the reception of holy stones from Jerusalem in Croatia and Georgia, respectively, and on the way they became emblematic for the recipient society. Trška discusses two stone relics from the Holy Land sent to the Republic of Ragusa (present-day Dubrovnik in Croatia) by the guardian of Mount Zion, Fra Bonifacio de Stephanis da Ragusa (Lopud, ca. 1500–Timişoara, 1581), tracing their reception through archival documents from 1558 until the late eighteenth century, examining their veneration in Ragusa and doubts about their authenticity. Zaza Skhirtladze focuses on a holy stone linked with the self-identity of the Georgian Orthodox Church, the Stone of Grace, traditionally brought from Jerusalem by Saint David, one of the so-called thirteen Syrian Fathers active in Georgia in the mid-sixth century. He traces the legend of how the stone was acquired along with two others in Jerusalem and their presentation and reception in Georgia, which led to the monastery being recognized as a pilgrimage site.

The interpretation of materials in their new locations depended on many factors, one of which was quantity. Lucy Donkin explores the qualities of *terra sancta* when measured in sacksful or shiploads rather than handfuls as described in texts from the Middle Ages and the beginning of the early modern period. She traces the dual representation of Rome and Jerusalem in cases where earth from the Camposanto Teutonico in Rome (that originated in Akeldama) was taken for the foundation of cemeteries elsewhere.

Using *spolia* or imitating the architectural spaces that marked the loca sancta made it possible to transfer sanctity and evoke their memory elsewhere. The third section, "Used materials, new locations," further documents the complexity of the relationship between matter and place by considering two cases in which stone was used to transfer meaning to a new site as *spolia* or as a narrative construction. Yamit Rachman-Schrire explores the tradition and translation of the Stone of Unction, which marked the place where Christ's body was anointed. Thought to have been transferred from Jerusalem to Ephesus in the first century and then to Constantinople in 1169, a slab identified as the Stone of Unction was displayed at the entrance to the Holy Sepulchre Church in Jerusalem toward the second half of the thirteenth century. She argues that in Constantinople the historical narrative of the stone's physical transfer gave it meaning, but that in Jerusalem, allegedly in its original location, the stone manifests the transmission of traditions and iconographical schemes from Constantinople and the Latin West to the Holy City.

Lawrence Nees examines the columns of the Dome of the Rock, noting that their different colors, origins, and arrangement inside the building could have been perceived as references to different sites in Jerusalem. All of the column shafts appear to be Roman *spolia*, removed from one or more Roman buildings. Nees postulates that the prominent red-colored columns axially aligned with the north, east, and south doors, and forming a pair at the west door, probably the main entrance to the Dome of the Rock (then and now), were either taken from the Nea Ekklesia or made reference to its columns, which were described as "flame-colored." The arrangement of the columns in the Dome of the Rock might have been a reference to Solomon's Temple, which stood on the same site.

The fourth section, "Installation and enactment," focuses on the framing of and interaction with transported materials. The way materials are displayed has a profound effect on viewers' engagement and perception of meaning. As noted above, materials could be presented in a reliquary; in a collection with other Holy Land relics; or juxtaposed with various other materials, pictures, or inscriptions. The next three chapters analyze the associations of natural materials with images, texts, and architectural spaces, and the effect of such installations on their beholders.

The box of stones from the Sancta Sanctorum chapel in the Vatican, discussed by Elina Gertsman and Asa Mittman, is perhaps the best-known example of combining stones, wood, cloth, and earth translated from loca sancta with identifying inscriptions and images in a wooden box. Their interest lies in the performative ways that viewers would have engaged with the wooden box and its content. They explore the mnemonic and affective thaumaturgical potential of the rocks and their painted environment.

Brad Hostetler expands the discussion to personal devotion in the Middle Byzantine period. Focusing on a twelfth-century reliquary now in the collection of the Protaton (Mount Athos, Greece), he argues that the relics, iconography, and inscriptions interact to transform it into a locus sanctus where the donor, whose image appears on the lid, ensures his tangible and perpetual presence at the site and hour of the crucifixion.

Nadine Mai focuses on the sculptured Calvary constructed by Anselm Adornes in his Jerusalem church in Bruges by considering the way that this centerpiece referred to its architectural frame. The rocky surface of the sculptured Calvary, which displays arbitrarily distributed *Arma Christi* and almost vanished horses' hoofprints, expresses a unique visualization of the passion drama via its location, the Golgotha Hill. Moreover, as the sculpture also houses natural materials from the Holy Land, it is also a reliquary, functioning as a statement of Christ's mundane permanence, which could be likewise replicated by walking through the surrounding architecture. Simultaneously preserving tools, sites, and relics and allowing liturgical re-enactment, the Bruges installation created an inspiring link to the original sites of the passion and transformed the chapel into a multifaceted synopsis of a holy place.

A final paper in the form of a postscript by Kobi Ben-Meir considers a contemporary re-enactment of the practice of collecting natural material relics in the work of the London-based artist Susan Hiller. Starting in 1969, Hiller collected water from various *loci* around the globe considered to be sacred, a prolonged action/creation that was finally sealed in 2009. She visited pilgrimage destinations for believers of different faiths and collected and bottled water that was considered to have magical powers to heal, protect, and provide heavenly blessing. While acting as a pilgrim collecting holy matter, Hiller, a former anthropologist, stored the holy water in cabinets that could simultaneously be perceived as (religious) reliquaries and as their secular-modern equivalents – first-aid cabinets or anthropology museum vitrines.

Notes

1 "Et hic quidem possedit agrum de mercede iniquitatis et suspensus crepuit medius et diffusa sunt omnia viscera eius; et notum factum est omnibus habitantibus Hierusalem ita ut appellaretur ager ille lingua eorum Acheldemach hoc est ager Sanguinis." Vulgate, Acts 1.18–19, translated in the Douay – Rheims and King James versions as: "And he indeed

hath possessed a field of the reward of iniquity, and being hanged, burst asunder in the midst: and all his bowels gushed out; And it became known to all the inhabitants of Jerusalem: so that the same field was called in their tongue, Haceldama, that is to say, The field of blood."

2 Denys Pringle, *Churches of the Crusader Kingdom of Jerusalem: A Corpus* (Cambridge: Cambridge University Press, 1993–2009), vol. 3, 222. Akeldama appears in the sixth-century Madaba map and on most of the round maps of Jerusalem from the twelfth and thirteenth centuries. Milka Levy, "Maps of Jerusalem," in *Book of Jerusalem: The Crusader and Ayyubic Periods, 1099–1250*, eds. Joshua Prawer and Haggai Ben – Shammai (Jerusalem: Yad Ben Tzvi, 1991), 2, 449–505. Its location in the Hinnom Valley was attested to from the twelfth century on: see Pringle, *Churches*, 223.

3 Adrian Boas, *Jerusalem in the Time of the Crusades: Society, Landscape and Art in the Holy City under Frankish Rule* (London: Taylor and Francis, 2001), 185.

4 Jacobus da Verona described the absence of a stench in the burial ground, and later writers continued to testify that they sniffed the air in the presence of newly buried bodies and did not detect a foul odor. Amit Re'em, *The Crusader Charnel House at Acheldama and the Church of the Holy Sepulchre as Case Studies Regarding Death and Burial in Crusader Jerusalem* (unpublished MA thesis, Hebrew University of Jerusalem, 2004), 16. Ludolph of Sudheim, who stayed in the Holy Land from 1336 to 1341, wrote that bodies buried within the soil there are consumed within three days. See Ludolph of Sudheim, "De itinere Terre sancte," in *Enchiridion Locorum Sanctorum: Documenta S. Evangelii Loca Respicientia*, ed. Donato Baldi (Jerusalem: Franciscan, 1955), 579. Another trait associated with the soil is the deferment of the need of those buried there to pass judgment. See Anon Graecus, *Patrologiae Cursus Completus: Series Graeca CXXXIII*, ed. J.P. Migne (Paris: Garnier, 1844–1861), col. 997.

5 Explicit references to the removal of soil from Akeldama are found in accounts from the fourteenth century on. Early evidence of this habit, from the first decade of the twelfth century, might be inferred from the wording used by the Russian abbot Daniel (visiting Akeldama between 1106 and 1108), who wrote that "nothing is allowed to leave this holy place because it was bought with the blood of Christ," suggesting that there was immanent threat of such removal. *The Pilgrimage of the Russian Abbot Daniel* (London: Palestine Pilgrims Text Society, 1897), vol. 4, 38.

6 Ilaria Toesca, "A Majolica Inscription in Santa Croce in Gerusalemme," in *Essays in the History of Art Presented to Rudolf Wittkower*, eds. Douglas Fraser, Howard Hibbard, and Milton J. Lewine (London: Phaidon, 1969), 102–105; Sible de Blaauw, "Jerusalem in Rome and the Cult of the Cross," in *Pratum Romanum; Richard Krautheimer zum 100. Geburtstag*, eds. Renate L. Colella, Meredith J. Gill, Lawrence A. Jenkens, and Petra Lamers (Wiesbaden: Dr. Ludwig Reichert Verlag, 1997), 55–74; Alexander Nagel, *Medieval Modern: Art out of Time* (London: Thames & Hudson, 2012), 119.

7 These were often called Campo Sancto, a synonym for Akeldama itself: Marjetta Wis, "'Gottesacker' und 'Campo Santo': Spuren mittelalterlicher Glaubensvorstellungen im deutschen und italienischen Wortschatz," *Neuphilologische Mitteilungen* 58 (1957): 71–108. Our thanks to Lucy Donkin for making us aware of Wis' work and for sharing her copy of this important publication. For the trilateral relationship among the transported earth, the common name, and the architectural copy in the Camposanto in Pisa, see Neta Bodner, "Earth from Jerusalem in the Pisan Camposanto," in *Between Jerusalem and Europe: Essays in Honour of Bianca Kühnel*, eds. Renana Bartal and Hanna Vorholt (Leiden: Brill, 2015), 74–93.

8 The earth itself discerned between foreigners deserving of burial in it and locals, who were expelled from it miraculously and found above ground again. J. Brefeld, "Account of a Pilgrimage to Jerusalem," *Zeitschrift des deutschen Palästina-Vereins* 101 (1985): 138. Felix Fabri, writing between 1480 and 1483, also highlighted the retained magical qualities of the earth when it was taken to Pisa and Rome: *Wanderings of Felix Fabri (circa 1480–1483 A.D.)*, trans. Aubrey Stewart, 2 vols, Library of the Palestine Pilgrims' Texts Society, 7–10 (London: Palestine Pilgrims' Texts Society, 1892–1893), vol. 1, part 2 (1896), 537.

9 From the time that Christians began making pilgrimages to the Holy Land they took samples of such materials from the *loca sancta* to serve various ends: see Holger A. Klein,

"Eastern Objects and Western Desires: Relics and Reliquaries between Byzantium and the West," *Dumbarton Oaks Papers* 58 (2004): 283–314; Georgia Frank, "*Loca Sancta* Souvenirs and the Art of Memory," in *Pèlerinages et lieux saints dans l'antiquité et le moyen âge: mélanges offerts à Pierre Maraval*, eds. Béatrice Caseau, Jean-Claude Cheynet, and Vincent Déroche (Paris: Travaux et Mémoires: Association des amis du Centre d'histoire et civilisation de Byzance, 2006), 193–201.

10 For a discussion of Holy Blood and Holy Milk, two bodily categories that are outside the scope of this volume, see Willy Jansen and Grietje Dresen, "Fluid Matters: Gendering Holy Blood and Holy Milk," in *Things: Religion and the Question of Materiality*, eds. Dick Houtman and Birgit Meyer (New York: Fordham University Press, 2012), 215–231, and Caroline Walker Bynum, *Wonderful Blood: Theology and Practice in Later Medieval Northern Germany and Beyond* (Philadelphia, PA: University of Pennsylvania Press, 2007).

11 Ivory is not included in the list of natural materials examined in this book. For a discussion of ivory as a category of pilgrimage art, see Henry Maguire, "Ivories as Pilgrimage Art: A New Frame for the 'Frame Group'," *Dumbarton Oaks Papers* 63 (2009): 117–146.

12 Bruno Reudenbach, "Holy Places and Their Relics," in *Visual Constructs of Jerusalem*, eds. Bianca Kühnel, Galit Noga-Banai, and Hanna Vorholt (Turnhout: Brepols, 2014), 197–206.

13 Julia M.H. Smith, "Relics: An Evolving Tradition in Latin Christianity," in *Saints and Sacred Matter: The Cult of Relics in Byzantium and beyond*, eds. Cynthia Hahn and Holger A. Klein, *Dumbarton Oaks Symposia and Colloquia* (Washington, DC: Harvard University Press for the Trustees of Dumbarton Oaks, 2015), 41–60.

14 Ibid.

15 Reudenbach, "Holy Places," 203.

16 Mircea Eliade, *Patterns in Comparative Religion*, trans. R. Sheed (New York: New American Library, 1974); *Cosmos and History: The Myth of the Eternal Return*, trans. W.R. Trask (Princeton: Princeton University Press, 1991).

17 See the discussion in Sarah Hamilton and Andrew Spicer, "Defining the Holy: The Delineation of Sacred Space," in *Defining the Holy: Sacred Space in Medieval and Early Modern Europe* (Burlington: Ashgate, 2005), 1–23, esp. 3–5. A useful summary of current literature about space and place in medieval culture is Megan Cassidy-Welch, "Space and Place in Medieval Contexts," *Parergon* 27, no. 2 (2010): 1–12.

18 Robert Markus, "How on Earth Could Places Become Holy? Origins of the Christian Idea of Holy Places," *Journal of Early Christian Studies* 2, no. 3 (1994): 257–271, esp. 258.

19 Robert L. Wilken, *The Land Called Holy: Palestine in Christian History and Thought* (New Haven: Yale University Press, 1992), 88.

20 On the importance of tactile experience in early Christian pilgrimage, see Georgia Frank, "The Pilgrim's Gaze in the Age before Icons," in *Visuality before and beyond the Renaissance: Seeing as Others Saw*, ed. Robert Nelson (Cambridge: Cambridge University Press, 2000), 98–115.

21 Maurice Halbwachs, *La Topographie légendaire des évangiles en terre sainte: Etude de mémoire collective* (Paris: Presses Universitaires de France, 1941); Wilken, *Land Called Holy*, 90.

22 Markus, "How on Earth"; and P.W.L. Walker, *Holy City, Holy Places? Christian Attitudes to Jerusalem and the Holy Land in the Fourth Century*, Oxford Early Christian Studies (Oxford: Clarendon Press, 1990).

23 Halbwachs, *Topographie*; for Jerusalem and the concept of narrative, see Annette Hoffman and Gerhard Wolf, "Preface," in *Jerusalem as Narrative Space: Erzählraum Jerusalem*, eds. Hoffman and Wolf (Leiden: Brill, 2012), ix–xx.

24 Jonathan Z. Smith, *To Take Place: Toward Theory in Ritual* (Chicago: University of Chicago Press, 1992), 86.

25 For a summary of views on medieval concepts of space and place see Cassidy-Welch, "Space and Place", 1–12.

26 Robert G. Ousterhout, "Architecture as Relic and the Construction of Sanctity: The Stones of the Holy Sepulchre," *The Journal of the Society of Architectural Historians* 62 (2003): 4–23.

27 On portability and the cult of relics, see Julia M.H. Smith, "Portable Christianity: Relics in the Medieval West (c.700–1200)," *Proceedings of the British Academy* 181 (2012): 143–167.

28 Bianca Kühnel, "Virtual Pilgrimages to Real Places: The Holy Landscapes," in *Imagining Jerusalem in the Medieval West*, eds. Lucy Donkin and Hanna Vorholt (Oxford: Oxford University Press, 2012), 243–264.
29 Both of these are found, for example, in the twelfth-century "Jerusalem" in Bologna. See Robert G. Ousterhout, "The Church of Santo Stefano: A 'Jerusalem' in Bologna," *Gesta* 20 (1981): 311–321.
30 A contemporary example for this practice may be seen in the stone from the Holy Sepulchre in Jerusalem situated within a copy of the Sepulchre in the Franciscan "Holy Land Park" in Washington, DC: Ousterhout, "Loca Sancta", 117.
31 See recently Kühnel, Noga-Banai, and Vorholt, eds., *Visual Constructs of Jerusalem.*
32 Ann Taves, "Special Things as Building Blocks of Religion," in *The Cambridge Companion to Religious Studies*, ed. Robert Orsi (Cambridge: Cambridge University Press, 2005), 58–83.
33 Robert Orsi, *Between Heaven and Earth: The Religious Worlds People Make and the Scholars Who Study Them* (Princeton: Princeton University Press, 2005), 73–74.
34 Dick Houtman and Birgit Meyer, eds., *Things: Religion and the Question of Materiality* (New York: Fordham University Press, 2012).
35 Hans Belting, *An Anthropology of Images: Picture, Medium, Body* (Princeton: Princeton University Press, 2011); David Freedberg, *The Power of Images: Studies in the History and Theory of Response* (Chicago: Chicago University Press, 1989); Alfred Gell, *Art and Agency: An Anthropological Theory* (Oxford: Oxford University Press, 1998); W.J.T. Mitchell, *Picture Theory* (Chicago: Chicago University Press, 1994) and idem., *What Do Pictures Want: The Lives and Loves of Images* (Chicago: Chicago University Press, 2005).
36 See the introduction to Aden Kumler and Christopher R. Lakey, " 'Res et significatio': The Material Sense of Things in the Middle Ages," *Gesta* 51, no. 1 (2012): 1–13. For a list of recent works on materiality in medieval art, see "Notes from the Field: Materiality," *Art Bulletin* 95, no. 1 (2013): 10–37. See also Thomas Raff, *Die Sprache der Materialien* (Münster: Waxmann Verlag, 2008); Christy Anderson, Anne Dunlop, and Pamela H. Smith, eds., *The Matter of Art: Cultural Logics, 1250–1750* (Manchester: Manchester University Press, 2015); B. Buettner, "From Bones to Stones: Reflections on Jeweled Reliquaries," in *Reliquiare im Mittelalter*, eds. Bruno Reudenbach and Gia Toussaint (Berlin: Akademie Verlag, 2005), 43-59; J.J. Cohen, "Stories in Stone," *Postmedieval: A Journal of Medieval Cultural Studies* 1, nos. 1/2 (2010): 56–63; Beate Fricke, "A Liquid History: Blood and Animation in Late Medieval Art", in *Wet/Dry*, eds. Francesco Pellizi and Christopher Wood, *Res: Journal of Anthropology and Aesthetics* 63/64 (2013): 52–69.
37 Herbert Kessler, *Seeing Medieval Art: Rethinking the Middle Ages* (Peterborough: University of Toronto Press, 2004), 19–44.
38 Caroline Walker Bynum, *Christian Materiality: An Essay in Religion in Late Medieval Europe* (Cambridge, MA: Zone Books, 2011), 28–29.
39 Bynum, *Christian Materiality.*
40 Ibid., 20.
41 Jane Bennet, *Vibrant Matter: A Political Ecology of Things* (Durham, NC: Duke University Press, 2010).
42 Sara Ritchey, *Holy Matter: Changing Perceptions of the Material World in Late Medieval Christianity* (Ithaca, NY: Cornell University Press, 2014).
43 Kessler, *Seeing Medieval Art*, 19–42; Cynthia Hahn, *Strange Beauty: Issues in the Making and Meaning of Reliquaries, 400 – circa 1204* (University Park, PA: Penn State University Press, 2012).
44 Hahn, *Strange Beauty.*
45 An exception is Lucy O'Connor, "The Late Antique Wooden Reliquaries from the Chapel of the Sancta Sanctorum," *Bollettino dei Monumenti Musei e Gallerie Pontificie* XXXI (2013): 201–230.

I

Collecting and collections

1 Earth, stone, water, and oil

Objects of veneration in Holy Land travel narratives

Ora Limor

In his famous book *The Innocents Abroad*, Mark Twain ironically recounts how he and his fellow travelers collected stones known as periapts against barrenness at the Milk Grotto near Bethlehem:

> We went to the Milk Grotto, of course – a cavern where Mary hid herself for a while before the flight into Egypt. Its walls were black before she entered, but in suckling the Child, a drop of her milk fell upon the floor and instantly changed the darkness of the walls to its own snowy hue. We took many little fragments of stone from here, because it is well known in all the East that a barren woman hath need only to touch her lips to one of these and her failing will depart from her. We took many specimens, to the end that we might confer happiness upon certain households that we wot of.[1]

Mark Twain attests to an old and long-held custom of Christian pilgrims, as old as Christian pilgrimage itself, of collecting "specimens," to use his word, that they came across on their way and taking them home. Most of these specimens were of lowly substances, such as geological or botanical fragments, but others were relatively scarce and precious, such as rare scents.[2] Some objects, such as the stones of the Milk Grotto, were story-related. They were collected in places where distinct episodes from the biblical past or the venerated history had occurred and had left their marks on the sites forever. Others carried miraculous traits without a specific narrative. It stands to reason that such objects were also connected to some old myth that was either lost to memory or not committed to writing by the informants, that is, the visiting pilgrims.

Unlike relics of saints or those of the foundation myth of Christianity, such as the Holy Cross or the crown of thorns, the objects discussed here were accessible, found in abundance, and usually less carefully guarded.[3] Yet, like the famous rare relics, their value was not defined by their material worth but rather by the meaning attached to them.[4] Moreover, whereas the rare relics could be and actually were transferred elsewhere, the abundant objects, even if some were taken away, were never diminished and never ceased to exist. By the seventh century, most of the important relics were in Constantinople, which consequently became a magnet for relic chasers of all kinds and the source of Western relics through donation, diplomacy, trade, or theft.[5] At the same time, while Jerusalem was shorn of her precious historical relics, the natural materials remained *in situ* and retained the Holy Land attraction for believers. They could never be removed entirely and it was their locality that gave them power and significance. Although fragments of these relics were constantly being taken away, the Holy Land itself preserved its former state, retaining a truth beyond words, the truth of the sacred revelation that occurred there. The power of the natural elements – earth, stones, water, oil, plants, and scents – was absorbed through the senses. These elements could be touched, eaten, and smelled.[6] As such, they stimulated memory and understanding and were effective transmitters of knowledge.[7] Rather than souvenirs, they were signs, symbols, and replacements for the holy places themselves, retaining in their fragmentary and lowly matter all the virtues and miraculous powers of those places.[8]

Pilgrims were not only the collectors but also the reporters and commentators on these objects. They described the process of their collection and explained their meaning. Writing was one way to preserve the pilgrimage experience and to share it with others; collecting objects was another. Both commemorated the visit and transmitted the essence of the holy places in their didactic, sentimental, and sensual senses.[9] This chapter traces early evidence of object collection as attested to by early Holy Land pilgrims.

Egeria's *eulogiae*

In her letter to her "beloved sisters" describing her pilgrimage to the East and the liturgy of the Jerusalem Church (the accepted date of her stay in the Holy Land is 381–384), Egeria tells of *eulogiae*, "blessings," that is, benedictional gifts that pilgrims received at the hands of monks residing in the holy places.[10] To be sure, exquisite relics did not escape Egeria's attention, as attested to in her famous story about the thief who bit off a piece of the True Cross. Consequently, the cross was closely guarded whenever it was exhibited in public.[11] However, ordinary pilgrims, even people of some means, such as Egeria herself had probably been, had to make do with more mundane objects, which nevertheless had some contact with holiness. The first such instance in Egeria's treatise relates to Mount Sinai. After guiding Egeria and her entourage throughout a detailed tour of the Mount and its surroundings and performing the pilgrimage rites for them, the presbyters of the place gave them some fruit that grew on the Mount as eulogiae, "blessings."

It is clear from Egeria's words that the significance of these eulogiae laid in the venerated site from where they were taken. She notes that the monks showed the pilgrims the huge plain across which they could see Sinai, the Mountain of God. This was the huge plain where the children of Israel camped when Moses went up on the Mount and stayed there for forty days and forty nights (Exod. 24:18). In this very same valley, the pilgrims were shown the place where the Golden Calf was made (Exod. 32). A large stone marked the spot.[12] After crossing the valley, the group arrived at the foot of the Mount itself, where they found several hermit cells and lodged there. At dawn, they started to climb the Mount, an arduous task then as it is today. At ten o'clock, the group arrived at the summit, "where the Law was given, that is, the place where God's glory came down on the day when the mountain was smoking."[13] There was nothing on the summit but "the cave of holy Moses" and a small church. The presbyter, "an ascetic, as they say here," came out to meet the pilgrims and with him all the hermits who lived near the mountain. They performed the liturgical rites for the pilgrims, which included readings of the biblical chapters pertaining to the site and the performance of the offering. When the pilgrims came out of the church, the presbyters of the place gave them "eulogiae, that is, some fruit which grows on the very mountain."[14]

In Egeria's description, a eulogia (*bracha* in Hebrew, *benedictio* in Latin) should be understood as an object, a material gift – fruit in this case – that was given together with an oral blessing that bestowed upon the pilgrims a wish for the favor of God. Egeria stresses the fact that although the Holy Mount was all stony, there was a little soil around its foot where the "holy monks"[15] planted shrubs and set out orchards or vegetable beds and also around their cells, "and it looks as if they gather fruit growing in the mountain soil itself, which however they grow with their own labor."[16] The last sentence is most telling. The fruit's uniqueness stemmed from its dual contact with the holy – holy men and holy soil. It grew in the very soil of the Holy Mount and was nurtured by the holy men who lived there. Whereas the holy ground was the source of the fruits' sanctity, the words uttered by the presbyters gave them the favor of God and turned them into eulogiae. The donation was part of a solemn rite that sealed the visit as an orchestrated drama, a ritual.

Egeria does not mention other holy objects that could be found in Sinai, such as stones from the Holy Mountain or holy earth from the place where Moses stood when he received the Law. Unlike earth and stones, fruit can be eaten. It enters the pilgrim's body and saturates it with holiness. Egeria could perhaps keep some of the fruit and dry it or at least dry its leaves and bring it home to her beloved sisters, as she did with the letter of Abgar that she acquired in Edessa.[17] In any case, her detailed description of the bestowal of the fruit is clearly an indication that a eulogia was much more than a memento.

Egeria mentions eulogiae in three other places: On their way from Jerusalem to Mount Nebo, she and her group visited the spring where Moses took water from the rock and gave it to the children of Israel to quench their thirst (Exod. 17:6; Num. 20:8). The ascetic monks living there received the pilgrims hospitably, held prayers with them, and gave them eulogiae "as they are used to giving to those whom they receive hospitably."[18] Egeria does not specify the nature of these eulogiae. In Aenon near Shalem, where John the Baptist baptized the faithful (John 3:23), the presbyter gave the pilgrims eulogiae from the orchard of Saint John the Baptist, and so did the holy monks who had cells (*monasteria*) in the garden.[19] Finally, six miles from Haran, the pilgrims visited the well where Jacob watered Rachel's flock (Gen. 29: 2–10), a center of Christian life with a large, beautiful church and cells of hermits. After inspecting the huge stone that Jacob rolled off the well's mouth and performing the customary rituals at the church, the pilgrims visited the monks, who welcomed them. After preaching to them, the monks kindly gave eulogiae to Egeria and to all those who were with her, "as is the custom of the monks to give to those they kindly receive in their cells."[20] Egeria does not specify the nature of these blessings.

Apart from these four instances, in Ramses in Egypt, Egeria mentions a sycamore tree that was said to have been planted by the patriarchs. "Though it is now extremely old, and thus small, it still bears fruit, and people who have something wrong with them pick its twigs, which do them good."[21] Finally, Petrus Diaconus, in his *Liber de locis sanctis* (twelfth century) based partly on Egeria's account, mentions the altar made of the stone near the Sea of Galilee, where Jesus placed the bread (probably the bread of the miracle of bread and fish [Mark 6: 33–44]). People took little pieces of stone from it for their health (*pro salute sibi*), and it benefitted many.[22]

It is clear, then, that from the very early days of Christian pilgrimage, benedictory gifts were part and parcel of pilgrimage customs and a trait of monastic hospitality. Monks guarded the holy places, guided the pilgrims, and performed the pilgrimage rituals for them. They sealed the visit with "blessings," both verbal and material. As Caroline Walker Bynum writes, although it was easier for the faithful to obtain a verbal blessing, "people wanted objects . . . not mere words."[23] The combination of words and matter was essential, even indispensable, in forming the eulogiae, movable segments from the immovable sites.

A pilgrim's collection

By the sixth century, collecting eulogiae had become one of the main pilgrimage goals. The anonymous Piacenza Pilgrim, who visited the Holy Land around 570, went home with a sack full of objects he had collected in the holy places.[24] Most of them were small specimens that could easily be carried a long way, others were probably quite heavy, and all were collected for a blessing, "*pro benedictione*," a phrase the pilgrim uses time and again. Some of the objects were closely connected with pilgrimage rites, just as in Egeria's description, and depended on the authorities of the local holy places, presbyters, or monks, who performed the rites and presented the blessings. Some were of special value when given at a certain time within a liturgical setting. All, however, were local, dependent on the place from where they were taken and were understood as its representatives. A list of the blessings, when given in sequence, tells the story of the journey.

As long and impressive as the list of natural objects collected by the Piacenza Pilgrim is,[25] it is unlikely that we can create an exhaustive list of such objects. Variability, mobility, and renewability are distinctive characteristics of the phenomenon, along with the traditional nature imprinted in it. Many objects recur regularly in the sources and are listed in relic collections, whereas others appear and disappear, depending, *inter alia*, on the creativeness and inventiveness of local orally transmitted traditions.

One noticeable distinction between the descriptions of Egeria and that of the Piacenza Pilgrim pertains to the process of collection. Whereas Egeria describes herself as a passive actor who received gifts when offered to her, the Piacenza Pilgrim appears to have been very active in accumulating as many objects as he could. Yet, it would seem that when it came to holy matter, such as earth from Jesus' sepulchre, oil from Golgotha, or water from the Jordan River, he too was dependent on the offerings and benedictions of church agents. Another distinction is the quest for remedial powers attached to many of the objects. Egeria mentions only one such object, that is, the twig from Ramses (although the missing parts of her letter might include others), but the Piacenza Pilgrim seems to have been rather obsessive about medicinal herbs and objects.[26] However, more than anything else, the Piacenza Pilgrim's list of objects reflects the collection impulse of the late antique Christian pilgrim and his enthusiasm for relics of all sorts, and we can assume that he was not unique in that. His list includes objects about which he explicitly notes that he took a sample back with him and others with no such indication; that is only one way of sorting his inventory. Other ways are according to places, narratives (e.g., Old Testament vs. New Testament), or liturgical uses. However, as the present volume revolves around natural materials, it seems reasonable to sort his list according to matter. The entire list, sometimes complemented with other descriptions, forms what can be termed, to paraphrase Julia M. H. Smith's definition, the portable Holy Land.[27]

Earth

In itself a nondescript item, earth from holy places or earth that touched the holy became a customary relic.[28] Abundant, even endless, it could be carried in small amounts, and as it was hard to distinguish from mundane, ordinary earth, it was also immune to confiscation and taxation. On the other hand, and for the same reasons, it was not easy to authenticate, and it had no validity without a label or an artifact that described and defined it.[29]

Earth was either taken from a specific holy place or from a place touched by a holy person, preferably by Jesus himself. In the *City of God*, Augustine writes about the ex-Tribune Hesperius, who received from a friend a sample of sacred earth taken from Jerusalem, "where Christ was buried and rose again on the third day. He had hung it up in his bedroom to ward off any evil from his own person." When the house was purified of the malignant spirit that attacked it, he asked that the sacred earth be buried somewhere and a place of prayer established on the spot. This was done, and the site became a place for prayer and miraculous cures.[30]

The Piacenza Pilgrim also brought back earth as well as oil from the holy tomb, both "as a blessing":

> In the place where the Lord's body was laid, there is a bronze lamp that back then was positioned at his head. It burns there day and night. We took a blessing from it (*ex qua benedictionem tulimus*), and put it back. Earth is brought to the Tomb from the outside, and those who go in take a blessing from it (*benedictionem tollent*).[31]

The Piacenza Pilgrim is careful to go into detail as to how the blessings were acquired. The earth was not integral to the tomb but was brought in from outside, probably from a nearby place, and thus it was close, perhaps even identical in its geographical and geological features to the original earth in which Jesus was interred. Through physical contact with the tomb, the external earth became one with it, and thus a true *eulogia*.[32] Gregory of Tours, in 587–588, recounts the following:

> Marvelous power appears from the tomb where the Lord's body lay. Often the ground is covered with a natural radiant brightness; then it is sprinkled with water and dug up, and from it tiny [clay] tokens are shaped and sent to different parts of the world. Often ill people acquire cures by means of these tokens. The fact that these tokens frequently deflect the approach of snakes is remarkable. But what do I rashly dare to say about them, since faith believes that everything that the sacred body touched is holy?[33]

According to Gregory, the earth was not collected by the devotees themselves, but rather by church functionaries who sprinkled water on it, collected it, and shaped tokens out of it to be sent far and wide. One can surmise that their involvement also included a verbal blessing, inseparable from the token or from any *eulogia*, as can be deduced from its name.

Oddly enough, the Piacenza Pilgrim fails to mention the earth that believers collected from the place of Jesus' ascension on the Mount of Olives, which famously retains his footprints. This failure may attest to the eclectic nature of any pilgrim's collection, as keen on blessings and virtues as he may have been. The ascension earth is described in detail by Sulpicius Severus early in the fifth century:

> Moreover, it is an enduring proof of the soil of that place having been trodden by God that the footprints are still to be seen; and although the faith of those who daily flock to that place leads them to vie with each other in seeking to carry away what had been trodden by the feet of the Lord, yet the sand of the place suffers no injury; and the earth still preserves the same appearance which it presented of old, as if it had been sealed by the footprints impressed upon it.[34]

A similar story is told by Adomnán, the learned abbot of Iona in the seventh century, whose information relied, among other sources, on the report of the pilgrim Arculf. The place of Jesus' footprints, writes Adomnán, could not be incorporated into a pavement, for the ground rejected whatever was laid upon it. This is why the center of the Ascension Church lay open to the sky, without roof or vaulting. Relying on Sulpicius Severus but also adding new information, Adomnán continues:

> Nay more, so lasting is the proof that the dust was trodden by God, that the imprints of the feet are visible; and, though crowds of the faithful daily plunder the earth trodden by the Lord, still the spot suffers no perceptible damage, and the ground goes on keeping the semblance, as it were, of footprints. . . . Also, at the western side of the structure, there is a sort of door always open, so that people entering by it can easily approach the place of the sacred dust, and take particles of it by stretching their hands through the open perforation in the circular structure.[35]

The earth then had a will of its own. It preferred to stay just as it was created, without any intervention by a human hand, artist or craftsman, which was the reason that pilgrims and visitors to the place were so keen to acquire it. Moreover, although the sacred earth was well protected by a circular structure around it, it was still accessible, but in an organized, structured way. The collecting of holy earth was an orchestrated action that included physical activity (kneeling, stretching, hands extended) and probably, although not explicitly written, also a verbal blessing.

Stones

Several remarkable stones could be seen and touched on Mount Zion.[36] In the basilica, the Piacenza Pilgrim saw the cornerstone that was rejected by the builders (Ps.118:22; Isa. 28:16; Matt. 21:42). The Bordeaux Pilgrim, more than 200 years earlier, had seen the cornerstone on the Temple Mount, the place where it was when Jesus related to it in his sermon.[37] The stone was related to Isaiah 28:16: "So this is what the Sovereign Lord says: 'See, I lay a stone in Zion, a tested stone, a precious cornerstone for a sure foundation; the one who relies on it will never be stricken with panic.' " In Christian tradition, the name Mount Zion, historically associated with the Temple Mount, was transferred to the Western Hill, which may explain the migration of several Temple traditions to the new site. The Piacenza Pilgrim describes the stone in some detail, probably transmitting what he heard (and perhaps misunderstood) from a local guide:

> When the Lord Jesus entered that church, which used to be the house of Saint James, he found this ugly stone lying in the middle, so he took it and placed it in the corner. When you hold it in your hands and lift it, and then put your ear in that corner, the sound you hear is like the murmuring of many people.[38]

Notably, the Piacenza Pilgrim ignores the stone's allegorical import. Indeed, as a rule allegorical interpretations are rare in his text. On the other hand, it is clear that he touched the stone and listened to the corner where it lay, although he does not mention taking any pieces from it.

At the Church of Mount Zion, the Piacenza Pilgrim also mentions the many stones with which Saint Stephen was stoned. Other sources offer variants of this tradition. Although the *Breviarius de Hierosolyma* (Short Description of Jerusalem), dated around 530, mentions only one stone and locates it in the *sacrarium* of the basilica, Adomnán in the late seventh century indicates "the rock on which Stephen died by stoning outside the city," rather than the stone by which he was killed.[39] The tradition is multifaceted, attesting either to modifications that occurred over the years, to different understandings by different informants, or to changes that can occur in the transition from the oral story to the written text. Saint Stephen's stone (or stones) was the means of his torture and death, the object that made him a martyr and thus was somewhat similar in function to Jesus' cross. Like many other objects, it was a contact relic.[40]

In the same basilica, the Piacenza Pilgrim also saw two columns. The first was the flagellation column, one of Jerusalem's most celebrated relics, mentioned in most descriptions of the Holy Land.[41] The column very clearly retained the marks of Jesus' hands, fingers, and chest as he held it while being whipped, so people took "measures" (*mensura*) from it for any kind of disease and, wearing them around their necks, were cured. These, then, were protective objects, amulets of sorts. According to Gregory of Tours, many believers came to the column. They tied cords around it and then got them back as a blessing (*pro benedictione*) that would help against various illnesses.[42] The words "they got them back" again imply the involvement of the church authorities who conferred the blessings on the cords. The other column mentioned by the Piacenza Pilgrim is a small column (*columnella*) that was the setting for the cross on which Peter was crucified in Rome. This column is not mentioned in other Holy Land descriptions, and its presence on Mount Zion is quite bizarre.

The Piacenza Pilgrim also tells about a marble column that he saw on the highway on the way to Diospolis. When Jesus was taken toward it to be scourged, it was lifted up by a cloud and moved to that place. It had no base but stood directly on the ground. People took lights and incense up to it, and individuals possessed by demons were cured there on behalf of Saint George.[43]

"Measures" (*mensura*) were also taken from the stone on which Jesus stood while being sentenced by Pilate. The stone was at the former Praetorium of Pilate, close to the Temple Mount, transformed to the Basilica of Saint Sophia. Jesus' footprints could still be seen on the stone. "From this stone where he stood," writes the Piacenza Pilgrim, "come many blessings (*virtutes*). People take 'measures' (*mensura*) from the footprints, and wear them for various diseases, and are cured. This stone is decorated with gold and silver."[44]

Although failing to mention some stones, such as the ascension stone on the Mount of Olives, the Piacenza Pilgrim does offer information on other powerful stones in the Holy Land. He tells about a small round rock on Mount Carmel, which although solid, made a noise when shaken. A woman or an animal to whom it was attached would never miscarry.[45] He does not say whether he took a piece of it, but he does note that he took a blessing from the "bed" of Cornelius at Caesarea (probably meaning his tomb).[46]

Stones were also good for expressing contempt. The Piacenza Pilgrim tells of a gesture made by all those who passed by the tomb of Goliath on the road to Gaza: throwing three stones at it. Therefore "there is not a pebble left for a distance of twenty miles."[47] According to a tradition recorded by Petrus Diaconus, possibly based on Egeria's treatise, a similar gesture was performed at the tomb of Jezebel in Jezreel.[48]

This list of stones, either in their natural raw form or shaped as columns or tombs, is long and diverse. Many of them acted against their nature and exhibited a remarkable liveliness. There were stones that could speak, like the murmuring stone on Mount Zion or the small rock on Mount Carmel that made a noise when shaken; there were stones that become soft when touched by the holy, such as the stones bearing Jesus' footsteps on the Mount of Olives and in the Praetorium or the scourging column on Mount Zion. All of these stones revealed and proved the constant immanence of the divine presence in this world, as Glenn Peers writes: "This mode of understanding geology is related to perceiving nature as in a constant, contingent states of divinization: God works himself out in the world, he is immanent in all of it, and he reveals that presence in answering stones."[49] In Peers's view, material objects of this sort were not "objects of devotion," because that definition would imply that they were only passive receivers, whereas in late antique understanding, they were animated, they could act and answer, and they were active in making the world "full of God."[50]

Oil

The unique natural and cultural attributes of oil, its association with light and warmth, and its traditional use for anointment and chrism made it one of the most significant materials used as blessings.[51] Pilgrims to Jerusalem brought home olive oil, known for its quality and durability, from olive trees growing around the city. Oil from the Anastasis Church is probably the best-known Holy Land *eulogia*, owing primarily to the ampoules that carried it, many of which are preserved to this day in various places in Italy and elsewhere.[52] According to the Piacenza Pilgrim, it was given to the pilgrims in two places – Jesus' tomb and Golgotha. As mentioned above, at the *Monumentum*, that is, Jesus' tomb, pilgrims were given oil from the very lamp that was put at his head when his body was placed there. "We took a blessing from it (*ex qua benedictionem tulimus*)," writes the Piacenza Pilgrim, "and put it back."[53]

The description pertaining to Golgotha's oil is more detailed:

> In the courtyard of the basilica is a small room where they keep the Wood of the Cross. We venerated it with a kiss. . . . At the moment when the Cross is brought out of the small room for veneration . . . a star appears in the sky, and comes over the place where they lay the Cross. It stays overhead whilst they are venerating the Cross, and they offer oil for blessing in little flasks (*et offertur oleum ad benedictionem, ampullas medias*). When the mouth of one of the little flasks touches

> the Wood of the Cross, the oil instantly bubbles over, and unless it is closed very quickly it all spills out.[54]

Ampoules with oil sanctified by the True Cross were carried home as amulets. The obverse of the Dumbarton Oaks tin-lead ampoule (Washington, DC: Dumbarton Oaks, 48.18) bears the inscription: "Oil of the Wood [Tree] of Life of the Holy Places of Christ," an inscription that evokes the contemporary passage from the Piacenza Pilgrim cited above.[55] The Golgotha oil was offered to the pilgrims for blessing in small flasks during the veneration of the True Cross on Easter Sunday as part of the liturgies of the day. Pilgrims such as Egeria and the Piacenza Pilgrim did their best to follow the events of Christian venerated history not only geographically but also chronologically. Both were at the Anastasis Church during Holy Week. The Piacenza Pilgrim was also at the baptism place at the Jordan River on Epiphany. His description of the presentation of the True Cross on Easter Sunday complements that of Egeria, who tells how the cross was venerated by all those attending, but was also well guarded from them.[56] As the True Cross oil was given to pilgrims on Easter Sunday, the day of the resurrection, it stands to reason that the oil from the Holy Sepulchre lamp was given to them on Friday, the day of the crucifixion and burial. Place and time together created these oils' special virtues.

The Piacenza Pilgrim also mentions other oils, which although not connected to a specific biblical story, were still remarkable for their wondrous quality, mainly their healing powers. The Christian women at Paran came to meet the pilgrims and anointed their heads and soles with radish oil, known for its remedial powers.[57] The pilgrim also took "for a great blessing" (*pro grande benedictione*) rock oil that dripped from a natural rock on an island near Clysma, which smelled of sulfur and was effective against all diseases and especially efficacious for those possessed by demons.[58]

Water

Similar to the stone "measures," water is also a strong protector, but in much more diverse ways. One can drink it for a protective blessing, wash in it, wash clothes in it, sprinkle it, or take it home as a *eulogia*. Pilgrimage accounts provide evidence of all of these uses.

Following the liturgical cycle, on the day of the Epiphany the Piacenza Pilgrim was on the bank of the Jordan River. He describes the event thus:

> The moment he [the priest] starts blessing the water the Jordan turns back on itself with a roar and the water stays still till the baptism is finished. All the shipowners of Alexandria have men there that day with great jars of spices and balsam, and as soon as the river has been blessed, before the baptism starts, they pour them out into the water, and draw out blessed water (*aquam benedictam*). This water they use for sprinkling their ships when they are about to set sail. After the baptism everyone goes down into the river to gain a blessing (*pro benedictione*). Some wear linen, and many other materials which will serve as their shrouds for burial. And after the baptism the water returns to its place.[59]

We may assume that pilgrims drew water from the Jordan on other days as well, but it had special value and holiness on the day of Jesus' baptism. The Piacenza Pilgrim describes a ritual during which the water was first blessed and then scented with precious spices and balsam. The water of the Jordan, just like the Golgotha oil, was an active participant in the liturgy. Whereas the Golgotha oil bubbled when it was blessed and touched the cross, the water of the Jordan, when blessed, turned back on itself with a roar and stayed still till the baptism ended. Further, the oil performed its miracle on the Day of Resurrection, and the special virtue of the water was revealed on Epiphany. Much earlier than the Piacenza Pilgrim, John Chrysostom (ca. 349–407), in his homily on baptism, tells of a custom to obtain water from the Jordan River on the day of Epiphany and of the virtue of this water:

> Why is not that day, on which the Lord was born, considered Theophany – but rather this day on which He was baptized? This present day it is on which He was baptized and sanctified the nature of water. Because on this day all, having obtained the waters, do carry it home and keep it all year, since today the waters are sanctified. And an obvious phenomenon occurs: these waters in their essence do not spoil with the passage of time, but obtained today, for one whole year and often for two or three years, they remain unharmed and fresh, and afterwards for a long time do not stop being water, just as that obtained from the fountains.[60]

The Jordan water conferred a blessing on both land and sea and both in this world and the next.[61] Although not explicitly stated, we can surmise that the Piacenza Pilgrim did not fail to take home a flask with holy water from the Jordan and perhaps also from the Spring of Elisha near Jericho.[62] Earlier, he probably acquired a flask of water from the water-pots of the wedding at Cana. At the spring in Cana, he washed so as to gain a blessing (*pro benedictione*), as he specifies.[63] We can assume that he also gained a blessing by washing in the famous hot springs at Gadara, where lepers were cleansed,[64] as well as at the pool of Siloam, which had two marble basins separated by a screen: "Men wash in one and women in the other, to gain a blessing."[65]

In Jerusalem, pilgrims drank water from the crucifixion sponge kept at the Holy Sepulchre and perhaps also from the onyx cup of the Last Supper.[66] As many other pilgrims, the Piacenza Pilgrim also drank water "for blessing" from the skull of the martyr Saint Theodota. The skull was kept in a golden chest adorned with gems in a women's monastery at Mount Zion.[67]

Notable also is the curative dew coming down from a cloud that ascends from the Jordan Valley in the morning and arrives over Jerusalem at sunrise, dropping down like a shower on its holy places. Sick people collected it and all the food in the city's hospices was cooked in it. Many diseases were cured in places where it fell. The Piacenza Pilgrim explains that this was the dew of which the psalmist sang: "Like as the dew of Hermon, which fell upon the hill of Zion" (Ps. 133:3).[68] Wondrous dew also came down from the sky over a valley in Sinai. Called manna, it solidified into something like lumps of gum. The monks at the monastery had casks full of it and they gave little flasks of it to pilgrims as blessings. They also gave it to them to drink as a liqueur.[69]

After he left Jerusalem and drew near to Rachel's tomb on the way to Bethlehem, the Piacenza Pilgrim saw a spring of sweet water coming from a rock, up to seven pints. "Everyone has his fill, and the water does not become less or more." People say that the water began to flow on behalf of Mary, who on her flight to Egypt became thirsty and stopped there.[70] Apparently, the Piacenza Pilgrim also had his fill. Another site of sacred water near Bethlehem that the Piacenza Pilgrim fails to mention is noted about 100 years later by Adomnán, based on the account of Arculf. Adomnán tells about a rock outside the city wall, over which the first washing water of baby Jesus was emptied. When poured from the wall, the water of "the sacred washing" found a channel in the rock beneath and since then, without fail, it is full of the purest water. "Our friend Arculf saw it with his own eyes and washed his face in it."[71]

Botanical specimens

As noted by Egeria, fruits and twigs were favorites of the early pilgrims. Like the other objects discussed here, some were unique, canonical objects, such as the rod (*virga*) mentioned in the *Breviarius de Hierosolyma* as one of the objects kept in the *sacrarium* of the basilica on Mount Zion.[72] The rod, probably the rod of Aaron, held a rich typological interpretation. It symbolized Jesus as well as Mary, and it was understood that the verse "and, behold, the rod of Aaron for the house of Levi was budded, and put forth buds, and bloomed blossoms, and bore ripe almonds" (Num. 17:8) related to her. Other fruits and twigs were free of such typological weight but still carried holiness, such as the fruit that Egeria received from the holy monks.

Pilgrims were most impressed by the size and beauty of the fruit that grew near Jericho. According to the Piacenza Pilgrim, this lushness was due to the spring that Elisha made sweet (2 Kings 2:19–24), which supplied Jericho with its water and irrigated its fields and orchards. He found dates there that weighed a pound. He took some home and gave one to a local nobleman named Paterius. He also saw grapes good for soothing fever, citrons that weighed forty pounds, and a vine that yielded full baskets of grapes. These grapes were on sale at the time of the ascension and the wine made of them was sold in great jars at Pentecost.[73]

Although the fruits the Piacenza Pilgrim tasted in Sinai and in Egypt were not connected to any biblical story, he still found them worthy of documentation, as part of the exotic and wondrous things he encountered on his journey. He tells of roots (*radices*) that the Saracens gave the pilgrims in Sinai, which smelled sweeter than any spice. He was astonished by their taste and also by the fact that the Saracens would take nothing for them because it was the festival season.[74] In Surandela, he saw a pepper tree (*arborem piperis*) and picked some fruit from it. In Clysma, he was given bright green nuts from India: "People think they are from paradise." One has only to taste them to feel satisfied.[75]

The list of the Piacenza Pilgrim's botanical specimens can be supplemented by Adomnán, who writes about the oak of Mamreh, called also the oak of Abraham "because once upon a time he entertained angels under it." Arculf saw the oak "with his own eyes." In his time, it was only "a truncated spur (*spurium*) rooted in the earth" and protected by the roof of the church. Adomnán elaborates on the custom of taking small pieces of it and explains its logic:

> Now this cropped spur is hewn about on every side by axes, little splinters being carried away to the divers provinces of the world, out of veneration and remembrance for the oak, under which, as was mentioned above, the famous and noteworthy meeting with the angels was once vouchsafed to Abraham the patriarch.[76]

Balsam

The Holy Land was transferred far and wide through matter, taste, and smell. Of special value was balsam, which carried an exquisite fragrance and also had typological meaning. Jerome writes that upon seeing the balsam vines of En Gedi and Zoar, Paula meditated on the verse, "A cluster of cypress my love is to me, in the vineyards of Engaddi" (Cant. 1:14), a verse understood to refer to Jesus.[77] The Piacenza Pilgrim, in his description of Epiphany recounted above, writes that on that day, as soon as the river was blessed and before the baptism started, people poured spices and balsam into the water and then drew out holy scented water.[78]

Saint Willibald, who toured the Orient in the 720s, managed, through a clever stratagem, to smuggle valuable balsam out of the Holy Land, a feat that he remembered vividly and in great detail for many years.[79] He hid the balsam in a flagon, concealed beneath a can of petrol-oil. In his biography (written more than 50 years after the pilgrimage itself by the nun Hygeburc), the balsam adventure is framed as a heroic story that combines the protagonist's resourcefulness with the unfailing help of God to his believer, the devout pilgrim. According to the text, had the Moslem authorities in Tyre found the concealed balsam, he would have suffered a martyr's death.[80]

Balsam was the most celebrated product of Judea during the Roman and Early Byzantine periods and the most valuable one.[81] Exquisite, exotic, and important for liturgical use, it intensified the sensory contact with the divine through the sense of smell.[82] Willibald was even ready to risk his life for its sake.

The power of natural objects

Pilgrims such as Egeria, the Piacenza Pilgrim, and Willibald transmitted knowledge about the Holy Land in various ways. They told stories about its marvels, counted its virtues, and elaborated on its sacred traditions and the liturgies performed there; they also brought home objects that could be accessed through the senses – touch, sight, hearing, taste, and smell. These objects were varied and came from all corners of the land. Their quantity, variety, and the breadth of the geographical space from where they were taken implied that wherever one tread, the country as a whole was blessed.

"In Christian tradition, a relic was material stuff that was not (just) what it was," writes Julia M. H. Smith.[83] To turn this material stuff from a mundane, indistinct fragment into an object with meaning needed words. Pilgrims' descriptions supplied not just words. They supplied narratives, shedding light on the process by which collections of objects came into being. They told how the objects were taken from their places of origin, were brought to new places, and became relics. Pilgrims' evidence was even more powerful than the labels attached to medieval relics, as they themselves had brought the objects back and guaranteed their authenticity. On the other hand, once the objects came to new places, they were detached from the pilgrimage accounts, thus losing their topographical link, and became subject to the risk of anonymity, neglect, and oblivion.[84]

The dependence of matter on words is salient in the *Breviarius de Hierosolyma* mentioned above. It elaborates on the holy relics that were held and displayed in the Holy Sepulchre complex and the Mount Zion Basilica, several of them in a special treasury room, the *sacrarium*. Visitors to the basilica could see the flagellation column, the stone with which Saint Stephen was stoned, and the crown of thorns. They could also see the lamp in the place where Jesus taught his disciples and the rod enclosed in a silver column.[85] Of these, the stone and the rod are of special interest in the present context, as both are mundane, natural objects that acquire meaning only through their location within the church and the words attached to them, orally or in writing by church interpreters. These words were the key to "creative viewing," to use Jaś Elsner's words, that is, the key to a true understanding of the biblical truth and of the meaning of sacred history.[86] Old Testament typology (the rod), New Testament revelation (the cornerstone), and the first steps of the Christian Church (Saint Stephen's stone) all became accessible through fragmentary, incomplete natural objects that became not only metaphors but synecdoches for places as well as for events.[87]

This mode of seeing and understanding was built upon the education and the expectations of the pilgrims themselves. Judging from the Piacenza Pilgrim's account, it would seem that not all viewers had the capacity for such hermeneutical, contemplative viewing, and not all were able to look at one thing and understand another. But for such pilgrims, too, objects could do what texts could not.[88] Like the holy places from where they were taken and the liturgy performed there, natural objects bearing a blessing intensified and prolonged the pilgrimage experience. Seeing them, touching them, hearing stories about their virtues, and especially collecting them were powerful motives for traveling and made enduring the hardship of travel worthwhile.

Notes

* This research was supported by the I-CORE Program (The Israel Science Foundation), Center for the Study of Conversion and Inter-Religious Encounters (No. 1754/12). I am grateful to Julia M. H.Smith for reading the article and for her important suggestions and to Moshe Yagur for his help.

1 Mark Twain, *The Innocents Abroad*, ch. 55 (New York: New American Library, 1966), 437. There are several versions of this tradition. See, for example, *Adamnan's De Locis Sanctis*, 2, 3, ed. Denis Meehan, Scriptores Latini Hiberniae, 3 (Dublin: The Dublin Institute for Advanced Studies, 1958), 74–77; Adamnanus, *De locis sanctis libri tres*, 2, 3, ed. L. Bieler, Corpus Christianorum Series Latina (CCSL), 175 (Turnhout: Brepols, 1965), 207.

2 Gary Vikan, *Byzantine Pilgrimage Art* (Washington, DC: Dumbarton Oaks, 1982), 42–43; Holger A. Klein, "Eastern Objects and Western Desires: Relics and Reliquaries between Byzantium and the West," *Dumbarton Oaks Papers* 58 (2004): 283–314; Cynthia Hahn, "Loca Sancta Souvenirs: Sealing the Pilgrim's Experience," in *The Blessings of Pilgrimage*, ed. Robert Ousterhout (Urbana and Chicago: University of Illinois Press, 1990), 85–96. See also Franco Cardini, "Reliquie e Pellegrinaggi," in *Santi e Demoni nell'Alto Medioevo Occidentale (secoli V – XI)*, Settimane di studio del Centro Italiano di Studi sull-Alto Medioevo, 36 (Spoleto, 1989): 981–1041.

3 Cynthia Hahn calls them "secondary relics." See Hahn, "Loca Sancta Souvenirs," 85.

4 Klein, "Eastern Objects," 283–284.

5 Ibid., 299; 313.

6 Glenn Peers, "Object Relations: Theorizing the Late Antique Viewer," in *The Oxford Handbook of Late Antiquity*, ed. Scott Fitzgerald Johnson (Oxford: Oxford University Press, 2012), 970–993 (esp. 984).

7 Averil Cameron, "The Language of Images: The Rise of Icons and Christian Representation," in *The Church and the Arts*, ed. Diana Wood, Studies in Church History 28 (Woodbridge, Suffolk: Boydell & Brewer, 1995), 1–42 (at 16).

8 Hahn, "Loca Sancta Souvenirs," 92.

9 See Vikan, *Byzantine Pilgrimage Art*, 13; For memory, see Georgia Frank, "'Loca Sancta' Souvenirs and the Art of Memory," in *Pèlerinages et lieux saints dans l'antiquité et le moyen âge: mélanges offerts à Pierre Maraval*, eds. Béatrice Caseau, Jean-Claude Cheynet, and Vincent Déroche (Paris: Travaux et Mémoires. Association des amis du Centre d'histoire et

civilisation de Byzance, 2006), 193–201. Frank bases her conclusions on Mary Carruthers, *The Craft of Thought: Meditation, Rhetoric, and the Making of Images, 400–1200* (Cambridge: Cambridge University Press, 1998), 7–21.

10 A. Stuiber, "Eulogia," *Reallexikon für Antike und Christentum* 6 (1966): 900–927; B. Kötting, *Peregrinatio religiosa* (Münster: Regensberg, 1950), 403–413; Vikan, *Byzantine Pilgrimage Art*, 10.

11 *Itinerarium Egeriae*, 37.1–3, eds. A. Franceschini and R. Weber, CCSL, 175, 80–81; English Translation: John Wilkinson, *Egeria's Travels* (London: S.P.C.K, 1971), 136–137. See also Theodosius, *De situ Terrae Sanctae*, 31, ed. P. Geyer, CCSL, 175, 124.

12 *Itinerarium Egeriae*, 2.

13 Exod. 19:18; *Itinerarium Egeriae*, 3.2–3; English translation: Wilkinson, *Egeria's Travels*, 94.

14 *Itinerarium Egeriae*, 3.6.

15 " 'Sancti monachi'," Egeria's frequent label for the monks she encountered en route.

16 *Itinerarium Egeriae*, 3. 6: "quasi ex ipsius montis terra aliquos fructus capiant, quos tamen manibus suis elaborasse videantur."

17 Ibid., 17–19.

18 Ibid., 11.1.

19 Ibid., 15, 6.

20 Ibid., 21.3.

21 Ibid., 8.3; English translations: Wilkinson, *Egeria's Travels*, 102.

22 Petrus Diaconus, *Liber de locis sanctis*, V.3, ed. R. Weber, CCSL, 175, 99.

23 Caroline Walker Bynum, *Christian Materiality: An Essay on Religion in Late Medieval Europe* (New York: Zone Books, 2011), 128.

24 *Itinerarium Antonini Placentini: Un Viaggio in Terra Santa del 560–570 d.c.*, ed. Celestina Milani (Milan: Vita e pensiero, 1977); *Antonini Placentini Itinerarium*, ed. P. Geyer, CCSL, 175, 127–174, English translation: John Wilkinson, *Jerusalem Pilgrims before the Crusades* (Warminster: Aris and Phillips, 2002), 129–151 (Hereafter: Piacenza Pilgrim).

25 See below.

26 Among other things, he mentions an antidote against poisoning (*tiriaca*) made from serpents that were found on both banks of the Jordan River below the mountains: Piacenza Pilgrim, 12.

27 Julia M.H. Smith, "Portable Christianity: Relics in the Medieval West (c.700–1200)," *Proceedings of the British Academy* 181 (2012): 143–167 (esp. 158).

28 Vikan writes that the word "eulogia" (in the pilgrim sense) appears nearly a dozen times in the long, contemporary life of Symeon the Younger and in each instance it refers to a substance rather than to an action of "blessing." Most often, Saint Symeon's eulogiae came in the form of reddish earth or "dust" from near the base of the column – it is the "eulogia of [his] dust": Vikan, *Byzantine Pilgrimage Art*, 28.

29 Smith, "Portable Christianity," 151.

30 Augustine, *De civitate dei*, 22, 8. English translation: *City of God*, trans. Henry Bettenson (Harmondsworth: Penguin Books, 1972), 1039; see Blake Leyerle, "Pilgrim 'Eulogiae' and Domestic Rituals," *Archiv für Religionsgeschichte* 10 (2008): 223–237.

31 Piacenza Pilgrim, 18.

32 To this day, pilgrims attach personal objects such as jewelry to sacred sites such as Jesus' tomb, the tomb of Mary, the ascension stone, and the stone of unction in the belief that contact with the sacred channels sacredness to these objects and through them to their owners.

33 Gregory of Tours, *De Gloria Martyrum*, 6, ed. Bruno Krusch, MGH SRM 1, 2 (Hannover, 1885), 42; English translation: Raymond Van Dam, *Glory of the Martyrs*, 6 (Liverpool: Liverpool University Press, 1988), 27; see also ch. 5.

34 Sulpicius Severus, *Chronicon*, 2, 33, Sources Chrétiennes, 441, ed. Ghislaine de Senneville-Grave (Paris: Cerf, 1999), 300–303 (Migne, Patrologia Latina, 20, 148); See Paulinus Nolanus, *Epistulae*, 31, 4, ed. Guilelmus de Hartel, Corpus Scriptorum Ecclesiasticorum Latinorum (CSEL), 29 (Vienna: Tempsky, 1894), 271 (Migne, Patrologia Latina, 61, 327–328).

35 *Adamnan's De Locis Sanctis*, 1, 23; English translation by Meehan, 64–67.

36 Piacenza Pilgrim, 22.

37 *Itinerarium Burdigalense*, eds. P. Geyer and O. Cuntz, CCSL, 175, 15.

38 Piacenza Pilgrim, 22; English translation: Wilkinson, *Jerusalem Pilgrims*, 140 (with changes).

39 *Breviarius de Hierosolyma*, ed. R. Weber, CCSL, 175, 111; *Adamnan's De Locis Sanctis*, 1, 18.

40 Among the relics that Richard of Saint-Vannes brought back when returning from his Holy Land pilgrimage (1026–1027) was a stone thrown at him by the Saracens. See Hugo De Flavigny, *Chronicon*, ed. Georg Heinrich Pertz, MGH SS 8 (Hannover, 1848), 396 (Migne, Patrologia Latina, 154, 248); Hubert Dauphin, *Le bienheureux Richard, abbé de Saint-Vanne de Verdun*, Bibliothèque de la Revue d'histoire ecclésiastique, 24 (Louvain and Paris: Bureaux de la Revue, 1946), 293.

41 For example: *Itinerarium Burdigalense*, 16; *Itinerarium Egeriae*, 37, 1; *Breviarius de Hierosolyma*, 111; Theodosius, *De situ Terrae Sanctae*, 7. Adomnán does not mention the column but "the memorable rock on which according to the story the Lord was scourged," located outside the basilica, on the western side (*De locis sanctis*, 1, 18).

42 Gregory of Tours, *De Gloria Martyrum*, 6, MGH SRM 1, 2, 42.

43 Piacenza Pilgrim, 25.

44 Ibid., 23; English translation: Wilkinson, *Jerusalem Pilgrims*, 141 (with changes).

45 Ibid., 3.

46 Ibid., 46: "ex cuius lectu benedictionem tulimus."

47 Ibid., 31; English translation: Wilkinson, *Jerusalem Pilgrims*, 143.

48 Petrus Diaconus, *Liber de locis sanctis*, V.6.

49 Peers, "Object Relations," 987.

50 Ibid., 989.

51 Luigi A. Canetti, "'Olea sanctorum': reliquie e miracoli fra tardoantico e alto medioevo," in *Olio e vino nell'alto medioevo*, Settimane di Studio del Centro Italiano di Studi sull'alto medioevo, 54 (Spoleto, 2006): 1335–1415. I am grateful to Julia M. H. Smith for this reference.

52 The literature about the ampoules is vast. For the present purposes see, principally, Hahn, "Loca Sancta Souvenirs"; Frank, "*Loca Sancta* Souvenirs"; André Grabar, *Ampoules de Terre Sainte (Monza, Bobbio)* (Paris: C. Klincksieck, 1958).

53 Piacenza Pilgrim, 18.

54 Ibid., 18, 20; English translation: Wilkinson, *Jerusalem Pilgrims*, 139; See *Adamnan's De Locis Sanctis*, 3, 3 on the oil of the True Cross in Constantinople. See also ibid., 3, 5.

55 Vikan, *Byzantine Pilgrimage Art*, 22.

56 *Itinerarium Egeriae*, 37, 1–3.

57 Piancenza Pilgrim, 40.

58 Piacenza Pilgrim, 42.

59 Ibid., 11; English translation: Wilkinson, *Jerusalem Pilgrims*, 136.

60 Ioannes Chrysostomos, *De baptismo Christi*, Patrologia Graeca, 49, 365–366; English translation: Saint John Chrysostom, Archbishop of Constantinople, *Discourse on the Day of the Baptism of Christ*, Orthodox Church of America, Digital facsimile: http://oca.org/fs/sermons/discourse-on-the-day-of-the-baptism-of-christ.

61 Leyerle, "Pilgrim Eulogiae," 228–229.

62 Piacenza Pilgrim, 11. See below.

63 Ibid., 4.

64 Ibid., 7. One of his friends, Ioannes of Placentia, died there. The pilgrim elaborates about the procedure at the healing baths: The tank in front of the basin was filled in the evening, and the lepers were sent in with lights and incense and sat there all night. Those who saw a vision were cured and cleansed within a week.

65 Ibid., 24. English translation: Wilkinson, *Jerusalem Pilgrims*, 141.

66 Ibid., 20.

67 Ibid., 22.

68 Ibid., 9.

69 Ibid., 39.

70 Ibid., 28; English translation: Wilkinson, *Jerusalem Pilgrims*, 142. This is probably the site of the Kathisma Church.

71 *Adamnan's De locis sanctis*, 2, 3. English translation by Meehan, 74–77.

72 *Breviarius de Hierosolyma*, 111. See Julia M.H. Smith, "Eleventh-Century Relic Collections and the Holy Land" (in this volume). The *Breviarius* does not specify which rod he means. Although it is most probably the rod of Aaron, it could well be the rod of Moses. See Klein, "Eastern Objects," 307–308.

73 Piacenza Pilgrim, 14. On the abundance of the Lord's Field watered by the Spring of Elisha, see also Theodosius, *De situ Terrae Sanctae*, 20.
74 Piacenza Pilgrim, 36.
75 Ibid., 41.
76 *Adamnan's De Locis Sanctis*, 2, **11**. English translation by Meehan, 82–83.
77 "contemplata est balsami vineas in Engaddi et Segor": Jerome, *Epistula* 108, 11, ed. Isidorus Hilberg, CSEL, 55 (Vienna: Tempsky, 1912), 320; English translation: Wilkinson, *Jerusalem Pilgrims*, 86. A list of gifts given by the Byzantine emperor Alexios I to Henry IV included some *opobalsamum* among many other relics and exquisite objects. See Klein, "Eastern Objects," 291.
78 Piacenza Pilgrim, 11.
79 *Vitae Willibaldi et Wynnebaldi auctore sanctimoniali Heidenheimensis*, ed. O. Holder-Egger, MGH SS 15,1 (Hannover: Hahnsche Buchhandlung, 1887), 101; Regarding the recording of Willibald's pilgrimage experiences, see Ora Limor, "Pilgrims and Authors: Adomnán's 'De locis sanctis' and Hugeburc's 'Hodoeporicon Sancti Willibaldi'," *Revue Bénédictine* 114 (2004): 253–275.
80 "et si aliquid invenissent, cito illos punientes martyrizarent."
81 On its reputation in the Early Middle Ages, see the ninth-century text, Riccardo Quadri, ed., *Anonymi Leidensis de Situ Orbis Libri Duo* (Padova: Editrice Antenore, 1974), 74–75; see also Aryeh Grabois, *Le pèlerine occidental en Terre Sainte au Moyen Âges* (Paris: De Boeck & Larcier s.a., 1998), 121, note 15; Vikan, *Byzantine Pilgrimage Art*.
82 Peers, "Object Relations," 984; on other relics that Willibald might have brought from the Holy Land, see Smith, "Eleventh-Century Relic Collections."
83 Smith, "Portable Christianity," 151.
84 Hahn, "Loca Sancta Souvenirs," 91; Smith, "Portable Christianity," 154: "In a social perspective, meaning was nevertheless liable to be undermined by a loss of identity and relegation to anonymity. The failure of collective memory, the loss of documentation, or a change of storage receptacle might also obliterate any lingering association with the person(s) who first acquired them or the donor(s) whose generosity they commemorated. As a consequence, these tiny objects had fluid social, political, and historical meanings – or none at all."
85 *Breviarius de Hierosolyma*, 111.
86 Jaś Elsner, *Art and the Roman Viewer: The Transformation of Art from the Pagan World to Christianity* (Cambridge: Cambridge University Press, 1995), 4–5.
87 Smith, "Portable Christianity," 150; Peers, "Object Relations."
88 This is a paraphrase of Elsner, *Art and the Roman Viewer*, 119.

2 Eleventh-century relic collections and the Holy Land

Julia M. H. Smith

An eleventh-century manuscript now in Oxford includes a Latin list of the relics belonging to the Church of Saint Mary and Saint Peter in Exeter. After itemizing 146 relics, it concludes with a note that there were also many others whose identities were unknown because their names were not written down.[1] Although comments to this effect were common at the end of medieval relic lists, they were not a trope. The relics collected in large numbers by medieval churches were tiny, amorphous, and indeterminate objects that were rarely, if ever, directly visible. As a consequence, their significance and raison d'être were not self-evident.[2] Liturgical ritual, glittering receptacles, and precious silk wrappers did much to interpret and uphold their overall meaning, but rarely differentiated among them.[3] This chapter explores the ways in which the act of writing down the name of every particle in a relic collection provided its components with durable identities so that religious ritual and splendid settings could complete the task of endowing these paltry objects with massive and enduring significance.

In contrast to narrative delineations of the Holy Land in pilgrim itineraries, the relic lists presented here do not constitute geographically coherent accounts. Nor, unlike such discrete representations as the sixth-century painted box from the Sancta Sanctorum in Rome and the Holy Land theme parks of late medieval Europe, is their significance immediately obvious. Rather, prior to the First Crusade, objects from Jerusalem became subsumed into relic collections from which they have to be extracted with care. Whereas the presence of saints' relics under or inside altars took its cue from the Book of Revelation 6:9, there was no equivalent biblical rationale for placing objects from the Holy Land among them.[4] Their presence in medieval churches is the more notable for the absence of canonical or liturgical obligations to include them. How can we exploit the written documentation about relic collections in general to enhance our appreciation of the ways in which the Holy Land was represented, relocated, and made present in the Latin West?

Natural materials, the written word, and ecclesiastical ceremonial fused to relocate the earthly and heavenly Jerusalem into Europe's churches and, indeed, to elide them. To be sure, this was not new in the eleventh century, for the process can be glimpsed as far back as ca. 800.[5] In combination with the much more elaborate eleventh-century liturgical sources, however, the extensive documentation of Salian-era relic collections brings the dual significance of writing and ritual into focus by the turn of the millennium. Whereas ninth- and tenth-century evidence is mostly from monastic milieux, by the eleventh century it is possible to point to metonyms of the Holy Land in the context of episcopal interactions with their lay congregations. I demonstrate this by means of two complementary case studies. One is English and the other German: both concern bishops who worked hard to promote the reputation and resources of their dioceses in the third quarter of the eleventh century. They raise questions about how readily Holy Land relics were available, how they contributed to buttressing episcopal authority and identity, and whether Western churches shared a common experience of the Holy Land.

Leofric of Exeter (1046/50–1072) and Gundechar of Eichstätt (1057–1075) had much in common beyond their interest in relics. Both were courtier bishops appointed from within the ranks of their respective sovereign's household, yet neither came from the ranks of the highest nobility: having risen more through ability than their relatives' influence, both returned to their natal regions as bishops. Both also stewarded their sees through turbulent political times – in Leofric's case the Norman Conquest of England in 1066 and in Gundechar's the Saxon revolts against Henry IV.[6] Both presented themselves as members of the circle around Leo IX (1047–1054), the first of the eleventh-century "reforming" popes who, as bishop of Toul and then pope, fully shared their enthusiasm for relics.[7] Above all, both exploited the capacity of the written word to solidify episcopal ritual, fashion a persuasive narrative of their place in the history of their sees, and project it to their successors.[8]

***

Alongside the ordination of clergy, the consecration of churches was one of the most important tasks undertaken by an eleventh-century bishop.[9] Gundechar of Eichstätt certainly took it very seriously. In all, he consecrated 126 local churches and altars in the course of his pontificate, for every one of which he would have needed relics. Of direct concern here, however, are the altars within his own cathedral church. It fell to Gundechar to complete a cathedral whose rebuilding and enlargement had been under way since the pontificate of one of his predecessors, Heribert (1022–1042). In 1060, three years after his installation as bishop, he consecrated its high altar, and over the course of the next fourteen years, consecrated a further eleven altars in various locations within it, together with one in the episcopal abbey of Herrieden. It is through his disposition of more than 800 relics around these thirteen altars that we can track how Gundechar mapped the Holy Land onto the monumental fabric of his cathedral.

Owing to the large number of surviving books of episcopal rituals from tenth- and eleventh-century Germany, the ceremonials for consecrating an altar and for dedicating or blessing a church are well known. An advance announcement (*denuntatio*) of the deposit of relics urged everyone to attend, and on the day itself, after an elaborate consecration ritual, the relics were processed in a large portable casket (*feretrum*) into the church to the accompaniment of the antiphon "I have sanctified Jerusalem, says the Lord, and I shall give my saints the kingdom and the chosen tabernacle which I have prepared with the perfume of unguent. Alleluia."[10] Before the relics were sealed into the altar's *confessio*, the bishop returned to the church door and in the course of an address to the members of the congregation informed them "in whose honour the church was built and dedicated, and also the names of the saints who rested in it."[11] In this context, it became increasingly common to keep a written record of the dedication ceremony, sometimes including details of every single relic deposited.[12]

Gundechar stands out for his meticulous attention to this task. Archeological evidence from his cathedral's eastern crypt has revealed how inscriptions recorded these details for all to see on the walls immediately behind each altar, in large red lettering on white plaster (Figure 2.1).[13] These must have been composed before the relics

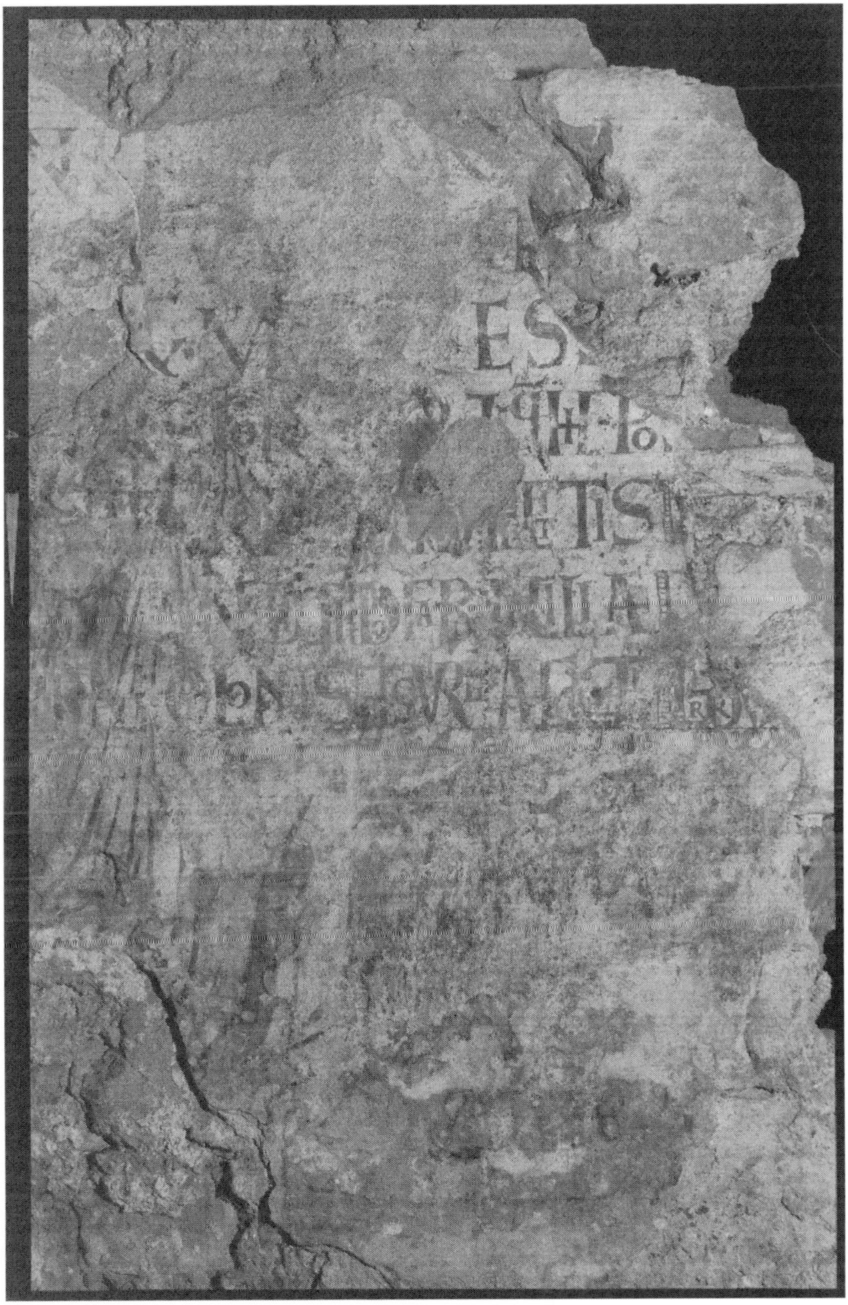

Figure 2.1 Painted relic inscription from the northern altar niche in the eastern, eleventh-century crypt of Eichstätt cathedral. The text corresponds very closely to the list of relics for this altar provided by Gundechar's *Liber Pontificalis*; traces of more recent overpainting visible. Photo credit: Bayerisches Landesamt für Denkmalpflege.

were cemented into the altars and compiled by copying the information from the tiny parchment labels customarily attached to every relic. In all likelihood, every altar had its own analogous inscription. The painted lettering thus made the invisible known.

In 1071/1072, Gundechar commissioned a magnificent *Liber Pontificalis* for his cathedral, now Eichstätt Diözesanarchiv MS B4.[14] This sumptuous presentation volume combines the full set of orders of ceremony for episcopal rituals now known as the "Romano-German Pontifical" with information on the history of his see, prepared with limited attention to historical accuracy.[15] Its purpose was to emphasize Gundechar's place in the unbroken succession of Eichstätt's bishops, which extended back to its first prelate, Willibald (d. 787), and to present the cathedral as the equivalent of a community of saints, the communion of all the living and dead members of the church. With this in mind, he placed it on the Willibald altar in the choir, where it emblematized the bishop's office and its pedigree.[16] He copied into it details of the dedication of all of the cathedral's altars, including comprehensive lists of the relics within them. By reproducing this information, Gundechar freed the relics from their hidden placement and their fixed inscriptions and gave them a new role as an integral part of the image of his pontificate that he presented to posterity.[17] His *Liber Pontificalis* thus allows us to explore the relics' contribution to Eichstätt's identity. Through careful analysis we can map the placement of Holy Land relics around the cathedral and deduce their significance (see Table 2.1).

Gundechar consecrated a total of thirteen altars. In the altar of the chapel in which he wished to be buried, he also placed his own pectoral cross, containing a further forty-five itemized relics.[18] As in other major eleventh-century churches, Eichstätt's multiple altars transposed the specific history of the diocese with its own patronal saints onto the universal narrative of Christian salvation. Typical of the age, it was a double-apsed church, and the altars' distribution in the choir, the nave and transepts, the eastern and western crypts, and the towering westwork created a three-dimensional liturgical space that served as the stage for carefully choreographed ritual enactments of priestly authority and the Christian salvation story.[19]

Gundechar began this effort in 1060, marking the day on which Willibald's priestly ordination was commemorated (22 July) by consecrating the altar in the choir dedicated to Willibald and for which, in due course, his *Liber Pontificalis* was designated. Only five of its seventy-four relics denoted the Holy Land. Then, chronologically second (28 October 1060) but transposed to first place in the manuscript sequence, came the original main altar, dedicated to Christ the Savior: "in honor . . . of his most holy birth, life-giving passion, most glorious resurrection and ascension, the descent of the holy spirit upon the apostles, the life-giving cross and Mary the holy mother of God, perpetually virgin, the nine orders of the spirits of the blessed, the patriarchs, prophets, apostles, martyrs, confessors, and virgins, but especially of those saints whose relics are here deposited." There follows a list of 156 relics, of which the first nine refer to Christ and his mother.[20] They constitute an eclectic selection, in no obvious narrative, geographical, or liturgical order. Although the altar of Saint Salvator enjoyed liturgical pre-eminence, others housed a higher percentage of Holy Land items. The variable numbers and proportions indicate that Gundechar did not simply divide a central relic set into evenly distributed parts. Not even two altars consecrated on the same day (10 July 1072; altars in the two tower chapels) contained matching relics. Some lacked any Holy Land or dominical relics altogether; most had four or fewer.[21]

Table 2.1 Gundechar of Eichstätt's altar consecrations

Date of consecration	*Sequence in Pontifical of Gundechar*	*Regesta number*[†]	*Altar dedication*	*Total number of relics deposited, according to Gundechar*	*Number of Holy Land relics deposited*	*Holy Land relics as proportion of the total*
27.vii.1060	2 (fol. 57r col. b – 57v col a)	224	St Willibald	74	5	7%
20.x.1060	1 (fol. 57r cols a – b)	227	St Salvator (high altar)	156	9	6%
5.xi.1060	3 (fol. 57v col. b – 58r col. a)	228	Sts Ulrich and Gunthildis	88	6	7%
17.x.1062	4 (fol. 58r col. a – 58v col. a)	231	Sts Mary and John the Evangelist	172[*]	14	8%
5.vi.1064	7 (fol. 59r col. a)	232	St Boniface, eastern crypt	26	3	7%
16.vi.1064	8 (fol. 59r col. b)	233	St Vitus, eastern crypt	24	0	0%
1.viii.1064	5 (fol 58v. cols a – b)	234	Sts Peter and Paul	54	1	2%
14.ix.1064	6 (fol. 58v col. b – 59r col. a)	235	Holy Cross	40	1	3%
1.x.1071	9 (fol. 59r col. a – 59v col. a)	240	St Salvator, Herrieden	49	9	18%
10.vii.1072	10 (fol. 60v col. a)	241	St Michael, south tower of westwork	[23][**]	0	0%
10.vii.1072	11 (fol. 60v cols a – b)	242	St Mary, north tower of westwork	[37][**]	2	5%
16.vi.1074	12 (fol. 60v col. b – 61r col. a)	247	St Willibald, western crypt	[35][**]	4	11%
8.vii.1074	13 (fol. 61r col. a)	248	St Kilian, western crypt	[20][**]	4	20%

Notes

[†]Franz Heidingsfelder, *Die Regesten der Bischöfe von Eichstätt* (Innsbruck: Wagner, 1915).
[*]Gundechar's total for this altar includes the 45 relics in his pectoral cross, which he deposited here.
[**]My totals; Gundechar did not supply the figure for the last four altars.

The largest absolute number, fourteen in total, was contained in the Chapel of Mary and John the Evangelist: fragments of the Lord's crib; the tree under which the shepherds were standing when the angels appeared to them; the palm that the Lord carried; the shroud in which his body was wrapped; his loincloth; the table of the Last Supper; the Holy Cross; the Lord's sepulchre; the rod of Aaron; the clothing of the Virgin; and the clothing of John the Baptist.[22] It was also here that Gundechar deposited his pectoral cross. This spot was central to Gundechar's self-identity, for he consecrated it on 17 October 1062, the fifth anniversary of the day he first occupied his episcopal throne, and also earmarked it for his own burial.[23] At the crucifixion, Mary and John the Evangelist had flanked Christ's cross: relics and dedication together made this chapel into Eichstätt's Golgotha. By developing it as his own funerary monument, Gundechar ensured his own presence at the foot of Calvary.

Although each of the thirteen altars had its own particular identity, their relics are always listed in accordance with a standard template: first come those of Jesus' life and passion, with those of Mary and John the Baptist (if any) tacked on at the end. There follow the apostles, then martyrs and confessors, and finally virgins. This classification had been widely (although not universally) used for relic lists since ca. 800, and effectively re-created the heavenly Jerusalem in the form of a list, exactly as litanies did.[24] As an inventory of the heavenly hierarchy, the format did not reference the earthly topography, and disentangling earthly Jerusalem from its celestial context is by no means straightforward. Although fragments *de ligno Domini* denote the cross of the crucifixion, they were so ubiquitous by the early eleventh century that their presence in seven of Gundechar's lists is of liturgical rather than geographical significance.[25] As the altar dedicated to the Holy Cross itself confirms, relics of the cross carried universal rather than topographical meaning: the piece of the *lignum Domini* in this altar was not accompanied by anything that might denote the Holy Land. Similarly, the six mentions of the *presepium Domini* need not necessarily refer to Bethlehem, for Santa Maria Maggiore in Rome had celebrated its crib relic since the seventh century. Also in this context, it is notable that the Holy Innocents are classified as martyrs, although their cult originated in the gospel narrative (see Table 2.2).

In several instances, however, we can be confident about the convergence of celestial and mundane topography. By far the most common dominical relic, present in ten of the lists, came *de sepulchro Domini*. Two other items, both only mentioned once, explicitly denote places: the mountain from which Christ ascended to heaven and "the place of Calvary." Although not identified topographically, several other objects refer to Jerusalem. The rod of Aaron (originally in the Temple's Holy of Holies) and the Column of the Flagellation were reported to have been in the church on Mount Zion by ca. 530, whereas of the various passion-related objects housed in the Basilica of Constantine, Eichstätt possessed the basin in which Jesus washed the disciples' feet and the sponge used at the crucifixion.[26] Still others – including the bread that Jesus blessed; the table of the Last Supper; the towel He used at the foot-washing; His shroud – are well known from elite collections of passion relics, such as those at Saint-Riquier, Oviedo, and the Pharos Chapel of the imperial palace in Constantinople.[27] Relics of John the Baptist, on the other hand, imply that some pilgrims had journeyed to Sebastiya, whereas in the sixth century, Mary's miracle-working clothes were reported to have been kept at Nazareth, although these geographical details were of no interest in Eichstätt.[28]

Table 2.2 Gundechar of Eichstätt's Holy Land relics, by frequency of mention

De sepulchro Domini	10
De ligno Domini	7
De presepio Domini	6
De ueste/uestimentis Sanctae Mariae	5
De sudario Domini	4
De mensa Domini	3
De linteo quo erat precinctus quando discipulis pedes lauit	3
De uirga Aaron	3
De linteo Domini	3
Sancti Iohannis Baptista	3
De fragmentis Domini	2
De palma quam dominus portauit	2
Sanctorum Innocentium	2
De arbore palme quam dominus portauit	1
De monte quo celum ascendit	1
De capillis et ueste Sanctae Mariae	1
De spongia Domini	1
De arbore sub quo pastores erant quando apparuit illis angelus	1
De sindone ubi corpus Domini inuolutum erat	1
De sancta cruce	1
De pane quem ipse Dominus benedixit	1
De columna ubi Dominus fuit ligatus	1
De pelue in qua Dominus pedes discipulorum lauit	1
De loco Caluariae	1
De uestimentis Sancti Iohannis Baptista	1
De pane quem ipse Dominus benedixit	1

Even the vegetation of Jerusalem and its environs had been incorporated into pilgrims' itineraries in a way that made it suitable for appropriation as relics: two items in the Chapel of Saints Mary and John the Evangelist had been fashioned from trees. Fronds from a palm that Jesus had carried presumably came from a tree somewhere along the Palm Sunday processional route from the Mount of Olives into Jerusalem, while a piece of "the tree under which the shepherds were when the angel appeared to them" was doubtless retrieved from the vicinity of the Monastery of the Holy Shepherds near Bethlehem. Unparalleled in other relic collections, this exceptional object perhaps derived from Willibald's own pilgrimage to the Holy Land in 724–726, for Hygeburg of Heidenheim reported that his itinerary included "the place where the angel appeared to the shepherds."[29]

The bishops of the Ottonian-Salian Empire commonly utilized relics among the many ritual props of their authority and manipulated them in status-enhancing ways. Although participation in the empire-wide networks of contact gave them ready access to large numbers of sacred particles of German and Italian origin, the channels through which Holy Land relics circulated are less easy to identify.[30] Various possible sources to which Gundechar had access can be suggested. In the first place, we may presume that when new altars replaced old ones in the previous cathedral, relic deposits were transferred from the old to the new. Indeed, in the eastern crypt, which was not rebuilt, Gundechar simply moved the two altars into different positions.[31] We have already seen a suggestion that at least one of the old altars contained relics that

Willibald had brought back from his pilgrimage to the Holy Land. Others might have gradually accumulated as secular and clerical benefactors donated relics and reliquaries to the church.[32]

Another source may well have been pilgrims returning from Jerusalem. We know that a large party of Germans had made its way, not without difficulty, to the Holy Land in 1065 and that it included Gundechar's own metropolitan, Archbishop Siegfried of Mainz. Furthermore, two of the three other bishops in the group presided over dioceses adjacent to Eichstätt – those of Regensburg and Bamberg.[33] It is hardly conceivable that none of them amassed relics during their journey, and so the normal gift exchange of relics among imperial bishops probably channeled recently acquired Holy Land items into Eichstätt. Indeed, the pattern of Gundechar's relic deposits suggests that an influx reached him part way through his pontificate: having had few, if any, Holy Land relics available for the four altars consecrated in 1064, he then deposited nine in Herrieden on 1 October 1071.

We know for certain of one pilgrim from Eichstätt who completed the round trip to Jerusalem. The calendar that Gundechar included in his *Liber Pontificalis* contains a selection of deceased persons to be commemorated liturgically, all of whom were important either in the history of Eichstätt or in Gundechar's own life. They include, on May 5, Meginward, pilgrim and "exile for God," who died a good death having returned from Jerusalem. Whether Meginward had joined the 1065 party cannot be ascertained, but the inclusion of his obit is a sure sign that he was somebody who mattered personally to his bishop. Although Gundechar himself never went on pilgrimage to the Holy Land, he experienced it at one remove through his encounters with those who had.

On balance, however, Gundechar evinced little interest in relics as metonyms of particular places in the Holy Land. They were a subset of his much larger relic supply, whose effect was cumulative and generic: he was evidently unconcerned about the particles' uneven distribution around the church or about obtaining a comprehensive assemblage representing as many biblical sites and events as possible. Rather, these sacred particles fused the earthly Jerusalem into its celestial and salvific counterpart. Their presence in his altars ensured that episcopal liturgy was at all times grounded in the framing narrative of the Christian story and that Jerusalem was present in Eichstätt and Eichstätt in Jerusalem.[34] Just as his *Liber Pontificalis* was not concerned with historical accuracy, so it was not bothered with topographic precision either. Instead, it projected a powerful message of Bishop Gundechar as the between mediator past, present, and future, as guarantor of the intercession of the saints of the heavenly Jerusalem on behalf of the deceased and living members of his flock. His designated burial place within his cathedral's Calvary Chapel enabled him to continue in this role after his own death: the miracles that occurred around his tomb earned him elevation to sainthood in 1309.[35]

* * *

Leofric of Exeter's self-fashioning was less ostentatious, probably because he could not afford lavish expenditure on either buildings or luxury manuscripts. His clerical training took place in Lotharingia, the western province of the German Empire, and clear traces of his continental networks are discernable in many of his activities in Exeter.

In 1046, Edward the Confessor appointed him to the small, rural see of Crediton, but four years later, Leofric secured Leo IX's permission to transfer his seat to Exeter, a flourishing walled city of Roman origin but with an impoverished mother church. His efforts to build up its landed endowments generated a series of spurious or interpolated documents, which make evaluation of his episcopacy difficult. Although, as we shall see, he had plentiful spiritual resources (in the form of relics), lack of financial means may explain why he continued to utilize the pre-existing church as his cathedral, which, at less than half the length and approximately one-third the width of Gundechar's cathedral, was a comparatively cramped liturgical space.[36] Instead of building activities, however, Leofric devoted much energy to stocking its library, acquiring large numbers of second-hand books as well as commissioning new workaday manuscripts; owing to the survival of a large number of them, Exeter's mid-eleventh-century liturgical and literary culture is exceptionally well documented.[37]

Among the many manuscripts he acquired, two stand out in this context. The first, Oxford Bodleian Library MS Bodley 579, known as the "Leofric Missal", served an analogous purpose to Gundechar's *Liber Pontificalis* but achieved its ends differently. Its core was a late ninth-century sacramentary probably produced in Canterbury, where substantial additions were made during the tenth century.[38] How Leofric acquired it remains unknown, but he personalized it by inserting supplementary liturgical and historical material of his own choosing and then presented it as a gift to his see for his successors' use. Just like Gundechar's *Pontifical*, this book confirmed Exeter's episcopal status, shaped its history, and commemorated Bishop Leofric, providing a coherent image of his place in the wider world.[39]

Among the notable additions to the Leofric Missal is the long relic list (fols 6r–6v) with which I opened this chapter (Figure 2.2). In Latin, it had been copied onto a blank leaf in the Mass book near the end of Leofric's episcopacy. Riddled with evidence of scribal carelessness and haste, it is essentially a functional document, one perhaps produced as part of the normal process of inventorying a bishopric's possessions when the incumbent died. After a brief prefatory note stating that it names Exeter's relics and that most of them had been gifted by King Æthelstan (924–939), it then catalogues the relics, but, unlike some of Gundechar's altar lists, no names are repeated. Beginning "First, from the blood of the Lord," it continues with "the wood of the Lord [i.e., the cross]" and then with another nineteen dominical or Holy Land relics, including the Holy Innocents, before proceeding to the apostles and on to martyrs, confessors, and finally virgins.[40] Although entries take the form of short, functional descriptions, analogous to those in many other eleventh-century relic lists, the Leofric Missal forms, in effect, a pedigree, a chain of succession from the Christian story's origin to Leofric's own day.

Exeter's relics anchored the heavenly Jerusalem to the cathedral, as in Eichstätt. Unlike Gundechar, however, Leofric lacked a sainted founder and predecessor whose body would have formed the point where his see's secular and sacred histories converged. Instead, the city claimed to have its own Roman-era martyr, Sidwell, or Sativola, who is listed, without further comment, four places from the list's end.[41] To this extent, the Exeter list, exactly like Gundechar's, is cosmology redacted as inventory, the specifically local version of the universal story of redemption.

But there the similarities end, for the relic list in the Leofric Missal cannot be evaluated without taking into account a second relic list also compiled in Exeter during the third quarter of the eleventh century (Figure 2.3). This is among the materials

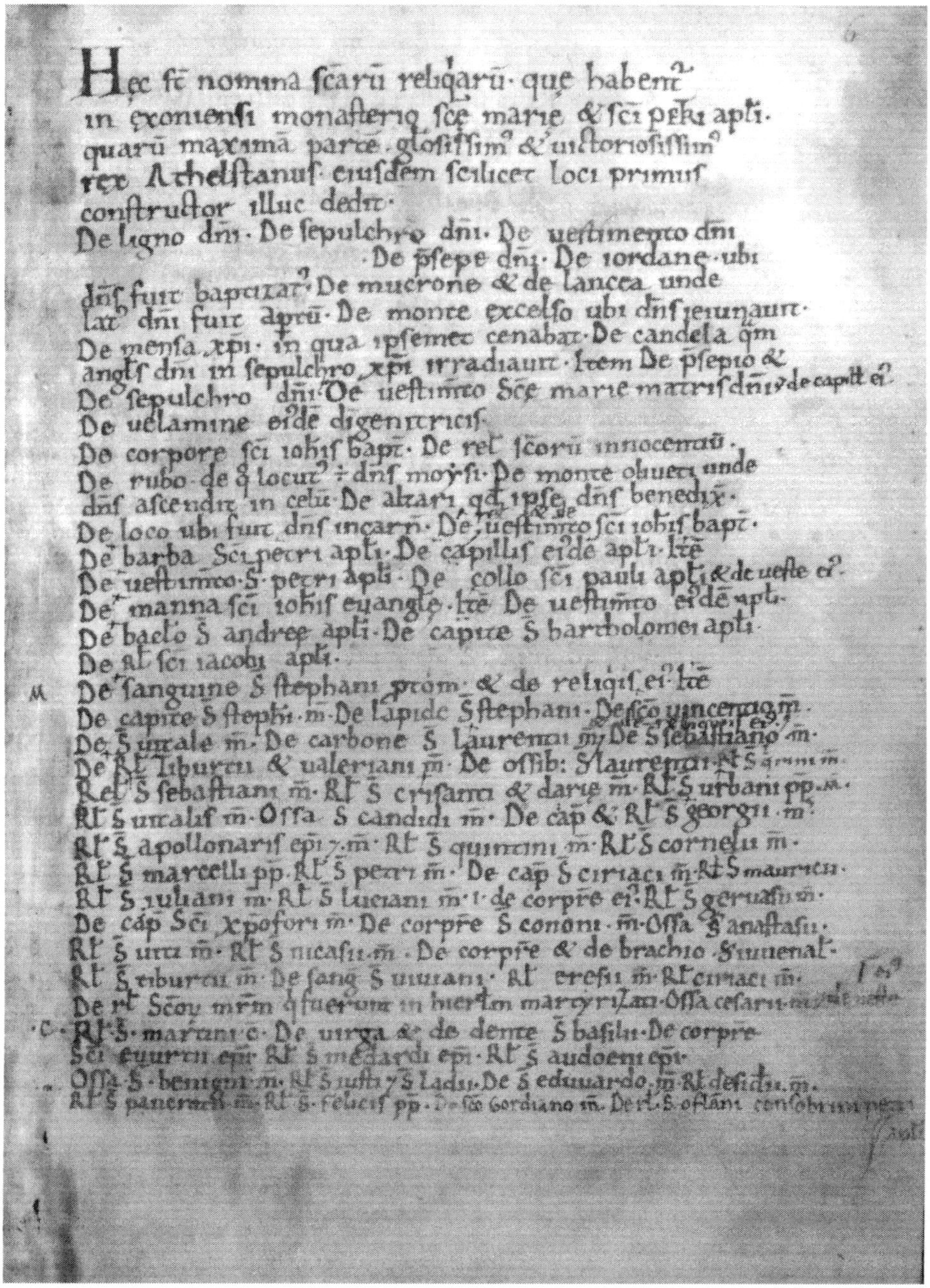

Figure 2.2 Latin relic list added to the Leofric Missal: Bodleian Library, MS Bodley 579, fol. 6r. More than half of Exeter's relics are listed on this page, including all the Holy Land and martyr relics. Photo credit: The Bodleian Libraries, The University of Oxford.

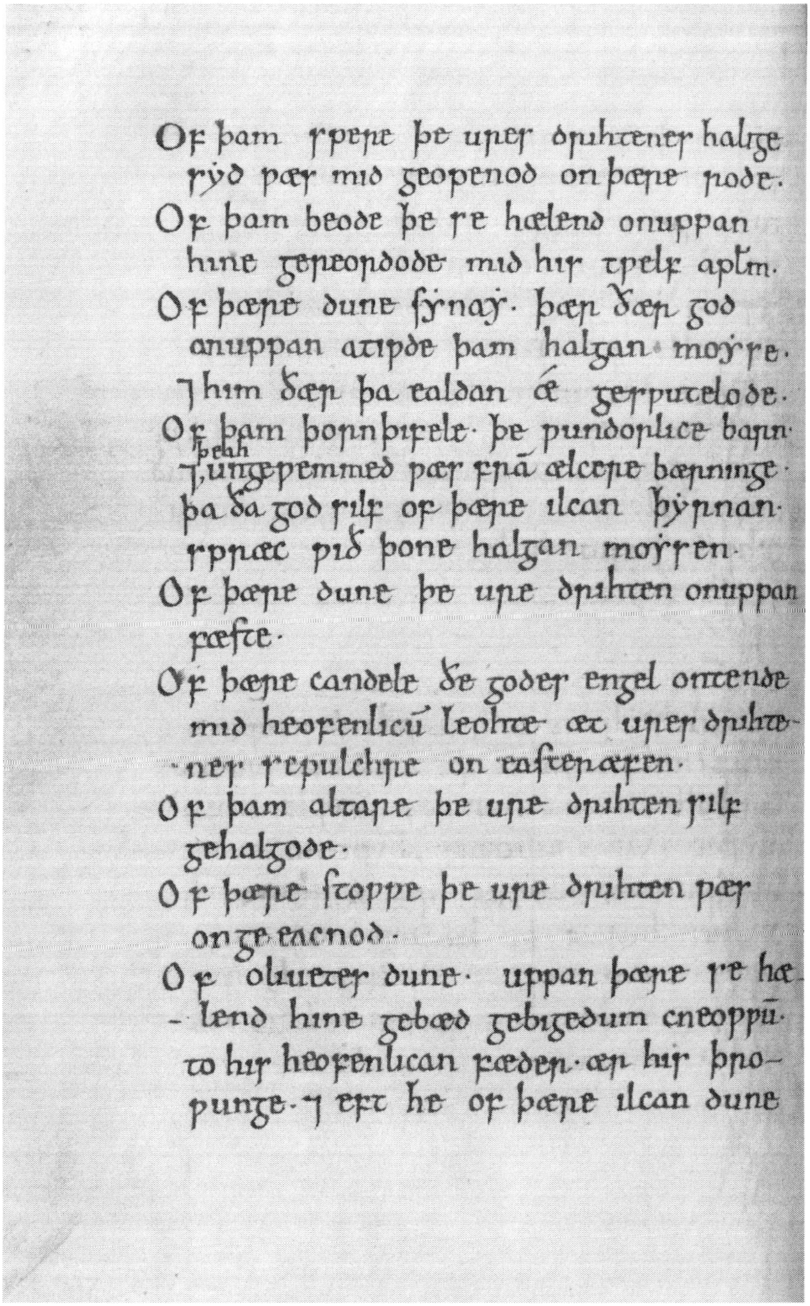

Of þam swere þe ures drihtenes halige
syð wæs mid geowenod on þære rode.
Of þam beode þe se hælend onuppan
hine gereordode mid his twelf aplm.
Of þære dune synay· þær ðær god
onuppan atiwde þam halgan moyse·
⁊ him ðær þa ealdan ǽ geswutelode·
Of þam þornþifele· þe wundorlice barn·
beah
⁊, ungewemmed wæs fram ælcere bærninge·
þa ða god silf of þære ilcan þyrnan·
spræc wið þone halgan moysen·
Of þære dune þe ure drihten onuppan
fæste·
Of þære candele ðe godes engel ontende
mid heofenlicū leohte æt ures drihte-
-nes sepulchre on eastereæfen·
Of þam altare þe ure drihten silf
gehalgode·
Of þære stoppe þe ure drihten wæs
ongeeacnod·
Of oliuetes dune· uppan þære se hæ-
-lend hine gebæd gebigedum cneowū·
to his heofenlican fæder· ær his þro-
wunge· ⁊ eft he of þære ilcan dune

Figure 2.3 Old English Exeter relic list prefaced to a ninth-century Breton Gospel book Bodleian Library, MS Auct D.2.16, fol 9v. The folio illustrated lists ten of Exeter's Holy Land relics. Photo credit: The Bodleian Libraries, The University of Oxford.

added to another ninth-century service book acquired by Leofric, the Breton gospel book that is now Oxford, Bodleian Library MS Auct. D.2.16.[42] Its second added booklet, written entirely in Old English (fols 8r–14r), includes a lengthy preamble explaining Æthelstan's generous donations followed by details of 138 relics. Linking preface to inventory is a direct first-person address emphasizing that the relics will be "explained" without any deceitfulness.[43] The list itself is divided into the same categories as in the Leofric Missal, but within each section the individual entries do not appear in the same order. There are also some notable omissions and additions.[44] It is not, however, simply a translation of the Latin list, for the two are independent but interdependent.[45] The *mise-en-page* of the Old English is spacious and calligraphic, but the page size is slightly smaller than that of the gospel book that it now prefaces: it may have been conceived as a self-contained booklet.[46]

Exeter's eleventh-century historical records note that upon his arrival in 1050, Leofric found the church to be "almost entirely despoiled" of its property and possessing only three books and one relic shrine.[47] Although the challenge of sorting through the contents of a large relic chest may perhaps explain the discrepancies between the lists, this cannot be the reason that the two versions of the inventory provide substantially different textualized reports of the objects it contained. These divergences, which extend to much more than just language and syntax, are most striking in the treatment of Holy Land objects.[48] For instance, the Old English list ignores the blood of the Lord and opens thus: "First, from that same precious tree, the holy cross, on which Christ suffered and thereon delivered us all from the power of the Devil."[49] Similarly, the Latin list's "From the crib of the Lord" is in the vernacular, "From the manger in which our Lord lay when he was born of St. Mary."[50]

The most striking contrast concerns the relics of John the Evangelist. The Latin gives "From the manna of St. John the Evangelist" immediately followed by "Again. From the clothes of the same apostle."[51] The Old English, on the other hand, reverses the sequence, placing the clothes first. But the object itself pales into insignificance, for the full entry reads: "From the clothes of St. John the Apostle and Evangelist, who was loved so much by our Lord that at his supper, He leaned upon his breast, and then when He suffered for us on the cross, He entrusted St. Mary, His dear mother, to the same John, His favourite, so that he should look after her."[52] Here, more clearly than anywhere else, the redactor of the Old English list turned the task of making an inventory into an opportunity for a didactic exposition of the gospel story.

What can we learn from the Old English list about conceptualizations of the Holy Land in Leofric's Exeter? In the first place, biblical topography and chronology both collapse. Relics of Moses (Mount Sinai; the burning bush) interrupt the dominical sequence, which itself is not structured to follow the order of events in the biblical narrative.[53] Nevertheless, there is a somewhat greater awareness of place than in Gundechar's lists, for the Jordan River, Mount Sinai, the mountain on which Jesus fasted, the place where He was conceived, and the Mount of Olives are all mentioned.[54] Evidently, Anglo-Saxon pilgrims had made extensive journeys beyond Jerusalem and its immediate vicinity. They also attended the annual ceremony of the paschal fire at the Holy Sepulchre, because Exeter possessed a relic "From the candle which God's angel kindled with heavenly light at our Lord's sepulchre on the Easter-vigil."[55] Unusually, Jerusalem is mentioned once by name, in the context of its post-biblical history: "There are also here the relics of many martyrs who were martyred at Jerusalem for Christ's name."[56]

Secondly, in contrast to Eichstätt's liturgical framing of its relics and the possible administrative purpose of Exeter's Latin list, the Old English inventory suggests a more didactic role. The preamble's emphasis on truthfully explicating the relics aloud has led to various suggestions about its purpose. Although the proposition that preamble plus list form the text of a sermon seems improbable, some kind of documentary prompt for oral communication does underlie parts of it.[57] An expository purpose would certainly help explain the use of the vernacular, perhaps also its origin as an independent booklet.

Several plausible occasions can be suggested. In the first place, we should consider Leofric's likely diocesan activities. A strong indication of these is provided by the two-volume collection of Old English homilies (now London British Library Cotton Cleopatra B. xiii fols 1–58 and London, Lambeth Palace Library MS 489) assembled for his use. Among them are four different sermons for the consecration of a church, which might have been used to supplement the relevant rituals in his Missal.[58] It is possible, indeed probable, that he undertook the consecration of new local churches within his diocese, exactly as Gundechar did, and in this context, the "explaining" of Exeter's relics may well refer to either the *denuntatio* or the address to the people demanded by the Romano-German Pontifical, which Leofric knew well.[59]

His homiliary also includes a sermon for the commencement of rogationtide, the days in the liturgical year when relics were carried around the fields to invoke divine blessing on the crops.[60] Rogation processions were essentially pleas for intercession and mercy by the bishop and his flock, in the presence of the assembled saints. They originated during the fifth century in times of distress and danger and constituted the principal occasions on which a bishop needed portable relics to carry from one place to another. Despite their routinization as part of the annual rhythm of the liturgical year, relic processions persisted as effective rituals in times of crisis. Early in 1068, William the Conqueror laid siege to Exeter, but the assault ended when "the flower of [the city's] youth, the older men and the clergy bearing their books and treasures" exited the gates and threw themselves on the king's mercy.[61] We can be sure that the clergy processed out carrying their relics, which the preface to the Old English list emphasizes were "the most precious treasures," and may suspect that Leofric's political acumen and mastery of liturgical pomp had brokered the cessation of hostilities.[62]

Finally, a third potential occasion – and surely the most significant – for proclaiming an exegesis of Exeter's relics was the formal transfer of the see from Crediton to Exeter, on 29 June 1050. On that day, in the presence of the entire royal court, King Edward the Confessor and Queen Edith linked arms with Leofric and, with the king on the bishop's right and the queen on his left, led him up the aisle of Exeter's ancient church, where they duly enthroned him "with great glory."[63] No standard ritual was available for such an exceptional event, and Leofric must have extemporized an appropriately splendid liturgy. The prestigious collection of relics that Æthelstan had gifted surely featured prominently for his successor to venerate, and an exposition of the biblical significance of selected items would have been a fitting way for the incoming bishop to take possession of Exeter's long and glorious Christian past. In a very real way, the relics became the foundation stones of his episcopate.

In sum, Exeter's relics were adaptable for various purposes. The characteristics of the written texts make it certain that they remained accessible, unlike at Eichstätt, and the various ritual uses posited here imply that a selection of objects might be extracted for different occasions. Although they always reinforced Leofric's episcopal authority, their exegesis might nevertheless have been adapted to fit the moment. Equally significant, the relics and the lists were both portable. Together with heavenly Jerusalem's saints, earthly Jerusalem's biblical topography could be transported into the nave of the cathedral, out of the city gates, and into the fields around the diocese. In Eichstätt, Gundechar had cemented the Holy Land into the architectural fabric of his cathedral, where it was hidden from sight but signaled by inscriptions, but in Exeter, Leofric and his clergy carried Jerusalem around for all to see. We have to envisage a stone from the top of Mount Sinai, a stub of candle wax from the Holy Sepulchre, a twig from the burning bush, and unspecified other objects (dust? scraps of altar coverings?) from biblical and Christian sites being held up to the eleventh-century populace as their significance was expounded. This was the Holy Land made real, yet nothing except the words on their labels would have distinguished them from ordinary stones, twigs, and candles. Priestly exposition and liturgical context supplied the exegetical framework, but the written word ensured that their significance was remembered, available to be read, recited, and explained. Without appropriate ritual, neither the earthly nor the heavenly Jerusalem could be conjured up in Europe, and without writing, natural objects could not function effectively as metonyms of the Holy Land.

Acknowledgement

I am very grateful to the Leverhulme Trust, whose award of a Major Research Fellowship gave me the ideal conditions in which to research and write this article.

Notes

1 Bodleian Library, MS Bodley 579, fol. 6v. This list was most recently edited by Patrick W. Conner, *Anglo-Saxon Exeter: A Tenth-century Cultural History*, Studies in Anglo-Saxon History, vol. 4 (Woodbridge: Boydell Press, 1992), 192–198; the colophon is §158.
2 I argued this in Julia M.H. Smith, "Portable Christianity: Relics in the Medieval West (c. 700–1200)," *Proceedings of the British Academy* 181 (2012): 143–167.
3 For the argument that the essential function of reliquaries was to explain and interpret their contents, see Cynthia Hahn, *Strange Beauty: Issues in the Making and Meaning of Reliquaries, 400–circa 1204* (University Park, PA: Penn State University Press, 2012).
4 Revelations 6:9 "And when he had opened the fifth seal, I saw under the altar the souls of them that were slain for the word of God."
5 François Héber-Suffrin and Anne Wagner, "Autels, reliques et structuration de l'espace monastique: L'exemple de Saint-Riquier," *Cahiers du Léopard d'Or* 12 (2011): 27–55.
6 Frank Barlow, "Leofric and His Times," in *Leofric of Exeter*, eds. Frank Barlow et al. (Exeter: University of Exeter, 1972), 1–16; Frank Barlow, "Leofric (*d.*1072)," in *Oxford Dictionary of National Biography* (Oxford: Oxford University Press, 2004), http://www.oxforddnb.com/view/article/16471; Tara Gale, John Langdon, and Natalie Leishman, "Piety and Political Accommodation in Norman England: The Case of the South-West," *Haskins Society Journal* 18 (2006): 110–131; Bruno Häuptli, "Gundekar II.," *Biographisch-Bibliographisches Kirchenlexikon*, 26 (Hamm: Traugott Bautz, 2006), 547–550; Stefan Weinfurter, "Sancta Aureatensis ecclesia: zur Geschichte Eichstätts in ottonisch-salischer Zeit," *Zeitschrift für bayerische Landesgeschichte* 49 (1986): 3–40.
7 Cyriakus Heinrich Brakel, "Die vom Reformpapsttum geförderten Heiligenkulte," *Studi Gregoriani* 9 (1972): 239–311, esp. 242–283. For Gundechar's closeness to Leo IX, see Weinfurter, "Sancta Aureatensis ecclesia," 35–36. Both Gundechar and Leofric included Leo IX among the select few whose obituary they added to their personal calendars: *Liber pontificalis eichstetensis*, ed. L.C. Bethmann, MGH SS VII: 248 (13. Kal. Mai.); Nicholas Orchard, *The Leofric Missal*, 2 vols (Published for the Henry Bradshaw Society by Boydell

Press, 2002), vol. 2, 60 (marginal addition to entry for XIII Kal. Mai) and see comments in vol. 1, 223, for the suggestion that this entry is in Leofric's own hand.

8 This article builds directly on the work of Sarah Hamilton, to whom I am greatly indebted. For her comparison of Gundechar and Leofric, see Sarah Hamilton, "The Early Pontificals: The Anglo-Saxon Evidence Reconsidered from a Continental Perspective," in *England and the Continent in the Tenth Century: Studies in Honour of Wilhelm Levison (1876–1947)*, eds. David Rollason, Conrad Leyser, and Hannah Williams (Turnhout: Brepols, 2010), 411–428. I am also grateful to Hiltrud Westermann-Angerhausen for help and advice throughout the research and composition of this chapter.

9 Louis I. Hamilton, *A Sacred City: Consecrating Churches and Reforming Society in Eleventh-century Italy* (Manchester: Manchester University Press, 2010); Henry Parkes, *The Making of the Liturgy in the Ottonian Church: Books, Music and Ritual in Mainz, 950–1050* (Cambridge: Cambridge University Press, 2015), 153, 161.

10 Cyrille Vogel and Reinhard Elze, *Le Pontifical romano-germanique du dixième siècle*, 3 vols, Studi e Testi, 226, 227, 269 (Vatican City: Biblioteca apostolica vaticana, 1963–1972), Ordo xxxviii (Denuntatio cum reliquiae sanctorum ponendae sunt in atrium); ordo xl (Ordo ad benedicendam aecclesiam) cl. 124, 127, 1: 123, 168, 169.

11 Ibid., ordo xl, cl 128, 1: 169. See cl. 132–137 (1: 170–171) for enclosing the relics. The placement of relics at the dedication of a church or an altar is discussed in detail with reference to continental texts by Helen Gittos, *Liturgy, Architecture, and Sacred Places in Anglo-Saxon England* (Oxford: Oxford University Press, 2013), ch. 6.

12 Philippe Cordez, "Gestion et médiation des collections de reliques au moyen âge: le témoignage des authentiques et des inventaires," in *Reliques et sainteté dans l'espace médiéval*, ed. Jean-Luc Deuffic, Pecia resources en médiévistique (Saint-Denis: Pecia, 2006), 33–63, esp. 39–41, 55–57.

13 Walter Sage, "Die Ausgrabungen in den Domen zu Bamberg und Eichstätt 1969–1972," *Jahresbericht der bayerischen Bodendenkmalpflege* 17/18 (1976): 178–238, esp. 225–226.

14 Andreas Bauch and Ernst Reiter, *Das 'Pontifikale Gundekarianum': Faksimile-Ausgabe des Codex B4 im Diözesanarchiv Eichstätt*, 2 vols (Wiesbaden: Ludwig Reichert Verlag, 1987).

15 The history of the Romano-German Pontifical has now been fundamentally revised: see Parkes, *Making of the Liturgy*; Brun Appel, "Die Altar- und Kirchenweihen der Bischöfe Gundekar und Otto," in *Das 'Pontifikale Gundekarianum': Faksimile-Ausgabe des Codex B4 im Diözesanarchiv Eichstätt*, eds. Andreas Bauch and Ernst Reiter (Wiesbaden: Ludwig Reichert Verlag, 1987), 148–174, notes the lack of attention to historical accuracy.

16 Parkes, *Making of the Liturgy*, 178.

17 Sage, "Die Ausgrabungen," 225, notes that painted inscriptions and manuscript copy are almost identical.

18 The relic lists were omitted from Bethmann's MGH edition of the *Liber Pontificalis* (MGH SS VII: 239–253); details in the accompanying tables are all taken from the facsimile edition (see note 14 above).

19 See, for example, Günter Bandmann, "Früh- und hochmittelalterliche Altaranordnung als Darstellung," in *Das erste Jahrtausend: Kultur und Kunst im werdenden Abendland an Rhein und Ruhr*, eds. Kurt Böhner, J. Hoster, and Victor H. Elbern, vol. 1 (Düsseldorf: L. Schwann, 1962), 371–411; *The White Mantle of Churches: Architecture, Liturgy, and Art around the Millennium*, ed. Nigel Hiscock (Turnhout: Brepols, 2003); Carolyn Marino Malone, *Saint-Bénigne de Dijon en l'an mil, totius Galliae basilicis mirabilior: interprétation politique, liturgique et théologique* (Turnhout: Brepols, 2009).

20 Bauch and Reiter, "*Pontifikale Gundekarianum*," fol. 57r.

21 See the helpful discussion by Appel, "Die Altar- und Kirchenweihen."

22 Bauch and Reiter, "*Pontifikale Gundekarianum*," fol. 58r.

23 The notice of the altar's dedication mentions that it took place "ordinationis vero suae anno 5, in qua etiam die ipse primum sedem intravit episcopalem," whereas the calendar in the *Liber Pontificalis* records "16. Kal. Nov. Dedicatio capellae sancti Iohannis euangelistae in qua domnus episcopus Gundechar secundus sepulturam suam preordinavit. . . . Gundechar episcopus sedem episcopalem primum intravit." (MGH SS VII: 246, 249).

24 Cf. Astrid Krüger, *Litanei-Handschriften der Karolingerzeit*, Monumenta Germaniae Historica: Hilfsmittel, 24 (Hanover: Hahnsche Buchhandlung, 2007); Michael Lapidge,

Anglo-Saxon Litanies of the Saints (Published for the Henry Bradshaw Society by Boydell Press, 1991).

25 There is only one instance of a relic labeled *de sancta cruce* rather than *de ligno Domini*.

26 *Breviarius de Hierosolyma* c. 3–4, CCSL clxxv, 110–111; see also John Wilkinson, *Jerusalem Pilgrims before the Crusades* (Warminster: Aris & Phillips, 2002), 365. I am grateful to Ora Limor for helping me with these objects.

27 Cf. Michele Bacci, "Relics of the Pharos Chapel: A View from the Latin West," in *Eastern Christian Relics*, ed. Alexei Lidov (Moscow: Progress-Tradition, 2003), 234–247; S. Berger, "Les reliques de l'abbaye de Saint-Riquier au IXe siècle," *Revue de l'Orient Latin* 1 (1893): 467–474; D. de Bruyne, "Le plus ancien catalogue des reliques d'Oviedo," *Analecta Bollandiana* 45 (1927): 93–96.

28 Wilkinson, *Jerusalem Pilgrims*, 331, 344–345.

29 Hygeburg, *Vita Willibaldi* c. 4, MGH SS XV/i: 98.

30 Brakel, "Die vom Reformpapsttum geförderten Heiligenkulte"; Eugenio Dupré Theseider, "La 'grande rapina dei corpi santi' dell'Italia al tempo di Ottone I," in *Festschrift Percy Ernst Schramm zu seinem siebzigsten Geburtstag von Schülern und Freunden zugeeignet*, eds. Peter Classen and Peter Scheibert (Wiesbaden: Franz Steiner Verlag, 1964), vol. 1, 420–432; Hedwig Röckelein, "Bernward von Hildesheim als Reliquiensammler," in *1000 Jahre St. Michael in Hildesheim: Kirche-Kloster-Stifter*, eds. Gerhard Lutz and Angela Weyer (Petersburg: Michael Imhof Verlag, 2012), 107–127; Anne Wagner, "Collection de reliques et pouvoir épiscopal au Xe siècle: l'exemple de Thierry Ier de Metz," *Revue d'Histoire de l'Eglise de France* 87 (1997): 317–341; Anne Wagner and Monique Goullet, "Reliques et pouvoirs dans le diocèse de Verdun au X – XIe siècle," *Revue Mabillon* n.s.10 (1999): 67–88.

31 Appel, "Die Altar- und Kirchenweihen," 155–159, for details.

32 This is widely attested in charters: see Julia M.H. Smith, "Material Christianity in the Early Medieval Household," in *Religion and the Household*, Studies in Church History, eds. John Doran, Charlotte Methuen, and Alexandra Walsham (Woodbridge: Boydell & Brewer, 2014), 23–46, esp. 42, and note 44.

33 *Annales Altahenses maiores*, a. 1065, ed. Edmund von Oefele MGH SSRG 4 (Hanover: Hahn, 1891), 66. See also *Vita Altmani episcopi Patavensis*, c. 3, MGH SS XII: 230.

34 Jerusalem was made doubly present when the Monastery of the Holy Cross and Holy Sepulchre was founded in Eichstätt nearly a century later: Diarmuid Ó Riain, "An Irish Jerusalem in Franconia: the Abbey of the Holy Cross and Holy Sepulchre at Eichstätt," *Proceedings of the Royal Irish Academy* 112C (2012): 219–270.

35 Häuptli, "Gundekar II," 549.

36 Christopher G. Henderson and Paul T. Bidwell, "The Saxon Minster at Exeter," in *The Early Church in Britain and Ireland: Studies Presented to C.A. Ralegh Radford*, BAR British Series, ed. Susan M. Pearce (Oxford: British Archaeological Reports, 1982), 145–175, plan on 157, dimensions on 163. Cf. Sage, "Die Ausgrabungen," 221–222, for dimensions of the eleventh-century cathedral at Eichstätt.

37 Versions of Leofric's own narrative of his acquistions and achievements were bound into two Exeter gospel books: for the text, see Conner, *Exeter*, 226–235. The fundamental study of Exeter's books remains Elaine M. Drage, "Bishop Leofric and the Exeter Cathedral Chapter, 1050–1072: A Reassessment of the Manuscript Evidence," unpublished D. Phil. dissertation (Oxford University, 1978). Conner's assessment of the creation of Exeter's book collection has been challenged by Richard Gameson, "The Origin of the Exeter Book of Old English Poetry," *Anglo-Saxon England* 25 (1996): 135–185. See also Joyce Hill, "Leofric of Exeter and the Practical Politics of Book Collecting," in *Imagining the Book*, eds. Stephen Kelly and John J. Thompson (Turnhout: Brepols, 2010), 77–98; Elaine Treharne, "Producing a Library in Late Anglo-Saxon England: Exeter 1050–1100," *Review of English Studies* 54 (2003): 155–172. The see's history and forged documentation is effectively appraised by Charles Insley, "Remembering Communities Past: Exeter Cathedral in the Eleventh Century," in *Cathedrals, Communities and Conflict in the Anglo-Norman World*, eds. Paul Dalton, Charles Insley, and Louise J. Wilkinson (Woodbridge: Boydell Press, 2011), 41–60.

38 Orchard, *The Leofric Missal.* Bibliography: Helmut Gneuss and Michael Lapidge, *Anglo-Saxon Manuscripts: A Bibliographical Handlist of Manuscripts and Manuscript Fragments Written or Owned in England up to 1100* (Toronto: University of Toronto Press, 2014), no. 585,

456–458. Digital facsimile: http://image.ox.ac.uk/show?collection=bodleian&manuscript=msbodl579.

39 Hamilton, "Early Pontificals."

40 Conner, *Exeter*, 192–193, §2–3.

41 Ibid., 198, §155. For further details on this obscure saint, see John Blair, "A Handlist of Anglo-Saxon Saints," in *Local Saints and Local Churches in the Early Medieval West*, eds. Alan Thacker and Richard Sharpe (Oxford: Oxford University Press, 2002), 495–565, esp. 554; Max Förster, *Zur Geschichte des Reliquienkultus in Altengland*, Sitzungsberichte der Bayerischen Akademie der Wissenschaften, Philosophisch-historische Abteilung, vol. 8 (Munich: Verlag der Bayerischen Akademie der Wissenschaften, 1943), 113–114.

42 Gneuss and Lapidge, *Anglo-Saxon Manuscripts*, no. 530, 423–424. Digital facsimile: http://image.ox.ac.uk/show?collection=bodleian&manuscript=msauctd216.

43 Conner, *Exeter*, 176–177, §1–6. I have commented on the historical implications of this preamble in Julia M.H. Smith, "Rulers and Relics, c.750–950: 'Treasure on Earth, Treasure in Heaven'," in *Relics and Remains*, Past & Present Supplement, ed. Alexandra Walsham (Oxford: Oxford University Press, 2010), 73–96.

44 Details: Förster, *Zur Geschichte*, 45–49. Förster's edition and discussion of the Old English list remains essential.

45 The inter-relationship of the Latin and Old English versions of the list is close but too complex to discuss here; suffice it to say that neither is a translation of the other.

46 Conner, *Exeter*, 174.

47 "Record of the Moving of the See of Devon," §6, ibid., 225. This tendentious narrative was placed at the front of the Leofric Missal and forms part of Leofric's self-fashioning: see Insley, "Remembering Communities Past."

48 Assessed by Förster, *Zur Geschichte*, 54–58. His assumption that this was necessary for the laity's sake is too restrictive.

49 Conner, *Exeter*, 178–179, §7. Except where noted, I cite Conner's translations of the Old English.

50 Ibid., 178–179, §10, cf. 193, §6.

51 Ibid., 193, §27–28.

52 Ibid., 178–179, §27.

53 Ibid., §14–15.

54 Ibid., §11, 14, 16, 19, 20.

55 Ibid.§17.

56 Ibid., 180–181, §71.

57 Ibid., 175 asserts that this is a work of literature belonging to the sermon genre and proposes that it was "read in the presence of the relics themselves."

58 Discussed by Gittos, *Liturgy, Architecture, and Sacred Places*, 220–221, 246–255.

59 Analyzed by Elaine Treharne, "The Bishop's Book: Leofric's Homiliary and Eleventh-Century Exeter," in *Early Medieval Studies in Memory of Patrick Wormald*, eds. Stephen Baxter et al. (Farnham: Ashgate, 2009), 521–537, with comments on the dedication sermons on p. 532. Leofric's knowledge of the Romano-German Pontifical is noted by Orchard, *The Leofric Missal*, vol. 1, 207.

60 Treharne, "The Bishop's Book," 525.

61 Orderic Vitalis, *Historia Ecclesiastica*, ed. Marjorie Chibnall, 6 vols (Oxford: Oxford University Press, 1969–1980), vol. 2, 212–213.

62 Conner, *Exeter*, 176–177, §4. I commented on this at some length in Smith, "Rulers and Relics."

63 Exeter narrative of the moving of the see, quotation from §6, Conner, *Exeter*, 225. See also Edward's charter uniting the dioceses: Christopher N.L. Brooke, Martin Brett, and Dorothy Whitelock, eds., *Councils & Synods, with Other Documents Relating to the English Church: 1, A.D. 871–1204*, part 1, 871–1066 (Oxford: Clarendon Press, 1981), 525–533. The date of June 29 was suggested by Frank Barlow, *The English Church 1000–1066*, 2nd ed. (London: Longman, 1979), 116–117, 154.

3 The popes and the loca sancta of Jerusalem

Relic practice and relic diplomacy in the Eastern Mediterranean after the Muslim conquest*

Manfred Luchterhandt

The idea that premodern treasure collections represented more than simply conglomerations of material, aesthetic, and spiritual values is a key insight in the field of culture studies, and relevant research can well be of value for the study of relics.[1] Particularly in cases where artifacts originating from different places and cultural spheres are assembled together at a political and religious focal point, the "cultural biography" of the objects and their perception cannot be understood through the prism of art history alone. As McCormick demonstrated for Sens and Chelles, a relic collection can also be viewed as a form of material communication, conveying information about the infrastructures in which people and things circulated – about trade, diplomacy, and pilgrimage.[2] Nevertheless, in his analysis of collections in the early medieval Mediterranean world, McCormick left out an important collection at the threshold between antiquity and the Middle Ages: the papal relics in the Lateran Palace in Rome.

The questions considered below relate to this collection, which is one of the most important in European history but thus far has been little researched in its entirety. When the medieval relic shrine of Leo III in the papal palace chapel Sancta Sanctorum (Figure 3.1) was reopened in 1903 after 400 years, its compartments were found to contain a unique ensemble of twenty-seven reliquaries, boxes, cases, and glass containers whose provenance from Byzantium, Jerusalem, Alexandria, and Sicily documents the extensive cultural relations of the various popes.[3]

Moreover, the containers were also found to hold one of the largest collections of relics from the medieval period. Their written labels, which provide information about persons and places in the Eastern Mediterranean, from Syracuse and Cyprus to the Sinai Peninsula, bear unique testimony to the international character of the sacred papal economy.

Unfortunately, at the present time, the potential of these authentics can be only partially utilized. The early transfer of the collection to the Vatican and its overhasty publication by rival scholars precluded its careful documentation before the relics, containers, and written labels were separated and archived. This situation continues to impede research because we can no longer identify the original containers for most of the relics, as Galland's 2004 edition of the *cedulae* and the excellent analysis carried out by Smith have demonstrated.[4] This not only inhibits questions concerning the history of the papal relic collection, but also hampers the investigation of many of the reliquaries, whose date, use, and provenance are difficult to reconstruct without knowing their content.[5]

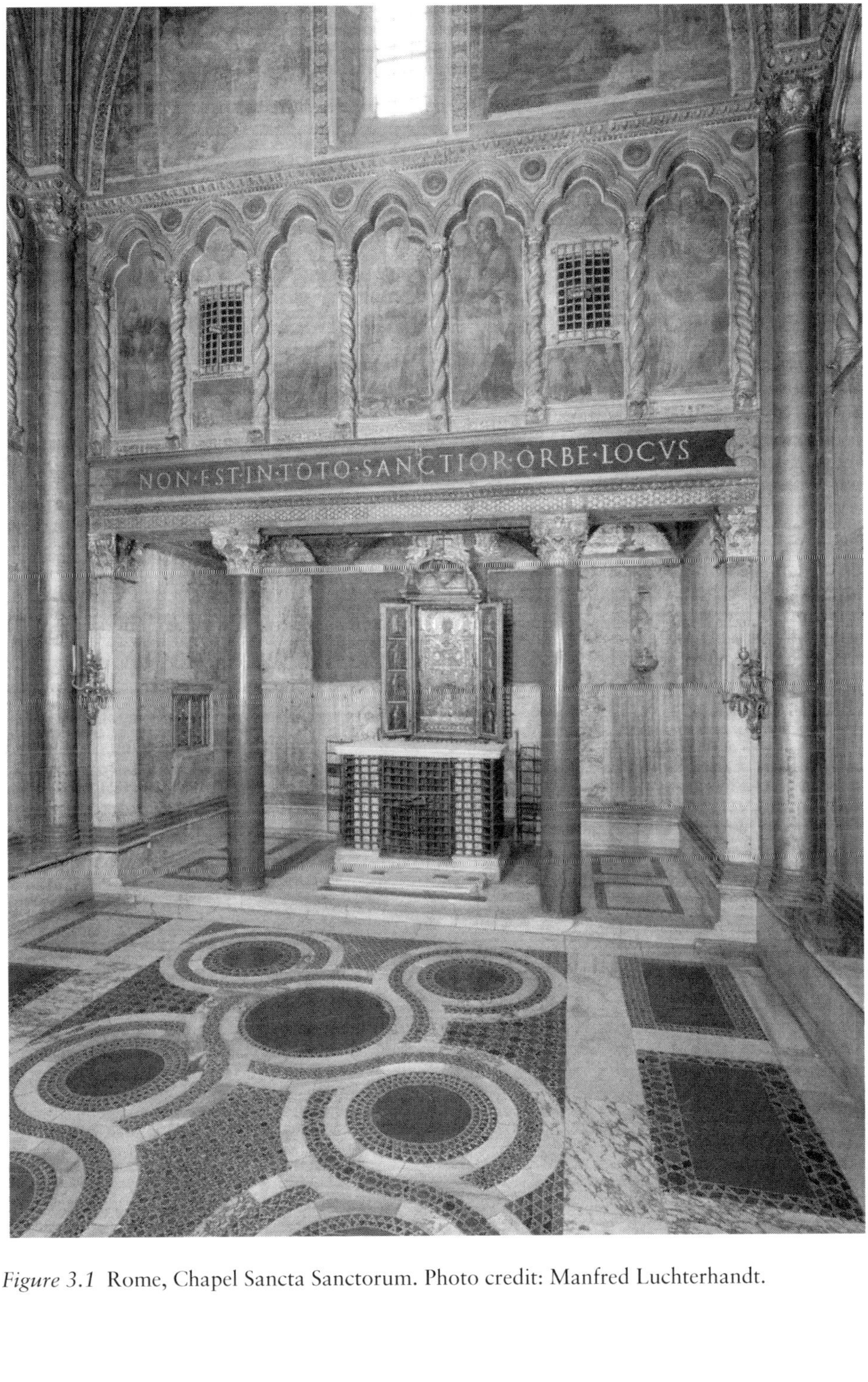

Figure 3.1 Rome, Chapel Sancta Sanctorum. Photo credit: Manfred Luchterhandt.

The consequences for the interpretation of this treasure can be demonstrated by just one example. Although the *cedulae* identify more than 50 relics from the Holy Land, discussion has been limited to a single object: the wooden reliquary box painted with scenes from the life of Christ (Plate 1), which has been one of the most frequently cited examples of pilgrimage art from Palestine.[6] Other wooden boxes (Figure 3.2) have been ignored, thus neglecting the opportunity to place this exceptional reliquary casket within a context of uses and practices.[7] Through the interpretation of the casket as "treasury art,"[8] the group of relics as an ensemble faded into the background, and questions of their geographical and political mobility between the papal court and other places of power were ignored. Even recent investigations of the casket from the perspective of visual culture have failed to consider how the reliquaries came to be in the Lateran Palace.[9]

I do not suggest that the following discussion reconstructs the history of the papal relic collection, as that would require the scientific investigation of many objects from which little art historical information could be gained. Rather, I proceed from a state of affairs that has only become clearly evident through the publication of the *cedulae*: that is, that even in its early days the papal palace possessed one of the largest documented assemblages of relics from Palestine from the Middle Ages and that these relics formed the historical core of the later collection.[10] Neither Theodelinda's treasure nor the relic collections of Charlemagne or of the imperial Pharos Chapel in Constantinople can be compared with the holdings of the papal collection in terms of the extent of their contemporary written documentation.[11]

The present study makes use of this potential by starting with the relics and investigating the objects and their sources in a way that also takes account of their places of origin. The relics are important pieces of evidence not only regarding Rome and the papacy but also in connection with the holy sites in Palestine, pilgrimage activity, and the political links between Rome and Jerusalem. Rather than dealing with the well-researched pilgrimage activity of late antiquity, I focus herein on the Islamic period up to about the year 900, as that is when the foundations of the papal relic collection were laid.

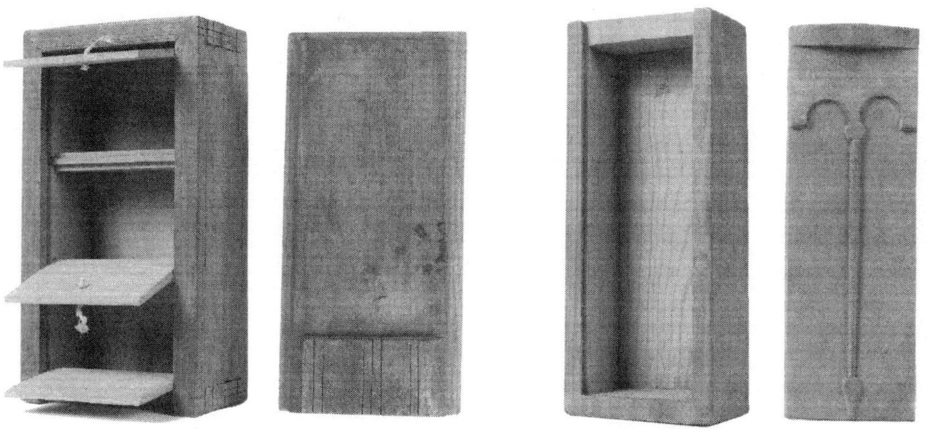

Figure 3.2 Wooden reliquary boxes, early medieval, probably Palestine, from the Sancta Sanctorum treasure, Rome, Vatican, Musei Vaticani, Museo Sacro. Photo credit: Vatican Museum.

Therefore this study is designed to answer the following questions: (1) What information does the Lateran treasure provide concerning the way relics were acquired in early Islamic Palestine and the practices and circumstances of post-antique pilgrimage? (2) What conclusions regarding their agents and bearers (pilgrims, ambassadors, monks, etc.) can be drawn from the quality, accessibility, and topographical distribution of these relics? (3) Through what ways and political channels did the relics come to Rome and to the papal court? (4) What were the prevailing political, social, and economic conditions that allowed for the high mobility of relics between Jerusalem and Rome in the early Middle Ages?

Relics dating from the early medieval period

One of the most important findings in Galland's study was that more of the Lateran collection was made up of relics from the loca sancta of Palestine than was evident from previously known medieval inventories. Fifty-seven of the 119 *cedulae* refer to locations extending from the Sea of Galilee to the Sinai Peninsula, and the handwriting on them dates from about the seventh century up to around the year 900.[12] Before that time the papal collection consisted primarily of relics relating to the life of Christ, establishing the papal palace and the adjacent *Basilica Salvatoris* as the focal point of a Roman cult of the Holy Land.

This historical profile is clearly evident if the two most important groups – the loca sancta relics and the relics from nonbiblical saints – are compared chronologically.[13] Although it was the cult of saints that distinguished Rome's historical identity, until the year 900 far more loca sancta relics than saints' relics were collected at the papal court, and even among the saints' relics few came from the West. The long-lasting refusal of the popes to permit the translation of Roman saints also had its effect on the collection in the papal palace. It was not until the tenth century that relics of Roman saints were more frequently found in the Lateran, and at the same time the tradition of relics relating to Christ came to an abrupt end (see Table 3.1).

The medieval inventories of the Lateran collection, which have so far dominated its interpretation,[14] do not reflect this history. They describe the Sancta Sanctorum after the changes in papal policy on relics, when the translation of *corpora* had become usual and the chapel was dominated by the cult of Roman saints.[15] However, very few relics that were at the center of attention at that time were in the chapel before the eleventh century. It is therefore necessary to redefine the historical identity of the papal relic collection for the early Middle Ages on the basis of the surviving relic labels.

Table 3.1 Chronology of relics held in the Sancta Sanctorum Chapel

Estimated date of cedulae	*Loca sancta*	*Saints*
Seventh/eighth centuries	14	11
Seventh century up to about 800	8	2
700–800	18	10
Eighth/ninth centuries	3	3
800–900	14	5
900–1000	–	4
1000–1300	–	39
Total	57	74

Eulogiae or relics? Changes in the quality and economy of post-antique pilgrimage objects

According to the evidence of the *cedulae*, the Lateran collection of relics from Palestine probably began in the seventh century, a time regarded as the "Byzantine phase" of the papacy, which was characterized by popes, officials, and monastic communities bringing their languages, culture, and religious practices from the East.[16] The new cults of saints that they brought to Rome also left their mark on the papal chapel: their authentics, which Galland did not always identify correctly, include surprisingly few Roman saints (see Table 3.2); rather there are many from Syria, Palestine, and Egypt – regions that were most strongly affected by the Arab conquest and consequently saw the flight of many refugees. For some Eastern saints, such as the Coptic Saints Amus and Arsenius, or the fathers of Judaean monasticism, Chariton and Theoctistus, the authentics provide the first evidence of a Roman cult.[17] This in itself makes the papal relic collection important evidence of the political upheavals that took place in the Mediterranean in the seventh century, as well as of the role of the church in these events.

As the largest group of objects apart from these saints' relics, the 57 loca sancta relics and their labels are invaluable sources of information regarding pilgrimage activity in early Islamic Jerusalem, for which very little material evidence has survived.[18] They not only confirm the continuity and intensity of interrelations between Rome and Palestine over a long period,[19] but also make it possible to conduct a statistical analysis of the forms and practices of the acquisition of relics to an extent not feasible on the basis of the written sources alone.

This applies above all to the custom of collecting solid materials as "site relics" in containers and recording their places of origin on written labels.[20] The most sophisticated product of this practice is the late antique wooden box that combines fragments of stone, wood, and textiles together with scenes from the life of Christ on the inner lid to create a reliquary made up of images and objects (Plate 1). Scholars regard this box as a typical product of the late antique pilgrimage industry around 600.[21] However, it only partly reflects the practices and customs of pilgrimage, as is true of other relics in the Lateran.

The only partial conformity between the objects in the papal treasure and the known sources concerning pilgrimage raises a question as to the extent to which pilgrimage activity changed its character under the economic, political, and religious conditions of Muslim rule. In their accounts from late antiquity, pilgrims reported that they collected primarily amorphous natural objects including oil, water, manna, wax, dust, leaves, and soil, as well as pictures, cloths, and *mensurae*, whose curative and protective powers were manifested in numerous miracles. As several scholars have demonstrated, it was not the value of these materials as memorabilia that was paramount, but their medicinal purposes and the protection they were thought to offer from dangers encountered during travel, such as storms, epidemics, and snake bites.[22] There is much less textual evidence relating to the stone relics in the box: pilgrims' reports refer to the stone table used to feed the multitudes, the pillars of Stylites, and fragments from the hill of Golgotha, that is, objects that were probably not assembled but rather chipped off a monument site.[23]

Table 3.2 The Sancta Sanctorum relics 650–900: groups, labels, places of origin

Gall. cat.	*Dating*	*Script group*	*Relic label*	*Objects (according to the 1906 catalog)*	*Saints, places of origin*
66	650–750	B	*de sepulcro sci maria*	–	Saint Mary, Valley of Jehoshaphat, Jerusalem
77	650–750	B	*de flumen iordane ubi dns pat(isatus)*	Two small stones	Baptism site, Jordan Valley
78	650–750	B	*de monte sina, de locu ubi muisis stetit quando tabulas recipit*	Earth	Mount Sinai
79	650–750	B	*de nota turra siloa*	–	Tower of City Wall near Siloam Pool, Jerusalem
28	650–750	C	*r(eliquias) sci aldigundae · uirginis*	–	Maubeuge, Belgium
29	650–750	C	*r(eliquias) de sco Isa[ac]*	Fragment of bone	(Identification not possible)
30	650–750	C	*r(eliquias) sco cosme. damiani*	Fragments of wax, sponge	Cyrrhus (Syria); Constantinople; Jerusalem
31	650–750	C	*r(eliquias) sco michahel*	White stone, sponge	Gargano (?)
41	650–750	D	*rel(iquias) sci amus*	–	Ammun, Monastery of Nitria, Egypt
44	650–750	D	*rel(iquias) sco caritto*	–	Tomb of Chariton, Tekoa, Palestine
20	650–750	E	*de sca lucia*	Fragments of bone, wood, fresco, small stone	Syracuse
21	650–750	E	*de cana gallilea uui dns ihs xps de aqua uino fecet*	Small stone	Cana, Galilee
22	650–750	E	*sci cibriani*	–	Cyprian of Antiochia
23	650–750	E	*de caluarie logo*	Small stone	Jerusalem, Church of the Holy Sepulchre
24	650–750	E	*de sco zaheo arbure*	Strip of linen	Tree of Zacchaeus, Jericho
25	650–750	E	*de sca sion*	Small stone	Jerusalem, Church of Holy Zion
10	650–750		*de petra supra qua corpus d(e)i genetricis labatus et myratus est in sca sion*	Small stone	Jerusalem, Church of Holy Zion
11	650–750		*de presepio xpi in bethleem*	Rubble	Bethlehem, Nativity Church
14	650–750		*de sanguine(m) dni n(ost)r(i) ihu xpi*	Small phial	Unknown
48	650–750		δε πετρο *scae calbariae*	Two small stones	Jerusalem, Church of Holy Sepulchre
65	650–750		*de arbore qui ipsi dns plantauit*	Fragments of cloth	Palestine
73	650–750		*de uultu domini incensum*	Charred substance	Unknown
90	650–750		*reliquias de sco constantino*	–	Constantinople; Rome (?)

(*Continued*)

Table 3.2 (Continued)

Gall. cat.	*Dating*	*Script group*	*Relic label*	*Objects (according to the 1906 catalog)*	*Saints, places of origin*
93	650–750		*de caluarie locu(m)*	Stone	Jerusalem, Church of the Holy Sepulchre
95	650–750		*de sepulchro xpi petra illa serratura qui serraui . . . illa*	(Stone)	Jerusalem, Church of the Holy Sepulchre
96	650–750		*saluatoris sce marta et sci michahel*	Small wooden tin	Unknown
97	650–750		*r(e)l(iquias) diuersas de lirino mon(asterio) id est palhiu(m) scae mariae nigro colore*	–	Abbey of Lérins, southern France
118	650–800	A	*terra de flumen iurdannis*	Kneaded earth	Baptism site, Jordan Valley
119	650–800	A	*cera de sepulcrus dni*	Wax drop	Jerusalem, Church of the Holy Sepulchre
120	650–800	A	*petra de presipet dni*	White earth	Bethlehem, Church of the Nativity
121	650–800	A	*terra de sepclus dni*	Kneaded earth	Jerusalem, Church of the Holy Sepulchre
122	650–800	A	*de petra de Calbarius locus*	(Stone)	Jerusalem, Church of the Holy Sepulchre
123	650–800	A	*spunnia dni*	Sponge	Jerusalem, Church of the Holy Sepulchre
124	650–800	A	*terra de spilunca Elisabeth cun Io[han]ne*	Kneaded earth	Ein Kerem, Grotto of Elisabeth
98	700–750		*. . . te sci sergi et bachi*	–	Resafa – Sergiupolis, Rome (?)
55	700–750		*d(e) sipunia dni*	Fragments of sponge, wood, cloth	Jerusalem, Church of the Holy Sepulchre
2	700–800	F	*primo de monte, ubi dns da Moisen locutus est*	Red stone	Mount Sinai
3	700–800	F	*de sepolcro dni*	White friable stone and powder	Jerusalem, Church of the Holy Sepulchre
4	700–800	F	*de presebio dni*	–	Bethlehem, Church of the Nativity
5	700–800	F	*de montem Caluarie*	Powder of white stone	Jerusalem, Church of the Holy Sepulchre
6	700–800	F	*de manna capo Iohanni de Emaecia*	(Manna)	Emesa (Homs), Church of Saint John
7	700–800	F	*de iordanne*	Friable stone and powder	Baptism site, Jordan Valley
8	700–800	F	*de colomna ubi dns flagellatus est*	Fragment of red marble (?)	Jerusalem, Church of Holy Zion
9	700–800	F	*de monte sinam*	–	Mount Sinai
63	700–800	G	*de lapide de ostio monumenti*	(Stone)	Jerusalem, Church of Holy Sepulchre
64	700–800	G	*de sco corni(lio) · de caesarea*		Caesarea Marittima

12	700–800	H	*de monte calbarie*	Two small stones	Jerusalem, Church of the Holy Sepulchre
13	700–800	H	*de bethlem*	Stone	Bethlehem
35	700–800		ΤΟΥ ΑΓΙΟΥ ΝΙΚΟΛΑΟΥ · ΕΚ ΤΗΣ ΚΟΡΙΦΗΣ	–	Myra, Asia Minor
45	700–800		*sci Pauli semlicus*	Fragment with hairs	Paul the Hermit or Paul *simplex*, Egypt
49	700–800		*reliquias de sco Stepha[no]*	Stones	Jerusalem, Church of Holy Zion
57	700–800		*rel(iquias) sci iuenii confes(soris)* (= Ivenius/Ivanius)	Brown substance	John Chrysostomos, Antiochia or Constantinople
60	700–800		*sca anasia* (= Anysia or Anastasia)	–	Thessaloniki or Rome
61	700–800		*de monte caluario*	Stone	Jerusalem, Church of the Holy Sepulchre
71	700–800		*rel(iquias) de sco arsenio*	Brown substance	Arsenius-Monastery, Egypt
74	700–800		*reliquias sci cruce(m)*	–	Unknown
75	700–800		*reliquias beator(um) martirs xiste . . . [pres]biteris* (uncertain)	–	Unknown
82	700–800		*reliquia . . .*	–	–
84	700–800		*rel(iquias) sci Lauranti*	Fragment of bone	Saint Lawrence, Rome
86	700–800		*rel(iquias) de colomna ubi dns flagellatus fuit*	Small stone	Jerusalem, Church of Holy Zion
87	700–800		*r(eliquias) de sepulchro dni*	White earth	Jerusalem, Church of the Holy Sepulchre
88	700–800		*de mensa apostoloru(m)*	Fragments of sponge and wood	Jerusalem, Church of Holy Zion or the Holy Sepulchre
89	700–800		*de sco teobtisto*	Stone	Monastery/Tomb of Theoctist, Judean Desert
94	700–800		*c rel(iquias) Pauli · apo(stoli)/sci Pauli ·*	Stripes of silk	Rome
42	750–850		*rel(iquias) de duodecim tronos*	–	Tabgha or Jerusalem
43	750–850		*rel(iquias) sci Zachariae*	–	Tomb of Zechariah, Eleutheropolis (Beth Guvrin)
46	750–850		*[s]co iordane*	–	Baptism Site, Jordan Valley
70	750–850		*de terra si [. . .] ome*	Earth	Unknown
76	750–850		*de sco epyfanio ep(iscop)o*	Fragment of cloth	Constantia, Cyprus

(*Continued*)

Table 3.2 (Continued)

Gall. cat.	*Dating*	*Script group*	*Relic label*	*Objects (according to the 1906 catalog)*	*Saints, places of origin*
83	750–850		*r(eliquias) sci domnini*	Fragment of cloth	(Identification not possible)
26	800–900	J	*reliquias de illa colomna dni*	Stone	Jerusalem, Church of Holy Zion
27	800–900	J	*reliquias de illo calice dni salutare*	Fragment of wood	Jerusalem, Church of Holy Sepulchre
47	800–900	J	*de sco caluario loco ubi xps crucifixus est*	Fragment of bone	Jerusalem, Church of the Holy Sepulchre
91	800–900	J	*reliquias de ligno saluatori ubi xpi postus fuit*	(Wood)	Bethlehem, Church of the Nativity; Rome (?)
99	800–900	K	*lapidem de spelunca ubi Elisaiuet cu(m) Ioh(anni) fugiens*	(Stone)	Ein Kerem, Grotto of Elisabeth
100	800–900	K	*lapidem ubi sed(erunt) angeli in monumentum*	Stone	Jerusalem, Church of the Holy Sepulchre
101	800–900	K	*lapidem de monte calbarie*	(Stone)	Jerusalem, Church of the Holy Sepulchre
102	800–900	K	*lapide d(e) fluuium Iordanem*	Kneaded earth	Baptism site, Jordan Valley
103	800–900	K	*terra d(e) sepulchrum dni*	(Earth)	Jerusalem, Church of the Holy Sepulchre
106	800–900	K	*lapide de presepie dni*	Stone	Bethlehem, Church of the Nativity
109	800–900	K	*cera de sepulchro dni*	Wax drop	Jerusalem, Church of the Holy Sepulchre
111	800–900	K	*oleo de sepulchro dni*	Oily parchment	Jerusalem, Church of the Holy Sepulchre
69	800–900	L	*pr praese(pio)*	Small stone	Bethlehem, Church of the Nativity
72	800–900	L	*calb[arie]*	Stones	Jerusalem, Church of the Holy Sepulchre
33	800–900		ΤΟΥ ΑΓ(ΙΟΥ) ΪΕΡΟΝΥΜΟΥ	Bone, sponge	Bethlehem, Church of the Church
56	800–900		*de sco georgio*	Fragment of wood	Diospolis – Lydda or Rome
59	800–900		*abdon (et) senn(en)·*	–	Rome
110	800–900		*de manica sci Petri*	Drops of blood (?)	Unknown

The preferred natural specimens that travelers took from the holy sites are also reflected in objects that have been preserved, from tokens made of earth to pilgrimage bottles made of glass, clay, lead, tin, or gold to hold oil from the holy sites or other substances.[24] These natural substances do not feature prominently in the Lateran collection, as one would expect. Predominant among the documented materials are solid substances (38), primarily stones (26), then textiles, wood, and sponge (12), with soil (8), dust, wax, or oil making up only 27 percent of the total. The most usual contact materials of late antiquity (oil, soil, manna, dust, textiles) account for only 30 percent, and there are very few liquids (see Table 3.3).[25]

The relative dearth of soil and liquids matches the absence of the well-known pilgrimage items such as are found in abundance in the relic collection of Queen Theodelinda (ca. 570–627).[26] In the Lateran collection, holy soil was no longer preserved in pressed medallions bearing stamped images, but rather was in small sacks or parchment; oil and wax were no longer in clay or metal pilgrim's bottles but in glass flasks sealed with parchment inserted in a wooden box or were trickled onto a sheet of parchment.[27] The period between the dating of Theodelinda's treasure and the Lateran *cedulae* thus reflects the transition from late antique to early Islamic Palestine, which was accompanied by a drop in the antique production of pilgrimage objects.[28]

Was this change the result of an economic decline? Or was there a shift in the market, with the collapse of the pilgrimage industry leading to a preference for solid objects that could be transported in any bag? Or were visitors less interested in curative substances than in obtaining permanent memorabilia for a church relic collection?

Unequivocal answers are impossible because economic aspects and shifts in religious practice are not mutually exclusive. Evidence of economic decline is provided by cases in which the pilgrim rituals did not change, but the objects did. For example, the Lateran holds several relics of *terra de sepulchrum domini*, which confirm that in the ninth century visitors to the Holy Sepulchre were still taking consecrated soil away with them, as they had done at the time of Gregory of Tours.[29] In the case of the *spelunca* of Elizabeth (Ein Kerem), however, simple lumps of earth or stone

Table 3.3 Materials of the Lateran loca sancta relics

Material	*Number*	*In percent*
Stones	26	48
Soil	8	15
Wood	5	9
Sponge	4	7
Textiles	3	5.5
Dust	3	5.5
Wax	2	4
Oil	1	2
Manna	1	2
Bones	1	2
Total	54	100%

relics take the place of pictorial medallions, such as have been preserved in Monza.[30] According to several accounts, oil from the Church of the Holy Sepulchre had not lost its importance as a medicinal remedy in early medieval times, although there is no longer any evidence of the production of *eulogiae*.[31] The change in the form of the objects at the same time that practices remained stable would seem to indicate that the market for pilgrimage articles did indeed alter in the early Islamic period.

It is more difficult to assess the influence of the cult of relics on the selection of objects. Pilgrims' reports seldom mention the collecting of stones as site relics, and standardized containers for transporting them have not survived. The medicine box adapted for that purpose in the Lateran collection (Figure 3.3) suggests that it made more sense to adapt an existing ornamental receptacle or, as in the case of the second stone reliquary (Plate 2), to craft a custom-made container.[32]

After all, relic collectives that were worth such effort and expense were probably the exception. The grouping of the labels according to their script indicates that the relics came into the Lateran in units of no more than four to eight objects, and this is about the number to be found in other church treasure collections.[33] With more than twenty pieces, the painted box (Plate 1) was thus an extremely extravagant individual item, rather than a mass-produced object "manufactured in large quantities for the pilgrimage market."[34] Furthermore, both stone boxes are considerably larger and heavier than the well-known devotional objects of late antiquity, which raises the question of whether they were indeed at all suitable for the traveling conditions encountered by pilgrims.

That the very construction of the containers took into account a certain practice of venerating relics is demonstrated by the cycle depicting the life of Christ on the interior lid of the painted box (Plate 1). Its location corresponds to that of later staurothekes and indicates that the relics were venerated in the opened reliquary, as was customary in the veneration of the cross as well as in other cases.[35] It is conceivable that the stone box was also intended for a relic of the cross but was later put to a different use. Its widely accepted pre-614 dating based on its representation of Christ's tomb falls during the time that the relic of the cross was still in Jerusalem.[36] The original use of the casket as a staurotheke is also suggested by the cycle of images, such as is also found on the Monza ampullae and on enkolpions and staurothekes, and particularly on objects containing relics of the cross.[37] The elaborate decoration of a liturgical object would doubtless be more in keeping with the status of a relic of the cross, since it is not usual in other objects from Jerusalem. Indeed, if that is the case, this relic box could be one of the oldest surviving staurothekes.[38]

Moreover, the size and design of the container is more similar to that of church reliquaries used in ceremonies than to lightweight pilgrim objects.[39] Was it this ritual use of relics that led to the preference for tangible, permanent objects? After all, stones and pieces of cloth were easier to work into a reliquary than oil or soil, which may explain their preferred use in the production of enkolpia.[40] The preference for solid objects might then be explained by the increasing transfer of such objects to altars and precious reliquaries for liturgical use, in which case the character of the relic as an amulet or medicinal remedy must have declined in importance.[41]

Provenance, geographical distribution, and modes of travel

If the geographical distribution of the relics (Figure 3.4) is taken as reflecting the mode of travel of the people who collected them, then a typical trend of the early Middle

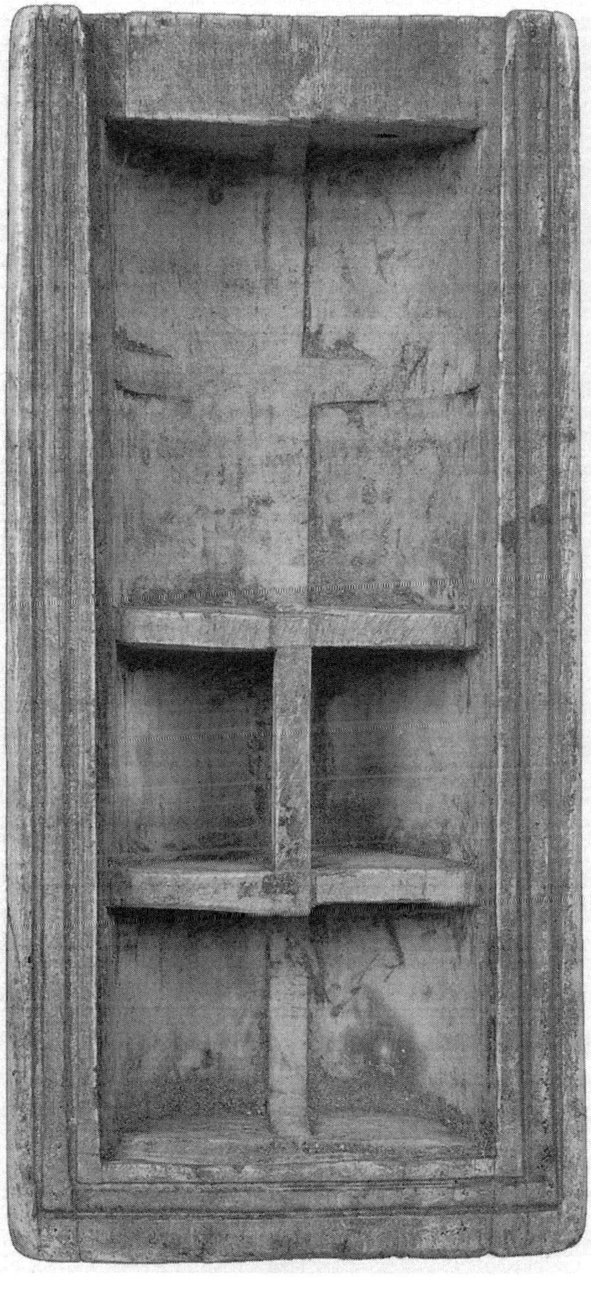

Figure 3.3 Medicine box, ivory, Egypt, fifth century (with medieval alterations), from the Sancta Sanctorum treasure, Rome, Vatican, Musei Vaticani, Museo Sacro. Image source: Christoph Stiegemann, Martin Kroker and Wolfgang Walter, eds., *Credo: Christianisierung Europas im Mittelalter*, vol. 2 (Petersberg: Imhof 2013), 132, figure 98.

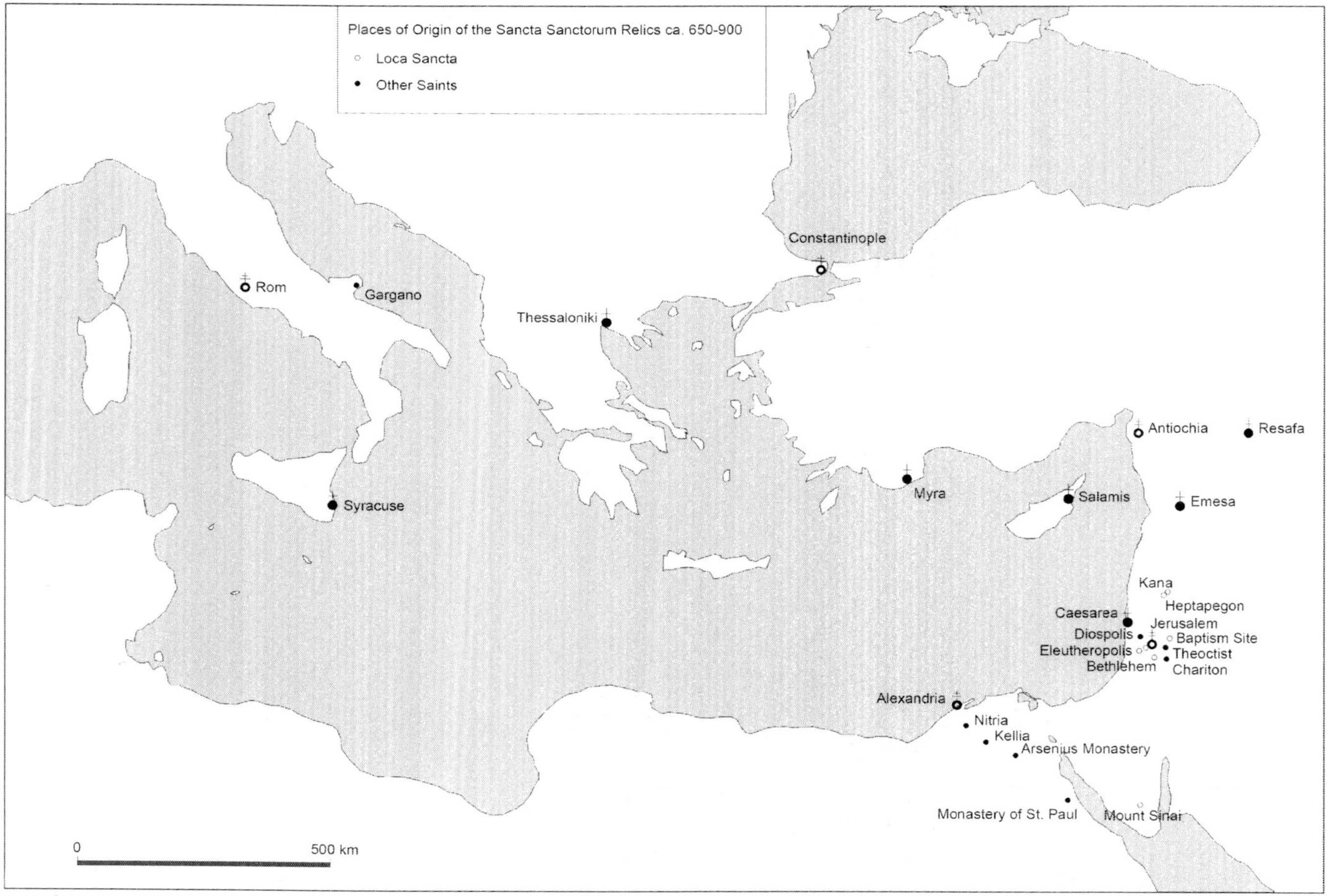

Figure 3.4 Eastern places of origin of the Sancta Sanctorum relics, ca. 650–900. Image credit: Manfred Luchterhandt.

Ages becomes apparent, namely, that interest in the Christian topography of Palestine in all its historical breadth declined, whereas the appeal of individual sites and relics increased.

Approximately 59 relics from 14 biblical sites were held in the papal collection, some of them in groups of four to eight objects, which in most cases probably came to the Lateran together, but without any discernible classification scheme.[42] Many multiples of certain relics suggest that they were not acquired according to any strategy. Rather, the collection of relics was accumulated successively through various missions whose destination was often Jerusalem. For instance, as many as nine relics of the rock of Golgotha came to the Lateran in this way (see Table 3.4).[43]

A map of their places of origin (Figure 3.4) shows that most of the relics came from the environs of Jerusalem; other regions were hardly visited. The most usual locations were no more than one day's journey from Jerusalem: Bethlehem (7), the Baptism site (5), Jericho (1), Ein Kerem (1), and Eleutheropolis (1). According to the relic labels few visitors were interested in the mountain of Moses in Sinai, and even fewer in the Christian Galilee.[44] If one also takes into account the places of origin of the saints' relics (Figure 3.4), other destinations were typical of the travel routes in the early medieval period: from southern Italy across the Mediterranean to the coasts of Asia Minor, Cyprus, and Syria or along the southern route via Alexandria, the Nile Delta, and northern Sinai.[45] The first route was taken in 721–726 by the Anglo-Saxon missionary Willibald, who traveled by sea from Syracuse via Ephesus and Cyprus and received his transit papers in Emesa.[46] The second route was taken in 867 by the Frankish pilgrim Bernard; setting out from Rome he traveled to Taranto and boarded the ship of a slave trader sailing to Alexandria, thereafter taking a well-developed pilgrimage route via Gaza to Jerusalem.[47]

Within Palestine, however, the places of origin of the Lateran relics do not at all correspond to the records that exist concerning late antique or early medieval pilgrims. For them, Galilee and the sites mentioned in the Old Testament were also important destinations, and they generally took more time to visit the periphery of the country.[48] The unknown collectors of the relics, by contrast, had little interest in sites that represented a Christian memorial landscape as defined by Halbwachs, an imagined Holy Land.[49] Cana and the site of the feeding of the multitude are only represented once,

Table 3.4 Relics according to the script groups of their labels (after Galland)

Group	*Date*	*Sites/relics*
B	ca. 650–750	Tomb of Mary, Jordan, Mount Sinai, Tower of Siloam
E		Cana, Golgotha, Zion, tree of Zacchaeus (Jericho), Cyprian, Lucia
A	ca. 650–800	Jordan, tomb of Christ (2×), Golgotha, Bethlehem, sponge, cave of Elizabeth
F	ca. 700–800	Tomb of Christ, Golgotha, Bethlehem, Jordan, Zion, pillar of the flagellation, Mount Sinai, John (Emesa)
G		Tomb of Christ, Caesarea,
H		Golgotha, Bethlehem
J	ca. 800–900	Golgotha, pillar of the flagellation, holy chalice, Bethlehem
K		Tomb of Christ (3×), Golgotha, Bethlehem, Jordan, cave of Elizabeth (possibly a duplication of group A)

Table 3.5 Provenance of the loca sancta relics

Jerusalem and Environs (33)
- Church of the Holy Sepulchre (23): tomb, angel's stone, Golgotha, sponge, chalice
- Basilica of Hagia Zion (6): pillar of the flagellation, stone of the anointment of Mary, Saint Stephen
- Grotto of Elizabeth (2)
- Saint Mary's tomb (1)
- Tower of Siloam (1)

Bethlehem and Church of the Nativity (7)
Jericho (6)
- Baptism site on the Jordan (5)
- Tree of Zacchaeus (1)

Eleutheropolis, Tomb of Zechariah (1)
Mount Sinai (3)
Galilee (3)
Emesa (1)

and Capernaum, Nazareth, and Mount Tabor not at all. Even in Jerusalem, the source of 60 percent of all the relics, their interest was one-sided: nearly all of them were from the Church of the Holy Sepulchre and the Basilica of Hagia Zion, the two main sites of pilgrimage (see Table 3.5). Other places were of little significance.

No internal logic can be discerned in the relic collection. The individuals who acquired these relics in Palestine preferred the most frequented pilgrimage sites, but they did not travel like foreign pilgrims. If the sites relating to saints are also taken into account (Figure 3.4), the region in which they moved corresponds more to the centers of ecclesiastical life around Jerusalem[50] and to the area of the Judaean monasteries, which looked after many of the holy sites.[51] The Church of the Holy Sepulchre and the Basilica of Hagia Zion, the most common places of origin of the relics, were the focal points of their activities; outside Jerusalem it was Bethlehem, the Monastery of Saint Gerasimus of the Jordan, and the more distant monasteries of Egypt and Sinai.[52] The fact that the relic collection more closely reflects the interests of Eastern monks than those of Western visitors is also confirmed by looking at the catalog of Epiphanius, the only Greek-language pilgrim guide from this period: it shows the same focus on Jerusalem, the Sinai, and the tombs of the fathers of monasticism, which counted more than Jerusalem in the monastic *peregrinatio*.[53]

As long as it remains unknown as to whether both groups of relics came to Rome with the same individuals, no conclusions can be drawn from this. It is evident, however, that Jerusalem was the primary destination of the travelers and that, unlike typical pilgrims, they had little time for or interest in the wider regions of the Middle East. The relics came with people who traveled directly between Rome and Jerusalem rather than those who explored the Holy Land with the curiosity of foreigners in an unfamiliar country.

The accessibility of relics and the social stratification of their collectors

Pilgrims' reports from the late antique period give the impression that the eulogiae received at the sites, from tombs and rivers, were available to most visitors.[54] However, this is not at all certain for many of the Lateran relics. Many of them suggest privileged access to places and authorities, which implies the existence of social stratification among those who conveyed them. Whereas soil, wax, and oil from the Holy Sepulchre were presumably still distributed generously as contact material, it is improbable in the case of the stone relics from the pillar of the flagellation or other sites. According to their *cedulae*, three such stones are from the "resurrection stone" of the Sepulchre and eight others are from the rock of Golgotha.[55] Given that the *Commemoratorium* of 808 notified custodians of relics from the tomb of Jesus, the clergy was unlikely to have given general clearance for the site to be plundered.[56] It is more probable that they themselves gave fragments of stone to suitable supplicants. Even Abbot Daniel (1104/07) reported that he received his stone from a custodian during a private viewing, "during which he made me swear not to tell anyone in Jerusalem."[57]

The two stone boxes in the Lateran (Plate 1 and Plate 2) provide indications that the members of the clergy themselves prepared and standardized the stone relics. They both show the same triangular carved stone, one bearing the inscription "from the life-giving site of the resurrection."[58] A similar form of control can also be presumed in the case of the hill of Golgotha. Although there is early evidence of inscribed stone relics from the hill, it is unlikely that the crucifixion site, which was surrounded by barriers and later walled in, was left open to pilgrims to help themselves.[59] That the clergy was concerned with this matter is demonstrated by the *Vita* of Martha, the mother of Simeon the Younger. In seeking to obtain a relic from the cross, Simeon first approached the staurophylax, who then also included stones from the rock in the golden cross.[60] The office of Custodian of the Cross was often held by a Sabaite monk,[61] so these monks presumably had access to many relics.

Such privileged access applied in particular to those major relics whose removal would have required official permission. Of the instruments of the passion and the Old Testament objects in the Church of the Holy Sepulchre, several (the holy sponge, the holy chalice, the *mensa apostolorum*) found their way to the Lateran, even though they were venerated in the relic chamber of the upper floor and other protected places.[62] According to the custodians of the holy chalice listed in the *Commemoratorium* of 808, this *sacrarium* was well guarded.[63] For whereas the Piacenza Pilgrim received only water to drink from the sponge,[64] Bishop Arculf was permitted to kiss the chalice through an opening in the shrine.[65] The high-ranking cleric was granted access to the original, but the crowd was given contact media. Insofar as the fragments of the chalice and sponge in Rome are not secondary relics, they would have been brought by a person of high status rather than in the baggage of a pilgrim.[66]

The same can be assumed regarding many relics from the Basilica of Hagia Zion, where only scraps of cloth from the tree of Zacchaeus recall a popular pilgrimage rite.[67] The removal of objects from the churches normally required the consent of the patriarch,[68] so it is likely that a large proportion of the relics came to Rome with the cooperation of the church authorities and via official channels.

Papal relic diplomacy? Communication and mobility

Many of the relics from Jerusalem probably came as diplomatic gifts.[69] Indications of this fact include not only the size of the containers and the exclusive nature of the sacred objects, but also the dignity with which they were treated (bags of silk/wool/linen, authentics). Generally speaking, the frequency with which relics were documented as diplomatic gifts indicates that such gifts would also have been customary between clerical institutions.[70] The correspondence of Gregory I contains references to thirty-seven relics in 854 letters culled from a much larger correspondence (at least two or three relics per annum).[71] Gregory once sent thanks to Jerusalem for a *benediction . . . de loco sancto* and several times to the patriarch of Alexandria for a *benedictio sancti Marci*.[72] Most of the relics were given in response to wishes expressed by the recipients. The papal chancery had form letters for this purpose, along with a procedure that was under the control of priests.[73] In view of the increasing veneration of relics in the early Middle Ages, later popes presumably sent relics even more frequently.[74]

That the patriarchate of Jerusalem also processed requests is confirmed by the *Vita Radegundis* and the *Life of Sainte Marthe, mother of Simeon*: a messenger took Simeon's request to Jerusalem, from where he received the relic at the hands of a church envoy enclosed in a golden cross.[75] The cross also contained fragments from the hill of Golgotha and a stone from the tomb of Christ. These two stone relics are also the most common types of objects in the papal collection. In the second box (Plate 2), the stone relics are supplemented by pieces of hardened soil and seals made of wood whose cross dies reflect motifs from pilgrims' tokens.[76] The *benedictionem et reliquias de sepulchro domini* taken to Aachen in 799 is an indication that this combination of relics and other holy presents was a typical gift from the patriarchs.[77] The number of relics in each consignment, namely four to eight, also corresponded to contemporary customs: the chancery of Pope Gregory usually sent four or five contact relics in response to requests.[78]

The details given in Gregory I's correspondence about travel routes are also informative.[79] His correspondence with Alexandria alone includes thirteen letters, which would have entailed even more journeys.[80] The envoys used the *navis alexandrina* between Egypt and Italy, which still supplied papyrus for the Curia in the ninth century.[81] In this way, at least five *benedictiones sancti Marci* were transported from this port city to the Lateran.[82] In regard to Jerusalem, it is known that papal officials brought a monetary donation and on another journey took back a relic.[83]

A further category of documents that testify to relic gifts is comprised of the synodal letters sent by newly elected patriarchs to their counterparts to affirm their orthodoxy: their tradition is also affected by many losses.[84] Of the four synodal letters sent by Gregory I, the register records only the one sent to Antioch: it contained a relic of Saint Peter, which the pope recommended as protection against sickness.[85] The emperor and patriarch of Constantinople made similar use of their monopoly on the relics of the cross: in 811 Nikephoros enclosed a precious staurotheke in a synodal letter to Leo III, and in 880 Photios sent a further one to the future pope Marinus.[86] As early as 638, Cyrus of Alexandria had received the Ekthesis along with a staurotheke.[87] It seems that the patriarchs of Jerusalem participated in these exchanges, as the patriarch of Constantinople also possessed relics from Jerusalem.[88] It is likely that only some of these gifts were placed in private chapels that were best suited to accommodate such tokens of friendship. Hence, the relics of the Sancta Sanctorum include those of bishop-saints from Antioch as well as of major saints from the region around Constantinople.[89]

If the relics received by the Lateran between 650 and 900 are taken as material sources to reconstruct the frequency of diplomatic letters between Jerusalem and Rome, then even the regular contacts via such synodal letters would have been sufficient for the accumulation of the collection. In the years 604–701 alone, during which there were twenty pontificates in Rome and additional terms of office in Jerusalem, some forty to fifty legations might have crossed the Mediterranean.

Diplomatic contacts between Rome and Jerusalem (634–880)

There is, however, a difficulty entailed in this explanation of the relics and reliquaries as diplomatic gifts. Contacts between Rome and early Islamic Jerusalem are insufficiently documented, even though patriarchs and monks were often among Rome's allies.[90] Only isolated sources have survived: evidence of how regularly the patriarch of Jerusalem corresponded with Western cities in 869 is provided by an envoy to Constantinople, who announced gifts and apologized for the irregularity of earlier letters.[91] On the Roman side, in 865 Nicholas I mentioned regular visitors from Alexandria, Jerusalem, and Constantinople.[92] The symposium to which the pope annually invited the monks of the other patriarchates was even better attended under his successors than previously.[93] *Negotiatores* was the term used in 869/870 to designate the monks who traveled between Rome and Jerusalem.[94]

The strongest evidence for these institutional contacts comes from documents concerning the Monothelite controversy (633–681), which indicate a high degree of mobility of persons, letters, and manuscripts.[95] Around 620, John Moschus fled from Bethlehem to Rome with his friend Sophronius and established a circle of pupils there.[96] When Sophronius brought the body of his master back to Bethlehem in 633/634, there were already close contacts between this group of exiles on the Aventin and the Curia.[97] Elevated to the patriarchate in 634, Sophronius in his struggle against Monotheletism immediately directed his diplomatic efforts toward the pope.[98] Up until 649, his legate Stephen undertook three journeys to Rome, the first two during the conquest of Bethlehem and Jerusalem.[99] The "Roman-Palestinian alliance" soon led to the election of Pope Theodore I (642–649), a native of Jerusalem,[100] and to the planning of the Lateran synod 649 under the leadership of the Eastern monks.[101] The pope appointed Stephen vicar for Jerusalem, an office that Martin I extended to Antioch. The amount of travel entailed in this "unprecedented interference" by the papacy in the Eastern churches[102] is evident from Stephen's report in 649, in which he wrote that during the Arab siege of Jerusalem Sophronius made him swear an oath at Golgotha to position Rome against the Monothelites.[103]

For the period after 649 there is a dearth of information regarding the authority of Rome in Palestine. According to Trombley, the vacancy of the patriarchate ended shortly after the council of 680/681, when the *topoteretes* Theodore was elevated to the rank of patriarch.[104] Theodore evidently ruled with papal approval and had good contacts with Rome, as his student Andrew of Crete was the first to report about Roman cult images.[105] The influence of the popes on Eastern church policy was demonstrated by the appointment of the papal legate as patriarch of Antioch in 681, when the former patriarch was exiled to Rome.[106] Contacts also seem to have existed at an early stage in regard to the iconoclasm controversy.[107] Such contacts were evident during the Lateran Council in Rome in 769, and they turned the meeting into a "*carrefour oecuménique*" in that controversy.[108]

The acts of 787 and a letter of Hadrian I quote a synodal letter from the patriarch that arrived in 767 and was read out in public after a copy had been sent to Pippin.[109] Paul I had already received a letter from Patriarch Cosmas in Alexandria, which was brought by a monk who traveled by ship.[110] These two documents suggest that envoys could pass the borders of the caliphate, which they probably did in fairly large numbers in connection with the Second Council of Nicaea in 787. For the ninth century, the regesta investigated by McCormick document an even greater degree of mobility.[111] As late as 879, John VIII wrote to the patriarch of Jerusalem confirming the arrival of monk-envoys and apologized for his simple reciprocal gift.[112] The response indicates that this exchange was nothing out of the ordinary.

From 634 on, the Jerusalem patriarchs had many reasons for requesting support from the popes.[113] Did they add weight to their entreaties, as at Aachen in 799, by giving precious relics as gifts? The way in which Gregory I received a *benedictio de loco sancto* in response to his intervention[114] implies that the sacred economy between the two churches was an established tradition, and it is unlikely that it would not have been continued by popes from Jerusalem or Syria. Even as early as 653, Martin I refuted the allegation of having collaborated with the Arabs by arguing that the matter in question only concerned alms for Christians.[115] In that case, Charlemagne's Eastern policy would have been preceded by an earlier assistance from Rome – perhaps on a smaller scale – which must have been a matter of course for a church with an ecumenical horizon and many refugees.

Cooperation with the caliphate probably also applied to pilgrimage activities. Frankish pilgrims such as Willibald, Frotmund, and Bernard received their travel papers in Rome and not in Byzantine Venice, which implies that there were recognized transport routes and agreement regarding the form of transit documents.[116] These contacts were beneficial to the Muslim side as well. It was preferable for them to assent to involvement by the popes or the Franks in Jerusalem than to agree to interference from their powerful neighbor, Byzantium.

Jerusalem relics in Rome in the seventh century

Some effects of these political contacts on the migration of relics can be seen very soon after the conquest of Jerusalem. No later than during the pontificate of Theodore I, the crib of Christ was placed in the Basilica of Santa Maria Maggiore, which has borne the epithet "*ad praesepe*" since the time of that pope.[117] The oratory for the relic was probably built before the end of the seventh century, and many popes donated precious gifts, including an icon of Mary.[118] As a new locus sanctus representing the grotto of the nativity in Bethlehem, this place was unique in Rome and was later included by the popes in their Christmas liturgy.[119]

That Theodore himself arranged the installation of the relic, as has often been postulated, is indeed plausible on chronological grounds: it was not until after the death of Honorius (638) that there was a rapprochement between Rome and the clergy in Palestine. Under Theodore I, it culminated in the appointment of Stephen as vicar, vested with wide-ranging competences in the patriarchate of Jerusalem. As *locum tenens*, only Stephen had the authority to release a relic of this rank: opportunities would have been offered by his visits to Rome up to 649. Was it a rescue mission, making the papacy take its share of responsibility? In his Christmas sermon of 634, Sophronius lamented the occupation of Bethlehem, which prevented the patriarch from venerating the relic of the crib.[120] In Rome in 649, Maximus again called for protection of the holy sites and Martin I compared the enemies of the earthly Jerusalem to those of the heavenly Jerusalem.[121]

The installation of the crib relic in Rome demonstrated the pope's determination to preserve the memory of Arab-occupied Palestine, an intention that is also evident behind the measures in respect to Santo Stefano Rotondo. A third case relates to the Lateran and the relic of the cross. As Sible de Blaauw has shown, relics of the cross are documented as early as in late antique Rome, although their whereabouts after the fifth century is open to question – in the *Basilica ad Hierusalem* in the oratory of the cross in the Lateran and in its counterpart in Saint Peter's Basilica.[122] All were associated with places of authority, but did not become objects of permanent veneration.

This changed when the festival of the Exaltation of the Holy Cross was taken over from Jerusalem. The earliest evidence comes from the references in the *Sacramentarium Gelasianum Vetus* and in the *Gregorianum Paduense* (658/681).[123] According to these documents, the veneration of the cross was already known in Rome, albeit perhaps limited to churches possessing a relic.[124] It was only under Sergius I (687–701) that veneration of the cross became an episcopal rite, following the discovery of an old silver casket in the *secretarium* of Saint Peter's, which contained a relic of the cross. This relic was thereafter venerated annually in the celebration of *Exaltatio Crucis* in the basilica of the Lateran.[125] The same rite is found shortly afterward in a further ceremony from Jerusalem – the *adoratio Crucis* on Good Friday.[126] On that day, the relic of the cross was carried from the Lateran to the Church *ad Hierusalem* and was venerated in an open casket by means of a kiss.[127] Despite contradictions in the written records, there is much to indicate that one of the reliquary crosses that was carried in these papal ceremonies was the lost jeweled cross that was found in the arca of Leo III in 1905.[128]

The transfer of the Holy Cross relic to the Lateran Palace suggests that in about 700 there must have been places where the popes kept their sacred objects. Since the oratory of Saint Lawrence was not mentioned until 768, its location is still problematic.[129] The earliest evidence of the papal treasure dates to 640, when troops of the exarch plundered the Lateran, stealing not only gifts given by rulers to Saint Peter's Basilica but also the *cymilia episcopii*.[130] The transfer of the Holy Cross relic to the palace thus followed a long established custom. Given its rank and function, it is more probable that Sergius I would have placed the papal cross in a chapel, perhaps in a *confessio*, such as existed in the *Oratorium Crucis* in the fifth century.[131] Was that the chapel in which the popes kept their relics from Jerusalem? It would be logical to assume so.

After all, this relic collection must have been built up mainly in the seventh century, when "Jerusalem" was a topical issue and the Lateran was faced with the loss of many treasures. An indication that a new start was probably made after 640 is that, with the exception of the *Capsella Vaticana* (610/641),[132] all reliquaries in the Sancta Sanctorum that were composed of precious metals and therefore attractive for plunderers date from after that time. Only a few years later, Theodore I built a new palace chapel and the *basilica Theodori*, which indicates that this construction work was carried out in preparation for the synod of 649.[133] As most of the *cedulae* were drawn up from the seventh century on suggests that the papal relic collection was established – or probably re-established – at that time as well.[134] One can imagine that it was the same pope of Jerusalem who turned a chapel in his palace into a focal point for commemorating the Holy Land.

Around 700, the Lateran collection would probably have still been a small ensemble of relics relating to places and to saints venerated by the Eastern diaspora, many of them brought by officials of popes or patriarchs, who operated as *negotiatores* between Rome and Jerusalem. A hundred years later, when Leo III donated the *arca* at the climax of new relations with Jerusalem, the identity of this treasure as a collection of relics relating to Christ was already clear. This was also respected when new objects were added, although the translation of Roman saints had by then become commonplace. The concentration on relics of Christ in the palace was an alternative model to the relic collections of the titular churches and probably facilitated the involvement of the chapel in the liturgy of the basilica.[135] However, the existence of the icon of Christ might also have motivated the addition of further such objects. For even before the pontificate of John X (914–928), a selection of loca sancta relics was placed on the icon's lower frame, so as to be carried in procession with the image through Rome.[136]

The basis for this further growth was probably again a bilateral diplomacy in which relics were exchanged for political support. In about 800, Patriarch Thomas sent his syncellus to Rome to ask for assistance in connection with Arab taxes.[137] Not only the capitularia of the Carolingian rulers but also the fate of the Sabaite monks who were martyred in Cordoba in 852 for collecting donations and the letters sent to the courts of Europe by Elias III of Jerusalem in about 880 document the Jerusalem Church's chronic lack of funds, which compelled their patriarchs to draw attention to the situation faced by the Christians there.[138] The papal collection of relics from Jerusalem is probably the most concrete result of these mutual relations, which were of benefit to both sides.

The cessation of these bilateral relations is less easily explained. As of the tenth century, no new relics from Jerusalem were added to the collection, despite the propaganda that was circulated in the High Middle Ages concerning the treasure in the papal chapel. The process coincides with chaotic conditions in the Curia and a sharp decline in long-distance trade and pilgrimage activity. Pope John VIII hinted at the background circumstances in 879 when he complained to Elias of Jerusalem that the Saracen threat prevented him from sending anything of value.[139] However, when trade relations were revitalized around 970/980, pilgrimages to Jerusalem resumed and soon exceeded the levels reached in earlier times.[140] Hence, there were subsequent opportunities to acquire new relics.

However, the papal relic collection was not to benefit from this new mobility. Its emergence had been based on relations that did not revive after the turn of the millennium, the primary reason being the successful recuperation policy of the Macedonians in Syria, which in the eleventh century led to a protectorate over the holy sites and control of the patriarchate. From at least 1027, the patriarchs were appointed by the imperial house, and their political home was Constantinople, not Rome.[141] Whereas the imperial armies brought relics relating to Christ with them from Syria to the Bosporus, which established the fame of the Pharos Chapel,[142] contacts with the Greek world in Rome were declining: recruitment problems in the monasteries and the conversion to Latin observance were only two aspects of this alienation.[143] As a place of refuge and a bastion of orthodoxy, the Roman papacy was no longer of any significance for Palestine, and there were fewer and fewer people in Rome for whom the old ecumenical world was still a biographical reality. Recollections of that ecumenical world gradually disappeared in the collective memory of the Lateran complex and were to develop a new, second life only in the High Middle Ages through the medium of legend.

Notes

* For their assistance and sharing of knowledge, I should like to thank Achim Arbeiter, Beat Brenk, Michael Grünbart, Bianca Kühnel, Julia M. H. Smith, and Hedwig Röckelein; for the translation Geraldine Schuckelt.

1 Patrick J. Geary, "Sacred Commodities: The Circulation of Medieval Relics," in *The Social Life of Things: Commodities in Cultural Perspective*, ed. Arjun Appadurai (Cambridge: Cambridge University Press, 2003), 169–191; Hedwig Röckelein, *Reliquientranslationen nach Sachsen im 9. Jahrhundert: Über Kommunikation, Mobilität und Öffentlichkeit im*

Frühmittelalter (Stuttgart: Thorbecke, 2002), 140–154; Lucas Burkart, Philippe Cordez, and Pierre Alain Mariaux, *Le trésor au Moyen Âge: Questions et perspectives de recherché – Der Schatz im Mittelalter: Fragestellungen und Forschungsperspektiven* (Neuchâtel: Institut d'Histoire de l'art et de Muséologie, 2005); Lucas Burkart, *Das Blut der Märtyrer: Genese, Bedeutung und Funktion mittelalterlicher Schätze* (Köln: Böhlau, 2009).

2 Michael McCormick, *Origins of the European Economy: Communications and Commerce A.D. 300–900* (Cambridge: Cambridge University Press, 2001), 283–318.

3 Hartmann Grisar, *Die römische Kapelle Sancta Sanctorum und ihr Schatz* (Freiburg im Breisgau: Herder, 1908), 1–9, 136–141; Mario Cempanari, *Sancta Sanctorum lateranense* (Rome: Tipografia Città nuova, 2003), 634–650; Bruno Galland, *Les authentiques des reliques du Sancta Sanctorum* (Vatican City: Biblioteca Apostolica Vaticana, 2004), 19–24; Kirstin Noreen, "Opening the Holy of Holies: Early Twentieth-Century Explorations of the Sancta Sanctorum (Rome)," *Church History* 80 (2011): 520–546.

4 Galland, *Authentiques*, 31; Julia M.H. Smith, "Care of Relics in Early Medieval Rome," in *Rome and Religion in the Medieval World*, eds. Valerie L. Garver and Owen M. Phelan (Farnham and Burlington: Ashgate, 2014), 179–205.

5 Lauer, *Trésor*, 97–104; on this problem Hedwig Röckelein, "Die 'Hüllen der Heiligen': Zur Materialität des hagiographischen Mediums," in *Reliquiare im Mittelalter*, eds. Bruno Reudenbach and Gia Toussaint (Berlin: Akademie Verlag, 2005), 75–88.

6 Charles R. Morey, "The Painted Panel of Sancta Sanctorum," in *Festschrift Paul Clemen* (Bonn: Cohen, 1926), 151–167; Bruno Reudenbach, "Reliquien von Orten: ein frühchristliches Reliquiar als Gedächtnisort," in *Reliquiare im Mittelalter*, eds. Bruno Reudenbach and Gia Toussaint (Berlin: Akademie Verlag, 2005), 21–41; Gary Vikan, *Early Byzantine Pilgrimage Art* (Washington, DC: Dumbarton Oaks, 2010), 19–20; Annabel J. Wharton, *Selling Jerusalem: Relics, Replicas, Theme Parks* (Chicago: University of Chicago Press, 2006), 22–24; Anja Kalinowski, *Frühchristliche Reliquiare im Kontext von Kultstrategien, Heilserwartungen und sozialer Selbstdarstellung* (Wiesbaden: Reichert, 2011), 67–69; Cynthia Hahn, *Strange Beauty: Issues in the Making and Meaning of Reliquaries, 400–circa 1204* (University Park, PA: Penn State University Press, 2012), 177–179.

7 Lucy O'Connor, "The Late Antique Wooden Reliquaries from the Chapel of the Sancta Sanctorum," *Bullettino dei Monumenti Musei e Gallerie Pontificie* 31 (2013 [2015]): 201–229, with some new observations, is an exception, although the author does not take into account the recent edition of the *cedulae*. O'Connor's paper came out after the completion of the present essay.

8 See Gabriele Mietke, "Wundertätige Pilgerandenken: Reliquien und ihr Bildschmuck," in *Byzanz: Die Macht der Bilder*, eds. Michael Brandt and Arne Effenberger (Berlin: Staatliche Museen zu Berlin Preußischer Kulturbesitz, 1998), 40–43; Martina Bagnoli, Holger A. Klein, C. Griffith Mann, and James Robinson, eds., *Treasures of Heaven: Saints, Relics, and Devotion in Medieval Europe* (Baltimore: Walters Art Museum, 2010), 69–81.

9 Beate Fricke, "Tales from Stones, Travels through Time: Narrative and Vision in the Casket from the Vatican," *West 86th* 21 (2014): 230–250.

10 Galland, *Authentiques*, 39–40, 87–89; Smith, *Relics*, 183, 203.

11 Recent Literature: John Elsner, "Replicating Palestine and Reversing the Reformation: Pilgrimage and Collecting at Bobbio, Monza and Walsingham," *Journal of the History of Collections* 9, no. 1 (1997): 117–130; Dennis E. Trout, "Theodelinda's Rome: Ampullae, Pittacia, and the Image of the City," *Memoirs of the American Academy in Rome* 50 (2005): 131–150; Philippe Cordez, "Karl der Große, die Dinge und das Reich: Reliquiensammlungen und Kirchenschätze," in *Karl der Große – Charlemagne: Karls Kunst*, eds. Peter Van den Brink and Sarvenaz Ayooghi (Dresden: Sandstein, 2014), 47–55; Paul Magdalino, "L'église du Phare et les reliques de la Passion á Constantinople," in *Byzance et les reliques du Christ*, eds. Jannic Durand and Bernard Flusin (Paris: Association des Amis du Centre d'Histoire et Civilisation de Byzance, 2004), 15–30.

12 On the problems of dating the handwritings, see Galland, *Authentiques*, 45–54; Smith, *Relics*, 183–184.

13 See Galland, *Authentiques*, 45–47. Biblical saints are classified as belonging to the loca sancta relics.

14 Lucas Burkart, "Die Aufhebung der Sichtbarkeit: Der Schatz der Sancta Sanctorum und die Modi seiner visuellen Inszenierung," in *Visibilität des Unsichtbaren: Sehen und Verstehen in*

Mittelalter und Früher Neuzeit, ed. Anja Rathmann-Lutz (Zürich: Chronos, 2011), 69–82; Daniela Mondini, "Reliquie incarnate: le 'sacre teste' di Pietro e Paolo a San Giovanni in Laterano a Roma," in *Del visibile credere: pellegrinaggi, santuari, miracoli, reliquie*, ed. Davide Scotto (Florence: Olschki, 2011), 265–296.

15 John McCulloh, "From Antiquity to the Middle Ages: Continuity and Change in Papal Relic Policy from the 6th to the 8th Century," in *Pietas: Festschrift für Bernard Kötting*, ed. Ernst Dassmann (Münster: Aschendorff, 1980), 313–324; Smith, *Relics*, 185–189.

16 Jean-Marie Sansterre, *Les moines grecs et orientaux à Rome aux époques byzantines et carolingienne* (Brussels: Palais des académies, 1983); Andrew J. Ekonomou, *Byzantine Rome and the Greek Popes* (Lantham, MD: Lexington Books, 2009).

17 Galland, *Authentiques*, no. 41, 44, 45, 71, 89; John Wilkinson, *Jerusalem Pilgrims before the Crusades* (Arminster: Aris & Phillips, 2002), 49, 209, 211, 282, 293, 340.

18 Cf. Janette Witt, *Werke der Alltagskultur: Vol. 1: Menasampullen* (Wiesbaden: Reichert, 2000), 61–66; Vikan, *Pilgrimage Art*, 31, 83–88.

19 Michael McCormick, "Les pélerins occidenteaux à Jerusalem (VIIIe – IXe siècles)," in *Voyages et voyageurs à Byzance et en occident du VIe au XIe siècle: actes du colloque international de l'Université de Liège (5 – 7 mai 1994)*, eds. Alain Dierkens and Jean-Marie Sansterre (Geneva: Droz, 2000), 289–306, esp. 296.

20 The term "site relic" is borrowed from Reudenbach, *Reliquien*.

21 Cf. note 6.

22 Sources in Harry Magoulias, "The Lives of Byzantine Saints as Sources of Data for the History of Magic in the Sixth and Seventh Centuries A.D.: Sorcery, Relics and Icons," *Byzantion* 37 (1967): 228–269; Bernard Kötting, *Peregrinatio religiosa* (Münster: Stenderhoff, 1950), 403–413; Brigitte Klausen-Nottmeyer, "Eulogien: Transport und Weitergabe von Segenskraft: Ergebnisse einer Zusammenstellung von Pilgerandenken," in *Akten des XII: Internationalen Kongresses für Christliche Archäologie* (Bonn, 22–28 September, 1991), ed. Ernst Dassmann (Münster: Aschendorff, 1995), 922–927, esp. 922–923; Josef Engemann, "Palästinensische Pilgerampullen im F.J. Dölger-Institut in Bonn," in *Jahrbuch für Antike und Christentum* 16 (1973): 5–27; idem, "Palästinensische frühchristliche Pilgerampullen: Erstveröffentlichungen und Berichtigungen," *Jahrbuch für Antike und Christentum* 45 (2002): 153–169; Gary Vikan, " 'Guided by Land and Sea': Pilgrim Art and Pilgrim Travel in Early Byzantium," in *Tesserae: Festschrift für Joseph Engemann*, Jahrbuch für Antike und Christentum, Suppl. vol. 18 (Münster: Aschendorff, 1991), 74–92; Andreas Hartmann, *Zwischen Relikt und Reliquie: Objektbezogene Erinnerungspraktiken in antiken Gesellschaften* (Berlin: Verlag Antike, 2011), 603–607.

23 Petrus Diaconus, *De locis sanctis*, ch. V. 4, ed. Georg Röwekamp, *Egeria, Itinerarium-Reisebericht*, vol. 2 (Freiburg: Herder, 2000), 338; Wilkinson, *Pilgrims*, 346; Georges Tchalenko, *Villages antiques de la syrie du Nord*, vol. 1 (Paris: Geunther, 1953), 245. Cf. note 60.

24 Cf. note 22; Mietke, "Pilgerandenken," 40–50; Vikan, *Pilgrimage Art*, 31–40.

25 Galland, *Authentiques*, 93–158. Several relic labels refer to more than one relic.

26 Except for some small phials found in a liturgical box in 1905: Lauer, *Trésor*, 99–100. Only one seal of earth, stamped with a cross and attached to a label "*de sancta Sion*" was documented by Lauer, *Trésor*, 126; Galland, *Authentiques*, no. 25, mentions a stone relic.

27 Galland, *Authentiques*, nos. 109, 111, 118, 121, 124; Lauer, *Trésor*, 107; O'Connor, *Wooden Reliquaries*, 218. O'Connor's late antique dating of the wooden box containing a glass phial is contested by the early medieval dating of the *cedula*, which sealed the glass phial.

28 Vikan, *Pilgrimage Art*, 83–88. For discussion of an economic change see Robert Schick, *The Christian Communities of Palestine from Byzantine to Islamic Rule* (Princeton: Darwin Press, 1995); Gideon Avni, *The Byzantine-Islamic Transition in Palestine: An Archeological Approach* (Oxford: Oxford University Press, 2014).

29 In glor. mart. 6 (MGH SSRM 1/2, 42); Galland, *Authentiques*, nos. 87, 103, 121.

30 Galland, *Authentiques*, nos. 99, 124; cf. Vikan, *Pilgrimage Art*, 21–22; for the site Schick, *Communities*, 293.

31 Life of Saint Stephen, ch. 52.1 – 12, in *The Life of Stephen of Mar Sabas* (CSCO 578: Scriptores arabici 51), ed. John C. Lamoreaux (Löwen: Peeters, 1999), 81–83; on later eulogiae, Witt, *Menasampullen*, 61–66; the most recent evidence was provided by Theodore of Paphos in 655: Anatole Frolow, *Le relique de la Vraie Croix: Recherches sur le*

développement d'un culte (Paris: Institut Français d'Etudes Byzantines, 1961), 174–175. See also William Anderson, "The Archeology of the Late Antique Pilgrim Flasks," *Anatolian Studies* 54 (2004): 79–93.

32 Lauer, *Trésor*, 102; *Credo: Christianisierung Europas im Mittelalter*, eds. Christoph Stiegemann, Martin Kroker, and Wolfgang Walter, vol. 2 (Petersberg: Imhof, 2013), 131–132 (C. Lega).

33 Galland, *Authentiques*, 49–52; regarding Sens cf. Maurice Prou and Eugène Chartraire, "Authentiques de reliques conservées au trésor de la cathédrale de Sens," *Mémoires de la Société Nationale des Antiquaires de France* 9 (1900): 142–143, 146.

34 Anton Legner, *Reliquien in Kunst und Kult* (Darmstadt: Wiss. Buchges, 1995), 60.

35 Gia Toussaint, "Schöne Schädel: Die Häupter der Heiligen in Ost und West," in *Knotenpunkt Byzanz: Wissensformen und kulturelle Wechselbeziehungen*, eds. Andreas Speer and Philipp Steinkrüger (Berlin: de Gruyter, 2012), 663–670; Holger A. Klein, "Brighter than the Sun: Saints, Relics, and the Power of Art in Byzantium," *Knotenpunkt Byzanz: Wissensformen und kulturelle Wechselbeziehungen*, eds. Andreas Speer and Philipp Steinkrüger, 635–654, esp. 649.

36 Morey, *Panel*, 153; Holger A. Klein, *Byzanz, der Westen und das 'wahre' Kreuz: Die Geschichte einer Reliquie und ihrer künstlerischen Fassung in Byzanz und im Abendland* (Wiesbaden: Reichert, 2004), 19–28.

37 Anna D. Kartsonis, *Anastasis: The Making of an Image* (Princeton: Princeton University Press, 1986), 101–105; Brigitte Pitarakis, *Les croix-reliquaires pectorales byzantines en bronze* (Paris: Klincksieck, 2006), 63–68.

38 I am preparing an essay on this topic.

39 Kalinowski, *Reliquiare*, 67–69.

40 Pitarakis, *Croix-Reliquaires*, 109–122.

41 Early examples: Achim Arbeiter, "Nahöstliche Terracotta-Eulogien aus einem Altardepot in Lusitanien," in *Syrien und seine Nachbarn von der Spätantike bis in die islamische Zeit*, eds. Ina Eichner and Vasiliki Tsamakda (Wiesbaden: Reichert, 2009), 199–213; Elsner, *Palestine*.

42 Most of the script groups refer to relics from the same origin, but not all, like group C, which combines relics from different countries: see Smith, *Relics*, 198.

43 Galland, *Authentiques*, nos. 3, 63, 87, 95, 103, 109, 111, 119, 121.

44 Galland, *Authentiques*, 108, probably corresponds to the Dodekathronon in the *Commemoratorium* of 808: Michael McCormick, *Charlemagne's Survey of the Holy Land: Wealth, Personnel, and Buildings of a Mediterranean Church between Antiquity and the Middle Ages* (Washington, DC: Dumbarton Oaks Research Library, 2011), 213.

45 Pierre Maraval, *Lieux saints et pèlerinages d'Orient* (Paris: Cerf, 1985), 163–167; Dietrich Claude, *Der Handel im westlichen Mittelmeer während des Frühmittelalters* (Göttingen: Vandenhoeck & Ruprecht, 1985), 239–249; McCormick, *Pèlerins*, 303–306.

46 MGH SS 15/1, 90–105: Rodney Aist, *The Christian Topography of Early Islamic Jerusalem: The Evidence of Willibald of Eichstätt (700–787 CE)* (Turnhout: Brepols, 2009), 31–34, 41–52; McCormick, *Origins*, 129–138.

47 Leor Halevi, "Bernard, Explorer of the Muslim Lake: A Pilgrimage from Rome to Jerusalem, 867," *Medieval Encounters* 4 (1998): 24–50.

48 Absent sites include Nazareth, Capernaum, Mount Tabor, Chorazin, Hebron, Mamre, Rachel's tomb, and Gilgal. On travel routes see Wilkinson, *Pilgrims*, 30–51 and Maraval, *Lieux*, 251–310.

49 Maurice Halbwachs, *La Topographie lègendaire des Èvangiles en Terre Sainte* (Paris: Presses universitaires de France, 1941).

50 McCormick, *Charlemagne*, 49–50.

51 Yizhar Hirschfeld, *The Judean Desert Monasteries in the Byzantine Period* (New Haven: Yale University Press, 1992), 10, 102–111; Bernhard Flusin, *Saint Anastase le Perse et l'histoire de la Palestine au début du VIIe siècle*, vol. 2 (Paris: Centre National de la Recherche Scientifique, 1992), 27; John Binns, *Ascetics and Ambassadors of Christ: The Monasteries of Palestine 314–631* (Oxford: Clarendon Press, 1996), 52–53; Vera von Falkenhausen, "Die Rolle der Wallfahrt nach Jerusalem im Leben der byzantinischen Mönche vor den Kreuzzügen," in *L'idea di Gerusalemme nella spiritualità cristiana del Medioevo, Atti del Convegno internazionale*, ed. Walter Brandmiller (Vatican City: Pontificio Comitato

di Scienze Storiche, 2003), 34–35; Phil Booth, *Crisis of Empire: Doctrine and Dissent at the End of Late Antiquity* (Berkeley, CA: University of California Press, 2014), 96–97.
52 Derwas J. Chitty, *The Desert a City* (Oxford: Blackwell, 1966), 168–178; Binns, *Ascetics*, 95–98, 156–161.
53 Herbert Donner, "Die Palästinabeschreibung des Epiphanius Monachus Hagiopolita," *Zeitschrift des Deutschen Palästina-Vereins* 87 (1971): 45–92; Andreas Külzer, *Peregrinatio graeca in terram Sanctam* (Frankfurt and Berlin: Peter Lang, 1994), 14–17; Flusin, *Anastase*, II, 26; Falkenhausen, *Wallfahrt*, 34–36.
54 Sources in Klausen-Nottmeyer, *Eulogien*, 922–923.
55 Galland, *Authentiques*, nos. 63, 95, 100; nos. 5, 12, 23, 48, 72, 93, 101, 122; Martin Biddle, *Das Grab Christi* (Giessen and Basel: Brunnen-Verlag 1998), 80–93.
56 McCormick, *Charlemagne*, 201.
57 Beat Brenk, "Der Kultort, seine Zugänglichkeit und seine Besucher," in *Akten des XII: Internationalen Kongresses für Christliche Archäologie (Bonn 1991)*, eds. Ernst Dassmann and Josef Engemann (Münster: Aschendorff, 1995), 102.
58 Lauer, *Trésor*, 102; Morey, "Panel," 151. A second, circular carved stone without inscription is repeated in the destroyed stone fragment in the second box, inserted in a circular compartment: O'Connor, *Wooden Reliquaries*, 213–214.
59 On the archeology see Bellarmino Bagatti and Emmanuele Testa, *Il Golgota e la Croce: Ricerche storico-archologiche* (Jerusalem: Franciscan, 1978), 41–67, 50; Gustav Kühnel, "Architectural Mise-en-Scène and Pictorial Turns in Jerusalem," in *Jerusalem as Narrative Space: Erzählraum Jerusalem*, eds. Annette Hoffmann and Gerhard Wolf (Leiden and Boston: Brill, 2012), 21–31.
60 Vie de Marthe c. 58–59, *La vie ancienne de S. Syméon Stylite le Jeune (521–592)*, ed. Paul Van de Ven, vol. 2: *Traduction et commentaire: Vie grecque* de *sainte* Marthe *mère de Syméon*, Subsidia Hagiographica 32 (Brussels: Soc. des Bollandistes, 1970), 302–303; Pitarakis, *Croix-reliquaires*, 110.
61 Magoulias, "Lives," 245–246.
62 Galland, *Authentiques*, nos. 27, 55, 88, 123. Breviarius, ch. 3 (CCSL 175, 110); Herbert Donner, *Pilgerfahrt ins Heilige Land: Die ältesten Berichte christlicher Palästinapilger (4. – 7. Jahrhundert)* (Stuttgart: Verlag Katholisches Bibelwerk, 1979), 235; Armenian guide, trans. Wilkinson, *Pilgrims*, 165; Epiphan. ch. 1: Donner, *Palästinabeschreibung*, 67, 82; Wilkinson, *Pilgrims*, 364–365.
63 McCormick, *Charlemagne*, 201.
64 Anon. Piac., ch. 20 (CCSL 175, 139).
65 De locis sanctis, ch. 7, 1–3 (CCSL 175, 191).
66 Epiphan. ch. 1: Donner, *Palästinabeschreibung*, 67, 83, mentions a greenish stone, whereas Arculf, De locis sanctis, 2.7.3, ED (CCSL 175,190), wrote about a silver stone. The latter may refer to the edging of the chalice mentioned by Epiphanius: Wilkinson, *Pilgrims*, 365.
67 Anon. Piac. 14 (CCSL 175, 140–141).
68 MGH SS 4, 44; MGH 15/1, 170; MGH SSRG 6, 108.
69 See Smith, *Relics*, 201–202.
70 R. Michalówski, "Le don de l'amitié dans la société carolingienne et les 'Translationes sanctorum'," in *Hagiographie, cultures et sociétés* (Paris: Ètudes Augustiniennes, 1981), 399–413; Peter Schreiner, "Diplomatische Geschenke zwischen Byzanz und dem Westen, ca. 800–1200: Eine Analyse der Texte mit Quellenanhang," *Dumbarton Oaks Papers* 58 (2004): 251–281, esp. 266.
71 John M. McCulloh, "The Cult of Relics in the Letters and Dialogues of Gregory the Great: A Lexicographical Study," *Traditio* 32 (1976): 145–184; idem., *Antiquity*, 314–316; Hack, *Codex Carolinus: Päpstliche Epistolographie im 8. Jahrhundert*, 2 vols, Päpste und Papsttum 35 (Stuttgart: Hiersemann, 2006–2007), 786–791, 1130–1139.
72 Reg. VI, 61; VII, 29, 37; VIII, 28, 33; X, 21; XIII, 26; XIII, 43: Pierre Maraval, "Grégoire le Grand et les Lieux Saints d'Orient," in *Gregorio Magno e il suo tempo* (Rome: Institutum patristicum "Augustinianum," 1991), vol. 1, 65–76; Piccirillo, *Gregorio*, 325–327.
73 McCulloh, *Antiquity*, 318–319.
74 For the seventh century only 62 letters have survived complete. Six of them mention gifts of relics (= 9.6 percent): Pietro Conte, *Chiesa e primato nelle lettere dei papi del secolo VII* (Milan: Vita e Pensiero, 1971), 35, 106–107, 406, 458–459, 484–485.

75 Vita Radegundis, ch. 14, MGH SSRM 2, 386; Vie de Sainte Marthe, ch. 52–70, ed. van den Ven, *Vie ancienne*, 296–312. See note 60.
76 O'Connor, *Wooden Reliquaries*, 213.
77 MGH SS 1, 186–187. See Michael Borgolte, *Der Gesandtenaustausch der Karolinger mit den Abbassiden und mit den Patriarchen von Jerusalem* (Munich: Arbeo-Gesellschaft, 1976), 61–62; Klaus Bieberstein, "Der Gesandtenaustausch zwischen Karl dem Großen und Harun ar-Rašid und seine Bedeutung für die Kirchen Jerusalems," *Zeitschrift des Deutschen Palästina-Vereins* 109 (1993): 152–173.
78 Conrad Leyser, "The Temptations of Cult: Roman Martyr Piety in the Age of Gregory the Great," *Early Medieval Europe* 9 (2000): 289–307; McCulloh, *Antiquity*, 316–317; Smith, *Relics*, 192.
79 Hack, *Codex Carolinus*, 795, 801–802.
80 Hack, *Codex Carolinus*, 805–809.
81 Reg. VI, 61; VII, 37; VIII, 28; X, 21; XIII; 43: Claude, *Handel*, 245.
82 Reg. V, 42; VII, 9; VIII, 2; XII, 1: Hack, *Codex Carolinus*, 729–730, 782–786, 826–827; Piccirillo, *Gregorio*, 326–327.
83 Reg. VII, 29; XIII, 26; Piccirillo, *Gregorio*, 325.
84 Conte, *Primato*, 319–322; Pauline Allen, *Sophronius of Jerusalem and Seventh-Century Heresy* (Oxford: Oxford University Press, 2009), 50–51.
85 Reg. I, 24–25. The accompanying letter cites all the patriarchs as recipients.
86 *Les Regestes des Actes Patriarcat de Constantinople*, eds. Venance Grumel and Jean Darrouzès, vol. 1, *Les Actes des Patriarches, Fasc. 2–3: les Regestes de 715 à 1206* (Paris: Institut français d'études byzantines, 1989), nos. 382, 515–516, 525, 539, 542–543, 554; Frolow, *Relique*, 205, 208, 214–215, 223, 226, 259–260, 261; Kartsonis, *Anastasis*, 118–119; Klein, *Byzanz*, 91, 181.
87 Frolow, *Relique*, 193.
88 Life of Saint Theodore of Sykeon, ch. 128, André-Jean Festugière, ed., *Vie de Théodore de Sykéôn* (Brussels: Soc. des Bollandistes, 1970), 103; Frolow, *Relique*, no. 88.
89 Galland, *Authentiques*, nos. 22, 57 (*Ivenius* = Syriac for John Chrysostomos).
90 Jean-Marie Sansterre, "Le monachisme byzantin à Rome," in *Bisanzio, Roma e l'Italia nell'Alto medioevo*, vol. 2, in book series: Settimane di studio del Centro italiano di studi sull'alto medioevo / 34, 1-2 (Spoleto: Presso la Sede del Centro, 1988), 701–746; Marie-France Auzepy, "Les Sabaites et l'Iconoclasme," in *The Sabaite Heritage in the Orthodox Church from the Fifth Century to the Present*, ed. Joseph Patrich (Leuven: Peeters, 2001), 309–310.
91 Mansi 16, 25ff, 314: Moshe Gil, *A History of Palestine 634–1099* (Cambridge: Cambridge University Press, 1992), 461.
92 Ep. 88, MGH Epp. 6, 478; Sansterre, *Monachisme*, 705–706.
93 LP II, 176–177; Sansterre, *Moines*, I, 141–143.
94 Mansi 16, 136E – 137D; McCormick, *Pèlerins*, 297.
95 Friedhelm Winkelmann, *Der monenergetisch-monotheletische Streit* (Frankfurt am Main: Lang, 2001), 45–184; Booth, *Crisis*.
96 Henry Chadwick, "John Moschus and His Friend Sophronius the Sophist," *Journal of Theological Studies* 25 (1974): 41–59; Allen, *Sophronius*, 15–23; Booth, *Crisis*, 90–139, 142.
97 Booth, *Crisis*, 106–107, 231–232; Robert Coates-Stephens, "San Saba and the Xenodochium de Via Nova," *Rivista di Archeologia Cristiana* 83 (2007): 223–256.
98 Mansi 11, 461–510; ACO II, 410–494; Booth, *Crisis*, 237–338.
99 Booth, *Crisis*, 242–244, 269–277.
100 Louis Duchesne, *Le Liber Pontificalis: texte, introduction et commentaire*, vol. 1 (Reprint; Paris: Boccard, 1955), 331, 333; Ekonomou, *Byzantine Rome*, 96–100; Erich Caspar, *Das Papsttum unter byzantinischer Herrschaft*, vol. 2 of *Geschichte des Papsttums von den Anfängen bis zur Höhe der Weltherrschaft* (Tübingen: Mohr, 1933), 541–553.
101 Erich Caspar, "Die Lateransynode von 649," *Zeitschrift für Kirchengeschichte* 51 (1932): 75–137; Rudolf Riedinger, "Die Lateranakten von 649, ein Werk der Byzantiner um Maximos Homologetes," *Byzantina* 13, no. 1 (1985): 517–534.
102 Caspar, *Lateransynode*, 124.
103 *Acta conciliorum oecumenicorum*, vol. 2/1, *Concilium Lateranense a. 649 celebratum*, ed. Rudolf Riedinger (Berlin and Leipzig: de Gruyter, 1995), 38–47; Christoph Schönborn,

Sophrone de Jérusalem: vie monastique et confession dogmatique, Théologie historique 20 (Paris: Beauchesne, 1972), 94–95.

104 Frank R. Trombley, "A Note on the See of Jerusalem and the Synodal List of the Sixth Ecumenical Council," *Byzantion* 53 (1983): 632–638.

105 Milka Levy-Rubin, "The Role of the Judean Desert Monasteries in the Monothelite Controversy in Seventh-Century Palestine," in *The Sabaite Heritage in the Orthodox Church*, ed. Joseph Patrich, 283–300, esp. 299; Ernst von Dobschütz, *Christusbilder: Untersuchungen zur christlichen Legende* (Leipzig: Heinrichs, 1899), 185*–186*.

106 Francis Xavier Murphy and Polycarp Sherwood, *Konstantinopel II und III* (Mainz: Matthias-Grünewald-Verlag, 1990), 246–247.

107 Marie-France Auzepy, "De la Palestine à Constantinople (VIII – IXe siècles): Étienne le Sabaite et Jean Damascène," *Travaux et mémoires (du) Centre de recherche d'histoire et civilisation de Byzance* 12 (1994): 183–218; idem, "Les Sabaites et l'Iconoclasme," in *The Sabaite Heritage in the Orthodox Church*, ed. Joseph Patrich, 301–314.

108 Jean Gouillard, "L'Église d'Orient et la primauté romaine au temps de l'iconoclasme," *Istina* 21 (1976): 25–54, esp. 31–33; Auzépy, *Sabaites*, 307–310.

109 Mansi 12, 1135–1146; MGH Conc. II/1, 89; MGH Epp. 3, 652–653; McCormick, *Origins*, 875, no. 173; Hans Georg Thümmel, *Die Konzilien zur Bilderfrage im 8. und 9. Jahrhundert: Das 7. Ökumenische Konzil in Nikaia 787* (Paderborn: Schöningh, 2005), 84–85; Hack, *Codex Carolinus*, 464, 979, 981.

110 MGH Epp. 3, 553: McCormick, *Origins*, 512, 874, no. 171.

111 McCormick, *Origins*, 852–972.

112 MGH Epp. 5, 143.

113 McCormick, *Charlemagne*, 7–8; Amnon Linder, "Christian Communities in Jerusalem," in *The History of Jerusalem: The Early Muslim Period 638–1099*, eds. Joshua Prawer and Haggai Ben-Shammai (Jerusalem: Yad Izhak Ben-Zvi, 1996), 139–142.

114 Reg. VII, 29: Piccirillo, *Gregorio*, 325.

115 Martinus, Ep. 14 (JE 2078), Migne PL 87, 199–200. See Wolfram Brandes, "Juristische Krisenbewältigung im 7. Jh.? Die Prozesse gegen Papst Martin I. und Maximos' Homologetes," *Fontes minores* 10 (1998): 141–212, esp. 172.

116 MGH SS 15/1, 92; MGH SS 15/1, 458; Gesta Sanctorum Rotonensium, III.8, in *The Monks of Redon: Gesta Sanctorum Rotonensium and Vita Convuoinis*, Studies in Celtic History 10, ed. Caroline Brett (Woodbridge: Boydell Press, 1989), 206–210; Karl Schmid, "Aachen und Jerusalem: Ein Beitrag zur historischen Personenforschung der Karolingerzeit," in *Das Einhardkreuz: Vorträge und Studien der Münsteraner Diskussion zum arcus Einhardi*, ed. Albert Hauck (Göttingen: Vandenhoeck & Ruprecht, 1974), 122–142, esp. 137.

117 LP 75, c. 2.

118 LP 92, c. 8; Sible de Blaauw, *Cultus et decor* (Vatican City: Biblioteca Apostolica Vaticana, 1994), 400–401; Dale Kinney, "The Praesepia in Santa Maria in Trastevere and Santa Maria Maggiore," in *Marmoribus vestita: Miscellanea in onore di Federico Guidobaldi, eds. Olof Brandt and Philippe Pergola* (Vatican City: Pontificio Istituto di Archeologia Cristiana, 2011), 777–795.

119 De Blaauw, *Cultus*, 430–436.

120 Hermann Usener, "Weihnachtspredigt des Sophronius," *Rheinisches Museum für Philologie* 41 (1886): 506.

121 *Acta conciliorum oecumenicorum*, vol. 2, no. 1, ed. Riedinger, 142; Mansi 10, 977–978; Booth, *Crisis*, 294–295.

122 *LP* 34, c. 22; *LP* 48, c. 3; *LP* 53, c. 7: Sible De Blaauw, "Translations of the Sacred City between Jerusalem and Rome," in *The Imagined and Real Jerusalem in Art and Architecture*, eds. Jeroen Goudeau, Mariette Verhoeven, and Wouter Weijers (Leiden: Brill, 2014), 136–165.

123 *Sacramentarium Gelasianum Vetus*, nos. 1023–1025, ed. Leo Cunibert Mohlberg, *Liber Sacramentorum Romanae Aecclesiae Ordinis Anni Circuli* (Rome: Casa Editrice Herder, 1960), 158; *Gregorianum Paduense*, no. 665, ed. Jean Deshusses, vol. 1, *Le Sacramentaire Grégorien: Ses principales formes d'aprés les plus anciens manuscripts* (Fribourg: Éditions Universitaires, 1992), 659; see Louis van Tongeren, *Exaltatio Crucis: Het feest van Kruisverheffing en de zingeving van het kruis in het Westen tijdens de vroege middeleeuwen*

(PhD dissertation, Tilburg University, 1995), 83–99, 117–118; De Blaauw, "Translations," 150–153.

124 Van Tongeren, *Exaltatio*, 119, surmises that it was introduced around 630.

125 LP I, 374; Van Tongeren, *Exaltatio*, 49–50; Eamonn Ó Carragain, *Ritual and the Rood: Liturgical Images and the Old English Poems of the Dream of the Rood Tradition* (Toronto: University of Toronto, 2005), 230–231.

126 *Ordo Romanus* 23, c. 9–22, ed. Michel Andrieu, vol. 3, *Les Ordines romani du Haut Moyen Age* (Leuven and Paris: "Spicilegium Sacrum Lovaniense" Administration, 1951), 270–272; Klein, *Byzanz*, 69–77.

127 Ò Carragain, Ritual, 148–150, 183–201.

128 Erik Thunø, *Image and Relic: Mediating the Sacred in Early Medieval Rome* (Rome: 'L'Erma' di Bretschneider, 2002), 17–23.

129 LP I, 469.

130 LP I, 328.

131 LP I, 242: De Blaauw, *Cultus*, 135–136.

132 Galit Noga-Banai, *The Trophies of the Martyrs: An Art Historical Study of Early Christian Silver Reliquaries* (Oxford: Oxford University Press, 2008), Cat. 15.

133 Manfred Luchterhandt, "Rom und Aachen: Die Karolinger und der päpstliche Hof um 800," in *Orte der Macht*, vol. 2 of *Karl der Große: Charlemagne (Catalogue of the Exhibition in Aachen 2014)*, ed. Frank Pohle (Dresden: Sandstein, 2014), 109.

134 Galland, *Authentiques*, 87–88.

135 Sible de Blaauw, "Il Patriarchio, la Basilica Laterenense e la liturgia," *Mélanges d'archéologie et d'histoire de l'Ecole Francaise de Rome, Antiquité* 116 (2004): 161–171, esp. 165–170.

136 Galland, *Authentiques*, 63–67; Gerhard Wolf, " 'Laetare filia sion: Ecce ego venio et habito in medio tui': Images of Christ transferred to Rome from Jerusalem," in *The Real and Ideal Jerusalem in Jewish, Christian and Islamic Art*, ed. Bianca Kühnel (Jerusalem: Centre for Jewish Art, Hebrew University of Jerusalem, 1998), 419–429.

137 Life of Michael Syncellus, ch. 6, Claudia Sode, ed., *Jerusalem-Konstantinopel-Rom: die Viten des Michael Synkellos und der Brüder Theodors und Theophanes Graptoi* (Stuttgart: Steiner, 2001), 202–204; Auzcpy, *Palestine*, 209. Concerning the finances of the patriarch see McCormick, *Charlemagne*, 7–22.

138 McCormick, *Charlemagne*, 184–196; Milka Levy-Rubin, "A Spanish Source on Mid-Ninth-Century Mar Saba and a Neglected Sabaite Martyr," in *The Sabaite Heritage*, ed. Joseph Patrich, 63–72; McCormick, *Pèlerins*, 299.

139 MGH Epp. 5, 143.

140 Francoise Micheau, "Les Itineraires maritimes et continenteaux des pélerinages vers Jérusalem," in *Occident et Orient au Xe siècle: Actes du IXe congrès de la Société des historiens médiévistes de l'enseignement supérieur public, Dijon, 2–4 juin 1978* (Paris: Les Belles Lettres, 1979), 79–104.

141 Linder, *Communities*, 128–129.

142 Magdalino, *L'église*, 23–25; Bernard Flusin, "Les reliques de la Saint-Chapelle et leur passé imperial à Constantinople," in *Le trésor de la Saint Chapelle*, eds. Jannic Durand and Jean-Pierre Laffitte (Paris: Ed. de la Réunion des Musées Nationaux, 2001), 20–31.

143 Sansterre, *Monachisme*; Bernard F. Hamilton, "The City of Rome and the Eastern Churches in the Tenth Century," *Orientalia Christiana Periodica* 27 (1961): 5–26.

4 Jerusalem refracted

Geographies of the True Cross in late antiquity

Laura Veneskey

In a letter dated sometime between 499 and 513, Avitus of Vienne petitioned Pope Symmachus to help him obtain a relic of the True Cross. Although his letter delicately articulates deference to the authority of Rome, Avitus is quite clear that his fragment should come from Bishop Elias in Jerusalem; what he wants from Symmachus is merely his episcopal leverage in obtaining it:

> [E]ven though we think that you have one of the relics of the Holy Cross in Rome, we still believe that this generous favor should be sought from the venerable patriarch of Jerusalem. In fact, by maintaining the true and inviolable purity of that sacrament within his jurisdiction of the pilgrimage-place, he is able to present us with a share in the desirable gift in such a manner as to free us from any hesitation and doubt.[1]

Avitus' preference for the Jerusalem relic was seemingly motivated by his anxiety regarding the authenticity of particles that had long been removed from Golgotha. Indeed, his concern was so great that it seems that he risked, albeit obliquely, impugning the pope's own specimen to achieve his goal. Although a Roman fragment would come with the implicit stamp of papal authority, from Avitus' perspective, the aura of authenticity attached to such a relic diminished after its removal from Jerusalem. Only wood obtained directly from the original source was above suspicion.

Surely, Avitus' concerns over the relic's authenticity were shared by at least some of his contemporaries, although his letter is unique in its frankness and insistence on unmediated translation. In all likelihood, most such exchanges were less direct, relying instead on a provenance of trusted hands to ensure authenticity. It was in just this way, for instance, that Paulinus of Nola had bequeathed an "almost indivisible particle of a small sliver" to Sulpicius Severus in the fifth century.[2] In his accompanying letter, Paulinus described the division and subdivision of this fragment, tracing its lineage to the ultimate source: "This goodly gift was brought to me from Jerusalem by the blessed Melania, a gift of the holy bishop John there."[3] Evidently, however, no earthly assurance – even that of the pope – was adequate for Avitus' particular purposes.

The rationale for Avitus' fixation on the material authenticity of his prize emerges in his letter of thanks to bishop Elias of Jerusalem, who seems to have been obliging regarding the request: by his gift of the wood, Avitus wrote, Elias had judged Avitus to be "not unworthy to share in the company of the earthly Jerusalem," indicating that his investment in the relic hinged at least partly on its ability to function as a topographical index of the Holy City.[4] It would seem, therefore, that he favored a

Jerusalem relic not only to ensure authenticity, but also in order to possess a piece of the city itself. Although not explicitly articulated, Avitus' pre-emptive rejection of such a gift from Symmachus raises the question of whether a relic from Rome would still be considered capable of translating this topography. Apparently, the desired relic effectively enacted a pilgrimage in reverse, bringing not only the wood of the cross, but the whole of its sacred milieu in Jerusalem. For Avitus, the bare wood was sufficient to accomplish such a translocation, but reliquaries from this period, which united pieces of that landscape with the True Cross, served as tangible demonstrations of this idea.[5]

Thus, it seems that a relic of the cross could transmit environs exceeding its own material substance, but likewise had the ability to accrue more than just Jerusalem's topography. Relics emanating from different sources differed with regard to both accessibility and desirability, and not all fragments were considered equal by their would-be possessors. The metonymy of the True Cross for Jerusalem was complicated by the seemingly endless division and distribution of its constituent splinters, slivers, and planks. Following its discovery in the fourth century, some portion of the relic remained in Jerusalem, but small particles changed hands as personal gifts, while larger portions underwent official translations, with the largest pieces coming to rest in Constantinople.[6] In a fourth-century rhetorical flourish describing the distribution of these relics, Cyril of Jerusalem intoned that "the whole world has since been filled with pieces of the wood of the cross."[7] His declaration is evocative of the paradoxical processes through which small particles, great in divine stature, were broken off and circulated, multiplying the original relic exponentially beyond its original volume and power.

In Cyril's description this transmission and diffusion is cast as overwhelmingly positive: the cross, though a finite object, possessed a property of seemingly infinite divisibility, multiplying in a manner after the loaves and fishes. However, a roughly contemporary passage from Egeria's *Itinerarium* tells of increased security measures around the cross in Jerusalem – the result of a relic-hungry pilgrim having once bitten off a segment of the wood – and indicates unease over the potential depletion of the primary relic.[8] The same fragmentation and circulation that increased the influence and presence of the cross in parts of Western Christendom also led to anxieties over both wholeness and authenticity. Perhaps partly as a result of increased worry over its depletion, access to pieces of the Jerusalem relic appears to have been increasingly restricted as time went on, as Egeria intimates. In the early fifth century, there are several textual mentions of particles being distributed from Jerusalem, but by the sixth century such references all but disappear.[9] The growing inaccessibility of the Jerusalem relic conflicted with the desire of devotees to possess such primary fragments, unmediated through other episcopal or imperial channels. In turn, the rarity of Jerusalem cross relics further distinguished them from their more readily available counterparts in Rome and Constantinople.

Such diasporic examples might differ subtly, but importantly, from the Jerusalem relic in that gifts from these stores would be colored by associations of papal or imperial power, as Avitus was acutely aware.[10] Recounting the long list of relic translations falsely attributed to Charlemagne, John McCulloh has observed that "medieval men seem to have felt that the quality of the donor added to the importance of the relics. . . . Such stories could also serve as 'proof' of the authenticity of the relics."[11] By extension, the location of a relic over time could influence its import as much as the status and identity of the people through whose hands it passed as a result of that context.

However, conceptual mutations of this nature were not universally positive, and might even be considered detrimental to the validity of relic fragments. Over time, the various translations and exchanges of the True Cross could affect the perceived valuation of its disparate pieces, and the longer a particular segment was away from its point of origin in Jerusalem, the more its perceptible authenticity was likely to erode. As Leslie Brubaker succinctly noted, "moving an object changes its meaning," and these mobile fragments could become naturalized to their new surroundings.[12] Thus, a relic of the cross, when detached and circulated, could potentially take on a host of layered associations.

This potential malleability is further demonstrated by two additional cases of relic acquisition in the early medieval West: the missions undertaken by the sainted Queen Radegund of Poitiers (569) and by Theodelinda, Queen of Lombard Italy (ca. 600). The circumstances of these acquisitions reveal an apparent shared interest in the cross as an object of pilgrimage and veneration – that is, one especially redolent of its associated sacred topography. Both queens received relics related to the cross, yet, in each case, the translations negotiated several centers of authority, including Rome and Constantinople, in addition to Jerusalem. Furthermore, although the question of authenticity was certainly at stake for all parties, the actions of Radegund and Theodelinda make it clear that authenticity need not be understood in a strictly material sense, but can relate equally to both pious experience and political authority.

Saint Radegund, queen of Gaul (569 CE)

On the surface, Radegund's ambitions were not dissimilar to those of Avitus in that she seems to have been motivated by a desire to possess a piece of Jerusalem's sanctity, yet the events surrounding her petition unfolded quite differently. A pious princess with a traumatic early history, Radegund became queen when she was forced to marry Clothar, the Frankish royal who had murdered her family.[13] Although Radegund retained her royal title and influence, she eventually left Clothar's side to found a monastery at Poitiers.[14] There she amassed a formidable relic collection, but above all she desired a piece of the True Cross, which she eventually received in 569 from Emperor Justin II and Empress Sophia in Constantinople, with the help of her nephew King Sigibert.[15] Radegund's monastery, originally dedicated to the Virgin, was reconsecrated in honor of the cross.[16] A relic of the True Cross remains at Poitiers to this day, though it is unclear whether this is the fragment obtained by Radegund, as oral tradition maintains.[17]

An intriguing detail from Radegund's story – one that her hagiographers have all but obscured – is that, like Avitus, she seems to have first sought the cross from Jerusalem; unlike Avitus, her petition apparently failed.[18] Whatever motivated Radegund's initial preference for the Jerusalem relic when a sliver from Constantinople was more readily at hand, her biographers would have us believe that her exceptional piousness spurred a desire for as direct as possible an experience of the holy places. Yet her choice to put the relics on public display, thus generating a surge in pilgrimage to Poitiers and augmenting her influence, demonstrates her political shrewdness as well.[19] What can be more certainly ascertained, however, is that Radegund's handling of the cross successfully evoked echoes of Jerusalem – both heavenly and earthly – in her homeland of Gaul. Her cloistered life and queenly status would have naturally made it impossible for her to make an eastward journey in person;[20] instead, through a combination of pious charisma and royal clout, she brought the embodied sanctity of the East to Poitiers. Gregory of Tours described Radegund's cross as being enclosed in a silver reliquary along with the bones of "the holy martyrs and confessors," also collected in "the East," a detail evocative of the cross's topographic significance in Poitiers.[21] The associated geography described here is not as specific as it was for Avitus, but is instead suggestive of a broader scheme in which the cross radiates a more generalized "Eastern" sanctity.

Baudonivia, another of the queen's biographers, concurred on this point, and she described the relic's context as being similarly Eastern, saying that Radegund held "the blessed wood of the Lord's cross enshrined in gold and gems and many of the relics of the saints that had been kept in the East living in that one place."[22] It would seem that again the cross was not simply cast as an instrument of salvation, but also as a metonym for the displaced sanctity of Palestine. Yet Baudonivia went further and spun the relic's Constantinopolitan origin into an advantage by capitalizing on the imperial association with Justin and Sophia. The metaphor of Radegund's relic was thus shifted, as she was compared to Helena at the *inventio* of the cross, as she fell to her knees in veneration before the holy wood:

> Thus, like Saint Helena, imbued with wisdom, full of the fear of God, glorious with good works, she eagerly sought to salute the wood where the ransom of the world was hung for our salvation that He might be snatched from the power of the devil. . . . What Helena did in oriental lands, Radegund the blessed did in Gaul![23]

In comparing Radegund to Helena, Baudonivia was following established convention,[24] but her analogy has the added dimension of also aligning Poitiers with Jerusalem, the site of the original discovery. Moreover, Radegund's receipt of the wood is not only portrayed as a new discovery, but the relic is described as undergoing its own personified passion upon its arrival in Poitiers, inscribing a restaged Jerusalem narrative into the Western landscape.[25]

Since the relic that sanctified Radegund's monastery in Gaul came ultimately from Constantinople, its relationship to the holy places had changed; its significance was thus articulated by alternative means. As Averil Cameron has argued, it necessarily carried implications of a political and religious alliance with Constantinople.[26] Such associations were seemingly not unwelcome to either side, since the Catholic Franks made an enticing prospective ally for the Byzantines, and a connection with Byzantium was particularly in line with King Sigibert's political ambitions.[27] The encounter saw Radegund walking a fine line between her courtly and pious personas, and the gift of the relic can therefore be seen not only as brought about by Radegund's devotion, but also as a reward for Frankish loyalty.

However, whereas the bequest bound Radegund to Justin and Sophia, Baudonivia emphasized instead the imperial analogy with Helena. She particularly stressed Radegund's own agency in physically drawing sacred matter to Poitiers, just as Helena had impelled the original discovery of such relics. Indeed, Radegund's faith is described as quite literally inspiring relics to motion.[28] Furthermore, such pious astral projection was not an isolated feat, in Baudonivia's telling:

> [Radegund] assembled a great multitude of saints through her most faithful prayers, as the East bears witness and North, South, and West acknowledge. From all sides,

> she managed to obtain those precious gems which Paradise has and Heaven hoards and as many came freely to her as gifts came in response to her pleas.[29]

The directional and spatial qualities of the cross' translation are pronounced in both Baudonivia's description of the event and Fortunatus' letter of thanks to the emperor, which, as noted by Cynthia Hahn, exploits the alignment of the four arms of the cross with the four cardinal directions.[30] Salvation was itself conceived of as directional, spanning not only the expanse between heaven and earth, but having also a cardinal orientation, firmly planted in the East.[31] Fortunatus emphasized the East to West motion brought about by Sophia's good will, but, as Moreira points out, "Baudonivia inverted the radiating configuration of this process to attribute to Radegund an active drawing in of salvation to Gaul. . . . Radegund . . . acted moreover as the channel or 'apostolic' vortex for the arrival of the symbol and substance of salvation in Gaul."[32] Thus, through her pious devotion (and political acumen), Radegund was able to overcome the distance between herself and the wealth of inaccessible holiness in the East: not as an "armchair pilgrim," spiritually projecting herself across Europe, but by actively relocating it to Poitiers in the form of the palpable sanctity of relics, renarrating their sacred history, and recontextualizing them on Gallic soil. Her relic may not have come directly from Jerusalem, as originally hoped, but its Constantinopolitan provenance added an imperial association, and Radegund's material encounter with Jerusalem was elevated through comparison with Saint Helena, the prototypical pilgrim queen.

Theodelinda, queen of Lombard Italy (ca. 600 CE)

In the case of the seventh-century court of Queen Theodelinda in Lombard Italy, there appears, again, to have been an interest in the cross as a topographic marker of pilgrimage. Although there is no record that the queen actively sought cross relics from Jerusalem, she did amass a collection of pilgrimage souvenirs sanctified by the holy wood. Theodelinda, a Bavarian Catholic, married into the Lombard royal family in 589.[33] The Lombards at that time comprised a mixture of pagans, Arian Christians, and Catholics, and although Theodelinda herself had demonstrated schismatic leanings, Pope Gregory the Great forged a significant political relationship with the queen, with whom he corresponded personally.[34] Just as Justin and Sophia had seen a prospective ally in Radegund and the Catholic Franks, Gregory viewed Theodelinda as a potential partner in the conversion of her subjects.[35] As part of her missionary activity, the queen established several Catholic churches in Italy, including the royal chapel of Saint John the Baptist in Monza, which were instrumental in the conversion of the Lombards to orthodoxy.[36]

Splinters of the True Cross came to Theodelinda's court in 603, when Gregory bestowed a pectoral reliquary cross upon her son, Adaloald, in acknowledgment of his Catholic baptism.[37] This object has not survived, but the cross relics it contained were necessarily reframed by their secondary origins in Rome and their role as a diplomatic gift: material proof of the faith to which the queen's son had been committed and a reward for his expected adherence to orthodoxy.[38] The circumstances of this exchange indicate that the significance of the relic was largely symbolic, a tangible sign of the Catholic faith and of Rome's relationship with the Lombard nobility. Although the Jerusalem origins of the wood and its role in the passion remained the ultimate source of its potency, these elements were also colored by the more immediate conditions of the bequest.

Thus, in a period when pieces of the Jerusalem cross were evidently difficult to attain, even for queens, it is interesting to observe the presence of these Roman relics alongside a group of roughly contemporary objects related to the Jerusalem relic at the Lombard court. This cache of thirty-six Palestinian pilgrimage ampullae was divided by the queen between her royal treasuries at Saint John the Baptist in Monza and the Monastery of San Columbano in nearby Bobbio. Little is known of their paths to these foundations, except that Theodelinda deposited them in her treasuries around the turn of the seventh century, making their donation roughly contemporaneous with Gregory's gift to Adaloald. As told earlier, these flasks were apparently filled with oil that had come into contact with the True Cross, likely during a ritual described by the Piacenza Pilgrim:

> At the moment when the Cross is brought out . . . a star appears in the sky, and comes over the places where they lay the Cross. It stays overhead whilst they are venerating the Cross, and they offer oil to be blessed in little flasks. When the mouth of one of the little flasks touches the Wood of the Cross, the oil instantly bubbles over, and unless it is closed very quickly it all spills out.[39]

Similarly, according to their inscriptions, these ampullae would have contained only sanctified oil, but emanating from the primary cross fragment in Jerusalem.[40] One major advantage of acquiring secondary relics of the type any pilgrim might possess was that Theodelinda had the option of sourcing her bequest directly from the Holy City without risking Gregory's ire or undermining his authority, as a request on the model of Avitus' might have done.

Like Radegund, however, Theodelinda was able to subvert an apparent handicap – in this case by conceptualizing the flasks as a complete collection. Circling the edge of each example is some variation on the phrase "Oil of the Wood of Life of the Holy Places of Christ" (Figure 4.1).

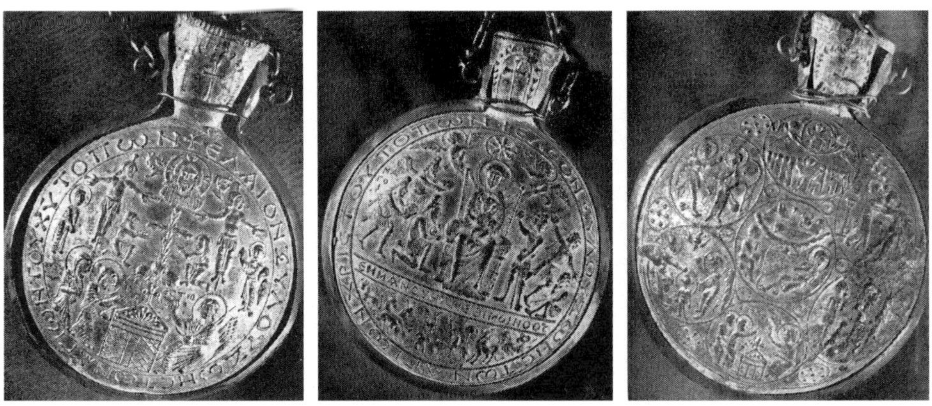

Figure 4.1 Three pilgrim ampullae: one with the Crucifixion and Women at the Tomb (left); one with the Adoration of the Magi (center); and one with loca sancta scenes (right). Byzantine Palestine, sixth–seventh centuries, tin-lead alloy, cathedral treasury of St. John the Baptist, Monza, Italy. Image source: Grabar, André and Denise Fourmont, *Ampoules De Terre Sainte (Monza, Bobbio)* (Paris: C. Klincksieck, 1958).

The iterative textual articulation of their common point of origin effected by their grouping as a collection underscores their common Palestinian origins, stressing the topographic quality that could not have been thus accentuated in the case of Adaloald's singular Roman relic. Furthermore, in describing the enclosed oil as ontologically derived from the "*places* of Christ" (emphasis added), the ampullae inscriptions do not serve merely as labels for their material contents, but rather extend to include the ever more absent topographic referents for which those contents had become stand-ins. The images circumscribed by the text affirm this metonymy and also provide examples: each flask bears not only an impression of the crucifixion or an aniconic cross, but also multiple other loca sancta scenes, most commonly the *Ascension*, the *Adoration of the Magi*, and the *Women at the Tomb*, framed around a schematic version of the contemporary Church of the Holy Sepulchre (Figs. 4.1 and 4.2).[41]

Two examples, one at Monza and the other at Bobbio, display full cycles of locus sanctus images, encompassing a number of Scriptural sites beyond Jerusalem, extending to include the *Visitation*, the *Annunciation*, the *Nativity*, and the *Baptism* (Figure 4.1, right). This expanded geographic network remained visibly embedded in the ampullae, regardless of whether they still contained sanctified oil by the time they arrived in Italy.

Jaś Elsner has called attention to these ampullae as containers for geographic relics that "evoke the sacred myth of their origins" through their unconventional assemblage as sets.[42] The Bobbio collection, which consists of twenty now fragmentary flasks, was discovered inside a wooden casket, itself enclosed within a Roman sarcophagus, buried in the basilica's crypt. These ampullae were accompanied by an

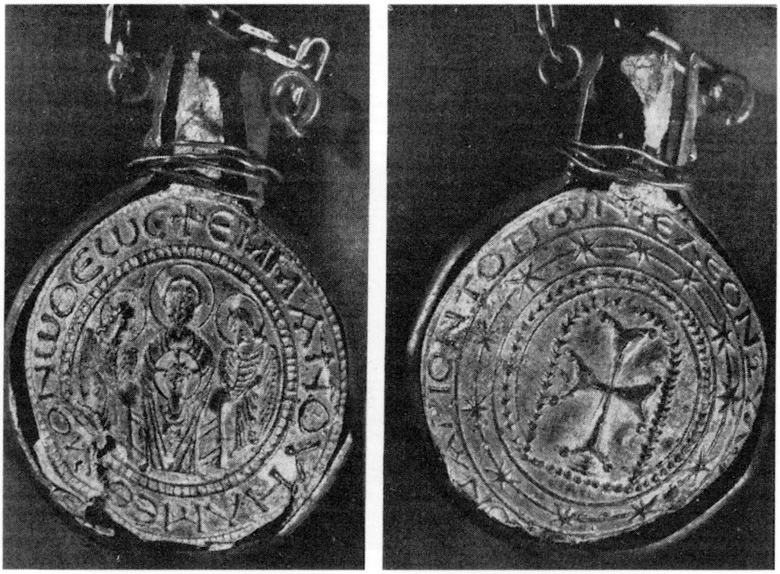

Figure 4.2 Two sides of a pilgrim ampulla with the Adoration and the Cross, Byzantine Palestine, sixth–seventh centuries, tin-lead alloy, cathedral treasury of St. John the Baptist, Monza, Italy. Image source: Grabar, André and Denise Fourmont, *Ampoules De Terre Sainte (Monza, Bobbio)* (Paris: C. Klincksieck, 1958).

assortment of pilgrimage objects, including several relics and eulogiae of Irish origin, and a number of clay tokens of Eastern extraction, featuring impressed images of the flight of Saint Elizabeth, the Transfiguration, Mary Magdalene, Golgotha, the Holy Sepulchre, and even Saint Symeon the Younger.[43] The Monza group, comprising 16 examples, had a comparably more visible occupation, residing continuously in the cathedral treasury. Although their base materiality is in contrast with the treasury's more opulent occupants, the ampullae are, as at Bobbio, accompanied by additional pilgrim souvenirs.[44] Notable among these objects are twenty-six glass vials of blessed oil from the Roman catacombs, gathered from lamps burning alongside the martyrial relics.[45] Owing to the very different contexts of these collections – one cloistered and the other conspicuous – Elsner identifies a different functional logic underpinning each. Each collection, he argues, lends authenticity to the authority of the foundation to which it was attached, but the former does so through burial and the latter through display: at Bobbio, the entombed ampullae and sanctified earth anchor a new saint's foundation in Palestine; at Monza, the visibility of Theodelinda's bequest could have bolstered both civic feeling and the queen's own political clout.[46] Common to both foundations is the mobilization of artifacts associated with a distant time and place in support of a contemporary center of authority.

There are two primary substitutions at work in these ampullae: the oil for the wood and the wood for the places. Through the combination of relic, image, and text, a single flask could encompass not only the cross, but any number of other holy places in the East, and a given pilgrim's direct experience of each. Therefore, beyond any essential metonymy through which cross relics might represent Jerusalem on their own, the ampullae present the sanctity of Palestine by integrating different aspects of these sacred places into a unified whole; as a result, a single example might be considered a discrete collection of locus sanctus relics. In the context of the treasury, the ampullae did not correspond to the experiences and memories of a particular pilgrim, but became instead collectively emblematic of a range of remote pilgrimage destinations.

The Palestinian flasks thus do not simply constitute one collection nesting within another, as Elsner has observed; matryoshka-like, each is an individual collection within a collection – within yet another collection.[47] The framing of these nested groups in Theodelinda's treasury exploits the metonymy of the cross to effect an additional substitution beyond the two previously mentioned: as a purposeful set, they stand not only for the cross and the holy places, but also for the very idea of pilgrimage *to* them. Whereas Radegund transported locus sanctus and sacred narrative by playing the supplicant, Theodelinda, in acquiring serialized souvenirs from Palestine and subsequently donating them to her church foundations, became less a symbolic pilgrim than the new custodian of these markers of topographic sanctity. For, the ampullae formed coherent collections of the sort that, outside Monza, could only be found in Jerusalem itself, ready for distribution to pilgrims. Thus, the Monza group, which was never distributed in this way, acted as a material sampling of a suspended, incipient pilgrimage narrative, maintained under the queen's care. In curating the treasury, Theodelinda acted as the local arbiter of this distant holiness; the ampullae, in turn, served to validate her pious authority, as well as the status of her church foundations as pilgrimage destinations in their own right.

The same can be said of the Roman vials with which the Palestinian examples were paired: in each instance, Theodelinda altered their trajectories by construing them as sets rather than as discrete objects with individual significance. The gift of a single ampulla from either city would have carried with it suggestions of its *locus sanctus*, but Theodelinda's choice to present her foundations with unbroken collections of these objects emphasized seriality and anonymity over irreplaceability and individuality. As collections of collections, they meta-narrate the prototypical act of pilgrimage to the pre-eminent holy sites of both the East and West. In the cases of both Bobbio and Monza, as at Poitiers, the relics were contextualized among assorted sanctified matter – sacred earth, cloth, and pilgrim tokens from various sources – that would have been broadly understood as remnants of a codified "Eastern" sanctity.[48] But whereas at Poitiers the importation of such Eastern exemplars appears to have been of paramount concern, at Monza and Bobbio these more remote sacred souvenirs were combined with relics of local, European derivation: at Bobbio, the remains of the monastery's founder, Saint Columbanus, and at Monza, the relics of Roman martyrs. These collections thus articulate an expanded notion of Christianized topography, uniting Scriptural with more recent sacred narratives.

At Monza, Rome was represented as a sacred site on its own terms, rather than one through which the sanctity of Jerusalem was filtered. Equally, Jerusalem was made present through the material echoes of relics that had remained rooted in the East. Like Avitus, Theodelinda evidently prioritized a Jerusalem derivation for her relics, and the ampullae communicated these authentic origins through both text and image, thereby bolstering both the queen and her churches. Although this concern did not manifest in the wood itself, the ampullae represented the idea of Jerusalem as a place of pilgrimage, with the True Cross as its heart. Situated within the unconventional context of the treasury, these smallest of fragments – mere traces of the sanctified wood – were no longer discrete containers indexing the memory of individual encounters; rather, they assumed a monumental role as a reordered and reassembled set of fragmentary pieces indexing the fullness of relic, place, and archetypal experience.

The cases of Avitus, Radegund, and Theodelinda represent only a few of the possible ways in which the geographic significance of the True Cross might have been harnessed in late antique Europe. The mobility of the cross relics allowed for them to accrue multiple associations across different locations and to be deployed not only as potent spiritual objects, but also as signifiers of power and authority, both religious and political. The relics continued to communicate aspects of Jerusalem's topography and the passion narrative across time and space, but in the context of late antique Europe, they provided an axis around which new sacred narratives could also turn. In this way, the treasuries of Theodelina and Radegund expanded on the model established by Avitus; these Western collections did not just bring together loca sancta within Jerusalem, they operated on a much broader scale, extending well beyond the immediate topographical setting of Jerusalem and bringing together sacred matter and sacred narratives from all over Christendom.

Notes

1 Avitus of Vienne, Epistola XVIII, *PL* 59: 236. Translation adapted from Danuta Shanzer and Ian Wood, eds. and trans., *Avitus of Vienne: Letters and Selected Prose* (Liverpool: Liverpool University Press, 2002), Epistola 20, 154.

2 Paulinus of Nola, *Sancti Pontii Meropii Paulini Nolani Epistulae*, ed. Guilelmus de Hartel, vol. 29, Corpus Scriptorum Ecclesiasticorum Latinorum (Vienna: Österreichische Akademie der Wissenschaften, 1999), Epistola 31.1, lines 6–15, 268; *Letters of St. Paulinus of Nola, Letters 23–51*, trans. P. G. Walsh, vol. 2 (Westminster, MD, and London: Newman and

Longmans Green, 1967), 125–126. See also a similar translation described by Gregory of Nyssa: Gregory of Nyssa, *Vie de sainte Macrine*, trans. Pierre Maraval (Paris: Éditions du Cerf, 1971), 238–243 (Migne, *PG*, 45–46).

3 Paulinus of Nola, *Epistolae*, Epistola 31.1, lines 6–15, 268; *Letters of Paulinus of Nola*, 125–126.

4 Avitus of Vienne, Epistola XXIII, *PL* 59: 239–240B; Shanzer and Wood, *Avitus of Vienne*, Epistola 25, 156. The association of the cross (as both relic and image) with Jerusalem extends beyond the obvious fact of the crucifixion. On the developing theology of the cross, Jewish and Platonic traditions, and their relationship to the topography of Jerusalem, see Stefan Heid, *Kreuz, Jerusalem, Kosmos: Aspekte frühchristilicher Staurologie*, Jahrbüch für Antike Christentum, Ergäzungsband, 31 (Münster Westfalen: Aschendorff, 2001), 13–105, esp. 61–76.

5 See the reliquary crosses of St. Symeon Stylites the Younger and Theodore of Sykeon: *Version géorgienne de la Vie de Sainte Marthe*, ed. Gérard Garitte (Louvain: Secrétariat du Corpus Scriptorum Christianorum Orientalium, 1968), chs. 58–59, 302–303; *Vie De Theodore De Sykeon*, vol. 1, ed. A.J. Festugière (Brussels: Société des Bollandistes, 1970), chs. 128, 103. Securely identifiable cross reliquaries from late antiquity are scarce, and the degree to which surviving texts accurately describe lost objects must necessarily remain ambiguous. What seems clear from these accounts, however, is that their authors understood the True Cross at least partly in terms of Jerusalem's sacred topography. One need only look at the Sancta Sanctorum reliquary to understand the associative geographies of relics from the holy places, including the True Cross.

6 The literature on the True Cross and the legends of its discovery is vast. See Anatole Frolow, *La Relique de La Vraie Croix: Recherches Sur le Développement d'un Culte* (Paris: Institut français d'études byzantines, 1961); Id, *Les Reliquaires de la Vraie Croix* (Paris: Institut français d'études byzantines, 1965). For a recent introduction, see (with accompanying bibliography) John Wortley, "The Wood of the True Cross," in *Studies in the Cult of Relics in Byzantium up to 1204*, ed. John Wortley (Farnham: Ashgate, 2009), 1–19; Holger A. Klein, *Byzanz, der Westen und das "wahre" Kreuz: Die Geschichte einer Reliquie und ihrer Künstlerischen Fassung in Byzanz und im Abendland* (Wiesbaden: Reichert, 2004).

7 Cyril of Jerusalem, *Catecheses*, 4, 10; *Cyrilli Hierosolymorum archiepiscopi opera quae supersunt omnia*, eds. W.K. Reischl and J. Rupp, vol. 1 (Munich: Libraria Literiana, 1848–1860), 1:100 (Migne, *PG*, 33:469).

8 John Wilkinson, *Egeria's Travels* (Warminster, UK: Aris & Phillips, 1999), 155. Wilkinson observes in note 5 that a large piece of the Jerusalem cross was thought to have been stolen by a Syrian.

9 See the roughly chronological progression of attested relics of the cross, which decrease precipitously over time, in Frolow, *Les Reliquaires De La Vraie Croix*, 155–188. The Piacenza Pilgrim describes what seems to be a well-established ritual for generating secondary relics of the cross around 570. John Wilkinson, *Jerusalem Pilgrims before the Crusades* (Warminster, UK: Aris & Phillips, 2002), 139; *Travels of the Piacenza Pilgrim*, 20. Whereas ordinary pilgrims could not have expected to receive primary portions of this pre-eminent relic, Gaul's Queen Radegund's failed petition to Jerusalem for a piece of the holy wood in the late sixth century is perhaps more telling about their increasing rarity.

10 This is not to suggest that a donation from Jerusalem would have come without the baggage of contemporary power relations, but rather that these would have had a quite distinct constellation.

11 John McCulloh, "The Cult of Relics in the Letters and 'Dialogues' of Pope Gregory the Great: A Lexicographical Study," *Traditio* 32 (1976): 148.

12 Leslie Brubaker, "The Elephant and the Ark: Cultural and Material Interchange across the Mediterranean in the Eighth and Ninth Centuries," *Dumbarton Oaks Papers* 58 (2004): 175–195. For more on the dynamics of circulation and on the "life cycles" of objects, see also the essays in Arjun Appadurai, ed., *The Social Life of Things: Commodities in Cultural Perspective* (Cambridge and New York: Cambridge University Press, 1986).

13 Jo Ann McNamara, "'Imitatio Helenae': Sainthood as an Attribute of Queenship," in *Saints: Studies in Hagiography*, ed. Sandro Sticca (Binghamton, NY: Medieval and Renaissance Texts and Studies, 1996), 63.

14 Ibid., 63–64.
15 Baudonivia, "Life of Radegund," in *Sainted Women of the Dark Ages*, trans. and eds. Jo Ann McNamara and John E. Halborg (Durham, NC and London: Duke University Press, 1992), 96–99; Gregory of Tours, *HR*, 9.40; *The History of the Franks*, vol. 2, trans. O.M. Dalton (Oxford: Clarendon Press, 1927), 413; Isabel Moreira, "Provisitatrix optima: St. Radegund of Poitiers' Relic Redemption to the East," *Journal of Medieval History* 19 (1993): 290; Annabel Jane Wharton, *Selling Jerusalem: Relics, Replicas, Theme Parks* (Chicago: University of Chicago Press, 2006), 31.
16 McNamara and Halborg, eds., *Sainted Women*, 98, note 119.
17 The authenticity of this relic notwithstanding, the reliquary that houses it likely dates to the eleventh or twelfth century. For a review of the dating controversy, see Robin Cormack, "Reflections on Early Byzantine Cloisonné Enamels: Endangered or Extinct," in *Thumíama ste mneme tes Laskarínas Mpoúra*, eds. Panagiotes L. Bokotopoulos Rena Andreade, Cyril Mango, Jean-Pierre Sodini, and Manoles Hatzedakes (Athens: Benaki Museum, 1994), 68–69. More recently, see Cynthia Hahn, "Collector and Saint: Queen Radegund and Devotion to the Relic of the True Cross," *Word & Image* 22, no. 3 (2006): esp. 268–269, 272; Lynn Jones, "Perceptions of Byzantium: Radegund of Poitiers and the Relics of the True Cross," in *Byzantine Images and Their Afterlives: Essays in Honour of Annemarie Weyl Carr* (Farnham: Ashgate, 2014), 123.
18 Isabel Moreira's analysis reveals that a relic mission purported by Radegund's biographer Baudonivia to have sought the finger of Saint Mamas from Jerusalem was, in fact, a failed petition for the cross. Her other biographers are silent on the alleged mission after Mamas' finger. Regarding what must have been a second attempt, Gregory of Tours simply stated that the cross was acquired after Radegund "repeatedly sent envoys to the East," with the oblique implication that a successful petition was made to Jerusalem. See Moreira, "Provisitatrix optima," 289–298.
19 Lynn Jones, in an article published while this essay was under final revision, treats the topic of the queen's increasing local authority in some depth. See Jones, "Perceptions of Byzantium," esp. 110–118.
20 McNamara, "Imitatio Helenae", 99, note 120.
21 Gregory of Tours, *Miracula*, 40.1ff.
22 Baudonivia, *Life of Radegund*, 2.16; in *Sainted Women*, 97.
23 Ibid., 99.
24 McNamara, "Imitatio Helenae," 51–80.
25 Moreira, "Provisitatrix optima," 302–303; Hahn, "Collector and Saint," 271. Hahn points out that in Radegund's narrative, the *inventio* is a "discovery of faith."
26 Averil Cameron, "The Early Religious Policies of Justin II," in *The Orthodox Churches and the West*, ed. Derek Baker (Oxford: Basil Blackwell, 1976), esp. 55–60. For more on the political (and economic) implications of the gift, see Wharton, *Selling Jerusalem*, 31–35.
27 Cameron, "The Early Religious Policies," 59; McNamara, "Imitatio Helenae," 65.
28 See, for instance, Baudonivia's account of the Mamas expedition, in which one of the saint's fingers separates itself from the hand in response to Radegund's faith: Baudonivia, *Life of Radegund*, 2.14, in *Sainted Women*, 95.
29 Ibid.
30 Hahn, "Collector and Saint," 271; see also Gerhard B. Ladner, "Medieval and Modern Understanding of Symbolism: A Comparison," in *Speculum* 54, ed. Kurt Weitzmann, et al. (1979): 231; idem., "St. Gregory of Nyssa and St. Augustine on the Symbolism of the Cross," in *Late Classical and Medieval Studies in Honor of Albert Mathias Friend, Jr.* (Princeton: Princeton University Press, 1955), 88ff.
31 Colin Morris, *The Sepulchre of Christ and the Medieval West: From the Beginning to 1600* (Oxford: Oxford University Press, 2005), 59. On the directional and cosmological associations of the cross's geometry, with numerous later medieval examples, see Bianca Kühnel, *The End of Time in the Order of Things: Science and Eschatology in Early Medieval Art* (Regensburg: Schnell und Steiner, 2003), ch. 3, esp. 136–142; Heid, *Kreuz, Jerusalem, Kosmos*, pt. 1, esp. 96–105.
32 Moreira, "Provisitatrix optima," 301–302.
33 Jan T. Hallenbeck, "Pavia and Rome: The Lombard Monarchy and the Papacy in the Eighth Century," *Transactions of the American Philosophical Society* 72, no. 4 (1982): 14.

34 Hallenbeck, "Pavia and Rome," 14; Robert A. Markus, "Gregory the Great and a Papal Missionary Strategy," *Studies in Church History* 6 (1970): 30. On the complicated religious pluralism among the Lombards in this era, see Steven C. Fanning, "Lombard Arianism Reconsidered," *Speculum* 56, no. 2 (April, 1981): 241–258.

35 Dennis Trout, "Theodelinda's Rome: Ampullae, Pittacia, and the Image of the City, *Memoirs of the American Academy in Rome* 50 (2005): 133; Robert A. Markus, *Gregory the Great and His World* (Cambridge: Cambridge University Press, 1997), 99–100.

36 Hallenbeck, "Pavia and Rome," 14.

37 Gregory the Great, *Registri*, IV, 12 (*MGH, Epist.*, II, 431).

38 A small cross in the Monza treasury roughly fits this description and has sometimes been assumed to be the seventh-century original. See Angelo Paredi et al., eds., *Il Tesoro del Duomo di Monza* (Milan: Amilcare Pizzi, 1966), 30; Roberto Conti, ed., *Monza: Il Duomo e i suoi Tesori* (Milan: Credito Artigiano, 1988), 22–23. However, its stylistic similarities to several other pectoral crosses indicate a late-ninth-century date. See Anna D. Kartsonis, *Anastasis: The Making of an Image* (Princeton: Princeton University Press, 1986), 94–125.

39 *Piacenza Pilgrim*, 20; Wilkinson, *Jerusalem Pilgrims before the Crusades*, 139.

40 According to Conti, the ampullae still contain dried remnants of this oil. See Conti, *Monza*, 34.

41 Gary Vikan, *Early Byzantine Pilgrimage Art, Revised Edition* (Washington, DC: Dumbarton Oaks Research Library and Collection, 2010), 37.

42 Jaś Elsner, "Replicating Palestine and Reversing the Reformation: Pilgrimage and Collecting at Bobbio, Monza, and Walsingham," *Journal of the History of Collections* 9, no. 1 (1997): 123.

43 G. Celi, "Cimeli Bobbiesi," *La Civilta Cattolica* 74, no. 3 (1923): vol. 2, 504–514; vol. 3, 37–45, 124–136, 335–344, 423–439; Arturo Calzona, "Oriente e Occidente a Bobbio," in *Medioevo mediterraneo: l'Occidente, Bisanzio e l'Islam: Atti del Convegno internazionale di studi Parma, 21–25 settembre 2004*, ed. Arturo Carlo Quintavalle (Parma: Electa, 2007), 604, 608–610; Elsner, "Replicating Palestine," 119–120.

44 Owing to a papyrus document (*notitia*) that describes the donation of the vials as taking place "during the time of" Pope Gregory, scholars have tended to assume that they were a gift from him to Theodelinda. For a full transcription of the papyrus, see A. Sepulcri, "I papyri della basilica di Monza," *Archivio Storico Lombardo* 19 (1903): 258. This assumption is sometimes extended to include the Palestinian ampullae as well, although there is no hard evidence to support either claim. See Trout, "Theodelinda's Rome," 133–134, esp. notes 20, 21, 24. Grabar, in his foundational work on the ampullae, understood the metal flasks to be Theodelinda's donation. André Grabar and Denise Fourmont, *Ampoules de Terre Sainte (Monza, Bobbio)* (Paris: C. Klincksieck, 1958), 15–16, 32.

45 Conti, *Monza*, 34–36.

46 Elsner, "Replicating Palestine," 119–123.

47 Ibid., 123. The same is true of the Roman flasks, which are described in accompanying papyrus labels (*pittacia*) as containing the oil of "many thousands of saints" ("*multa milia sanctorum*"). See Sepulcri, "I papyri della basilica di Monza," 260–262.

48 These objects include three petite lead caskets containing an assortment of relics, among which are numerous wooden fragments, pieces of stone, a quantity of earth, and fragments of additional ampullae. See Conti, *Monza: I Tesori*, 34–36.

II

Agents of translation

5 *Una processione da farsi ogni anno con una Messa Solenne*

Reception of stone relics from the Holy Land in Renaissance Ragusa

Tanja Trška

The history of the Republic of Ragusa (modern Dubrovnik, Croatia) was shaped in a significant way by its fidelity to the Catholic Church, which held sway throughout the centuries of the Republic's independence – from 1358, the year of its liberation from Venetian rule and formal institution of the Republic, to 1808, when the state ceased to exist.[1] Its political identity in relation to the larger European states and the Holy See was defined on the basis of the much repeated metaphor *antemurale Christianitatis*,[2] considering the Republic's geographical position between the "schismatics" (Eastern Orthodox Christians) and the infidels (the Ottoman Empire). A major element in the state's religious identity involved material objects of veneration – numerous relics carefully preserved in the city's cathedral treasury (Figure 5.1) and churches of different monastic orders, palpable demonstrations of the Republic's centuries-old loyalty to the Catholic Church.

The safekeeping of relics was organized in such a way that neither the archbishop nor the clergy had access to more than a hundred sacred fragments housed in the cathedral treasury: in 1433, the Senate issued a decree to the effect that a total of six keys to the treasury were to be kept by three treasurers and three *procuratori* and that at least two members of each group should be on hand for every opening and temporary removal of the relics for processions.[3] The treasurers and the *procuratori* were nominated from among the Ragusan nobility, so that the various archbishops – between the fourteenth and eighteenth centuries these were always foreigners[4] – would not have an opportunity to appropriate holy fragments and send them back to their own homelands or donate them to their princely friends.[5]

Serafino Razzi, author of the first printed history of Ragusa (1595), recorded that there were 111 relics carried in the procession on the feast of Saint Blaise and that many more were housed in the cathedral.[6] The number of relics in the cathedral treasury mentioned by Giacomo di Pietro Luccari at the beginning of the seventeenth century – more than 300 – is somewhat exaggerated,[7] although a considerable number were destroyed in an earthquake in 1667. Finally, eighteenth-century historiographer Seraphinus Maria Cerva enumerated 160 extant relics and another fifty-two that were lost,[8] whereas the number of relics currently in the Treasury is 161, with twenty-one items reported missing.[9] Razzi especially praised the Ragusans for never parting with their sacred objects, even when these were requested by much more powerful states: he noted a request made by the city of Florence for a relic of the hand of their patron saint, John the Baptist, to which the Ragusans responded that "the city of Ragusa, situated on the borders with the infidel, is in greater need of sacred relics and saintly assistance than the city of Florence, which feels no similar danger."[10] Nonetheless, the

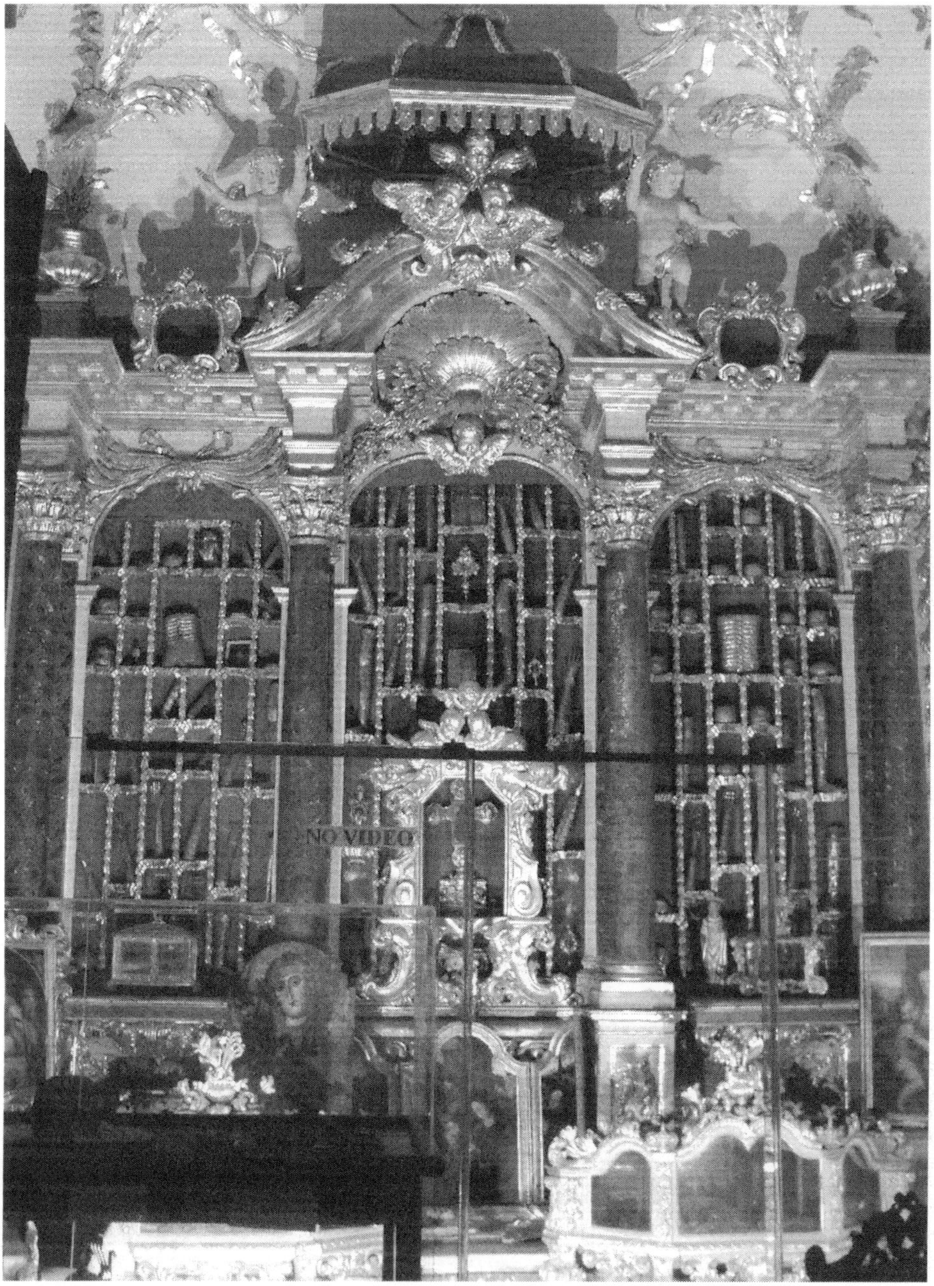

Figure 5.1 Dubrovnik, cathedral treasury, 1713–1717, attributed to Marino Gropelli.

need for political alliances gradually changed the Ragusans' attitude: in 1771, they ceded the relic of the hand of the Hungarian king Saint Stephen to the Empress Maria Theresa in return for her protection against a possible Russian invasion from the sea.[11]

The relics possessed by the Republic of Ragusa brought the state considerable prestige and were regularly mentioned in the accounts of pilgrims and city chroniclers. Apart from the relics of the hand and leg of Saint Blaise, patron saint of the Republic, fifteenth-century pilgrims Roberto di Sanseverino, Gabriele Capodilista, and Santo Brasca made special mention of a particularly venerated object from the Holy Land, the swaddling clothes in which Jesus was wrapped when he was presented to Saint Simeon in the Temple.[12] Moreover, these precious things were not overlooked in written histories: in his 1440 description of the city, Philip de Diversis mentioned "the relics of many saints encased in silver," among which "the completely white swaddling clothes in which Our Lord Jesus was wrapped when the prophet Simeon took him in his arms" stood out.[13] Churches of monastic orders in Ragusa – Franciscan, Dominican, and the Order of Saint Claire – owned other relics from the Holy Land: fragments of the True Cross, the Column of the Flagellation, and other relics related to the passion,[14] but their existence was not accorded any special attention, nor did they induce more elaborate ceremonies of veneration. Singular exceptions were the two stone fragments that arrived in Ragusa from the Holy Land in late 1558, the veneration of which immediately assumed the form of a state-inaugurated public procession.

The ties between the Republic of Ragusa and the Holy Land were strengthened around the middle of the sixteenth century, when a Ragusan, Fra Bonifacio de Stephanis (Lopud/Isola di Mezzo, ca. 1500 – Timişoara, 1581) was named to fill the distinguished post of Guardian of Mount Zion.[15] Fra Bonifacio entered the Franciscan order in Ragusa, where he completed the order's studies in theology; in 1543, he was sent to Paris to continue his studies in philosophy and theology. Before returning to Ragusa, he spent some time in Rome, where he maintained close relationships with many learned men, among whom was Felice Peretti, the future Pope Sixtus V.[16] Fra Bonifacio's service in the Holy Land began in 1551, after first being named Guardian of the Holy Land (he was re-elected in 1553, 1556, and 1563); during the years he spent in Jerusalem he restored many sacred structures, including the Holy Sepulchre, on which work was begun in 1555 (Figure 5.2).

He described the opening of that church in a letter written in Ston (a town and fort northwest of Ragusa) on May 13, 1570, some fifteen years after the event, in which he recalled the discovery of the fragment of the wood of the Holy Cross, which he divided into smaller pieces and sent to Rome and his homeland.[17] The retrieved fragment of the Holy Cross was initially broken into two pieces, of which one was left at the altar of the Holy Cross in the Chapel of the Apparition within the Holy Sepulchre, while the other was further divided and presented to Pope Pius IV, Cardinal Rodolfo Pio da Carpi, and Cardinal-Priest of Santa Maria in Ara Coeli (Clemente d'Olera). Fra Bonifacio kept the last fragment for himself and eventually donated it to the Diocese of Ston, where he was named bishop in 1564;[18] another piece of the Holy Cross was given to the Spanish king Philip II.

Fra Bonifacio's gifts were not limited to the relics of the Holy Cross: he sent fragments of the Column of the Flagellation to Pope Paul IV, the Holy Roman Emperor Ferdinand I, and the Republic of Venice.[19] His choice of recipients was mindful and diplomatic: he distributed the fragments to either the protectors of the Franciscan order (Minister-General of the Order of Friars Minor Clemente d'Olera and the

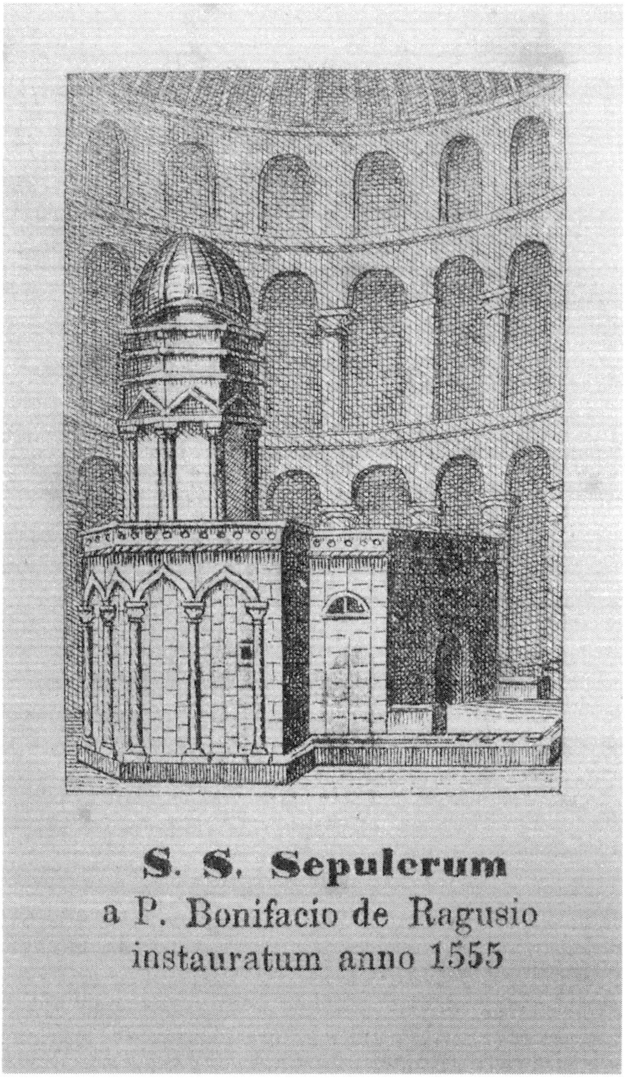

Figure 5.2 The Holy Sepulchre as restored by Fra Bonifacio de Stephanis, illustration in Bonifacio Stephano Ragusino, *Liber de perenni cultu Terrae Sanctae et di fructuosa eius peregrinatione*, 2nd ed., Venice: Ex Typ. L. merlo Ioh. Bapt. filii, 1875.

supporter of the Capuchins Rodolfo Pio da Carpi) and to sovereigns and political powers who could provide financial assistance for his efforts to restore the sites of the Holy Land – as did the Spanish king Philip II, himself a passionate collector of relics.[20] However, among the multitude of sacred fragments from the Holy Land that he distributed all over Europe, the relics that engendered the greatest veneration were the ones he brought to his homeland, the Republic of Ragusa: a fragment of the Holy Sepulchre and a particle of the Column of the Flagellation.

These two relics arrived in Ragusa on December 15, 1558, accompanied by a letter from Fra Bonifacio addressed to the Republic of Ragusa and dated July 24, 1558 (Figure 5.3).

He began by noting that many relics had passed through his hands during his service in the Holy Land and that he had donated precious objects to many churches, among which were the Basilica of Saint Mark in Venice and the Church of the Blessed Apostles Peter and Paul in Rome.[21] He went on to explain the origin of the two stone fragments he was giving to the Republic: the fragment of the Holy Sepulchre was detached during the 1555 restoration of the sanctuary and the piece of the Column of the Flagellation came from the red porphyry column preserved in the Chapel of the Apparition within the sanctuary of the Holy Sepulchre. The latter, originally kept in the Praetorium of Pilate,[22] was considered the true Column of the Flagellation ordered by Pilate before Christ's crucifixion, unlike the column of white-veined black granite preserved in the Church of Santa Prassede (brought to Rome by Cardinal Giovanni Colonna in 1223), identified as the one once situated in the house of Caiphas.[23]

The reception of the two stone fragments in Ragusa was immediate: on December 16, 1558, only one day after they arrived in the city, the Senate ordered the officials of the cathedral to place the relics in two separate containers and arranged for a general procession with a holy Mass in the Cathedral, which was to be held every year to commemorate the day of their arrival in the city. Moreover, the day of the procession – to be held *in perpetuum* – was proclaimed a nonworking day.[24] However, the decision of the Republic to accept the two relics and arrange for a public feast and procession was made without consulting the archbishop, despite the fact that the matter was clearly within the realm of the ecclesiastic authorities. It thus came

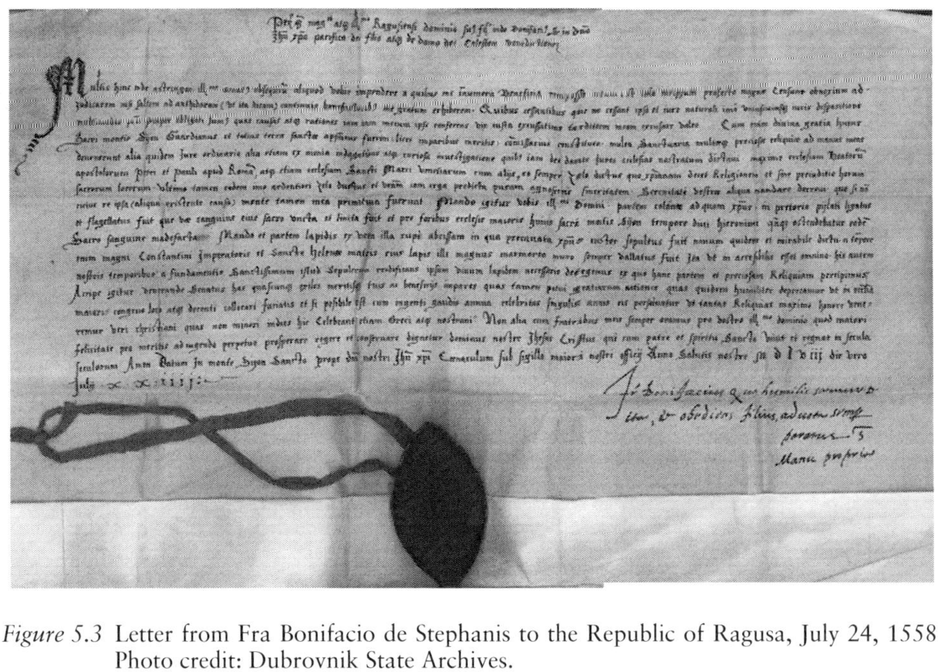

Figure 5.3 Letter from Fra Bonifacio de Stephanis to the Republic of Ragusa, July 24, 1558. Photo credit: Dubrovnik State Archives.

as a surprise to the reigning archbishop of Ragusa, Lodovico Beccadelli (Bologna, 1501 – Prato, 1572; archbishop 1555–1564),[25] who was anxious to restore order and improve the problematic state of the archdiocese. Beccadelli's efforts met with much resistance on the part of the Ragusan clergy and thus of the government, whose members were eager to protect the interests of their fellow noblemen.

One of the major difficulties was the interference of the state in ecclesiastical matters, a problem that was clearly reflected in the episode of the two relics from the Holy Land. Their arrival was followed by a completely autonomous decision on the part of the Republic to arrange a ceremony, which necessarily included the participation of ecclesiastical authorities, but it was not deemed necessary to get the approval of the latter when the relics arrived. Archbishop Beccadelli's cautious attitude in the matter is revealed in a letter dated January 19, 1559, addressed to the sacristan of the Apostolic Palace, Giovanni Giacomo Barba, regarding Fra Bonifacio's gift of "two fragments of stone in a box well sealed and enfolded," sent to the Republic to venerate them among the many relics they already possess.[26] The archbishop explained that he had informed the government that, though "all pious feelings are most praiseworthy, the introduction of new relics was not up to them," but remained the exclusive privilege of the Holy See. His greatest concern was to avoid departing from the "true rule" by allowing the acceptance of possibly inauthentic relics, especially after the Republic's decision "not only to venerate them, but to institute a procession with a solemn Mass to be held every year on the day of their arrival in the city, that is, on December 15."

Lodovico Beccadelli's prudence regarding the fragments from the Holy Land reflected his views regarding the ongoing efforts to reform the Catholic Church at the Council of Trent, which he followed closely. In fact, his conduct regarding Fra Bonifacio's gift anticipated the decree "On Invocation, Veneration and Relics of the Saints, and on Sacred Images," issued during the 25th session of the Council (December 3–4, 1563), which prescribed that all bishops "are first of all to instruct the faithful carefully about . . . reverence for their relics." The same conciliar decision demanded precisely what the archbishop of Ragusa requested from the Senate – mandatory consultation with ecclesiastical authorities: "Nor are any new miracles to be accepted, or new relics recognized, without the bishop similarly examining and approving them. And as soon as he learns of something of this kind, he should consult with theologians and other devout men and decide as truth and devotion suggest."[27]

The Republic's pronounced readiness to accept the relics from the Holy Land had its precedents in the arrival of other sacred fragments procured by Ragusan merchants, most of which did not have a "certificate of authenticity," such as the ones sent by the Guardian of the Holy Land: in 1550, the Republic accepted the arm of the Innocents, a thorn from Christ's crown, and the sponge and a part of the Virgin's veil and had them placed in the cathedral treasury.[28] However, less than a decade later, the importance accorded the two relics from Jerusalem was far greater, and their value must have been perceived in light of the fact that the geographical position of Ragusa ensured the city the status of a frequent stop for pilgrims traveling from Venice to the Holy Land.[29] The privilege of transporting pilgrims to the Holy Land, granted to the Republic by the Council of Basel (1433) and confirmed by successive popes,[30] ensured their presence in the city, so the existence of material objects from Jerusalem must have been viewed by their new owners as a sort of introduction to the sacred places the pilgrims were directed to as a *pars pro toto* preview of the Holy Sepulchre and the Column of the Flagellation available within the cathedral treasury. Their significance in the local context is revealed not only in the speed with which the Ragusan government arranged a public feast on the day of their arrival, but also in the rank assigned to the feast: a general procession and a holy Mass in the Cathedral – or, as Archbishop Beccadelli defined it, "*una messa solenne*" – corresponded to a Solemnity, a feast of the highest rank, which included the service of the Solemn or High Mass[31] in the principal church of the city and presupposed the participation of the highest ecclesiastical authority, the archbishop himself, who would be expected to conduct the procession.

The role played by the state in the case of the two stone relics from the Holy Land was in fact a reflection of the general attitude of the Ragusan government toward the ecclesiastical authority of its (foreign) archbishop, who was accepted and respected as long as he did not interfere in political issues, the Republic's internal affairs, or its long-established traditions – or, in this case, the traditions they had decided to create. Whereas most *translatio* narratives emphasize the participation of a local ecclesiastic, either as a commissioner or receiver of the newly acquired relics,[32] in Ragusa the archbishop was entirely disregarded by both parties, Fra Bonifacio as the sender and the Republic as the receiver. The authorization to accept and solemnize the two relics arrived only several months later: in his response to Beccadelli's letter, dated February 17, 1559, the sacristan of the Apostolic Palace noted that there were no obstacles to accepting the two stone fragments, since they were not relics of "new or nonsolemnized saints, but ancient relics venerated throughout the Christian world, if truthfully obtained."[33] At that point, the effect of such a response could have only been to give the archbishop peace of mind, since the secular authorities considered the matter long resolved. The issue of relics and processions in Ragusa would be brought to the attention of the highest ecclesiastical authorities several decades later: among the fifty-seven complaints regarding the Republic's intervention in ecclesiastical matters presented to Pope Gregory XIII in 1582, Archbishop Girolamo Matteucci (archbishop of Ragusa 1579–1583) noted that the Ragusans had too liberal an attitude toward relics and inaugurated processions as they pleased, and at the same time did not participate in those prescribed by the church.[34]

The practice of introducing state-inaugurated processions was certainly not uncommon during the Renaissance – for instance, the ducal procession in honor of Saint Theodore declared by the Venetian Senate in 1450[35] – but such a procession usually signified a revival of interest in a particular saint whose relics were already present in the civic and ecclesiastical traditions of a community. By contrast, the Ragusans' decision regarding the processional veneration of the two stone relics could not have been corroborated by any ancient tradition or common memory, but was evidently motivated by their origin, which transformed the two fragments into material reminders of the Holy Land. The importance ascribed to the relics only a day after their arrival becomes even more extraordinary in light of the fact that the cathedral treasury already held two fragments of the Holy Sepulchre, which had not been accorded similar treatment.[36] What evidently motivated the Republic to arrange an annual procession was not the fact that they were granted relics of the passion (others of which were already in churches of the monastic orders), but rather the circumstances of their arrival: in the eyes of the state, the accompanying letter from the Guardian of the Holy Land guaranteed their origin and therefore their authenticity – arguments that did not quite convince Archbishop Beccadelli. The statement that other relics had been sent to Rome and Venice certainly added value to the gift; it even assigned the two fragments a certain political significance, hidden in the notion that the Republic of Ragusa was considered an equal receiver of the authentic objects from the Holy Land as the objectively more powerful European rulers.

Despite Archbishop Beccadelli's objections, Fra Bonifacio's relics were recognized as authentic before the official confirmation from the Holy See, and the procession arranged by the state was included in the calendar of feast days officially celebrated by the Republic. In 1561, to honor the donor, the Senate further decided that the procession was to enter the Church of Saint Francis.[37] Public rituals, both secular and ecclesiastical, had a prominent place in the life of the Republic of Ragusa, while the century-long tradition of veneration of relics, accompanied by numerous processions (Figure 5.4), served as a demonstration of fidelity to the Catholic Church both inside and outside the Republic.

A special calendar of festivities (*Calendarium festorum celebrandorum secundum ordines Racusii*) introduced at the beginning of the fifteenth century singled out the feast days within the Christmas and Easter cycles that were of special significance for the state.[38] However, the institution of a feast day to commemorate the arrival of the two relics from the Holy Land was a case apart: no special dedication or festivity of a higher rank was prescribed for December 15,[39] so the day assigned to the solemn celebration of Fra Bonifacio's relics represented an addition to the feast days that were

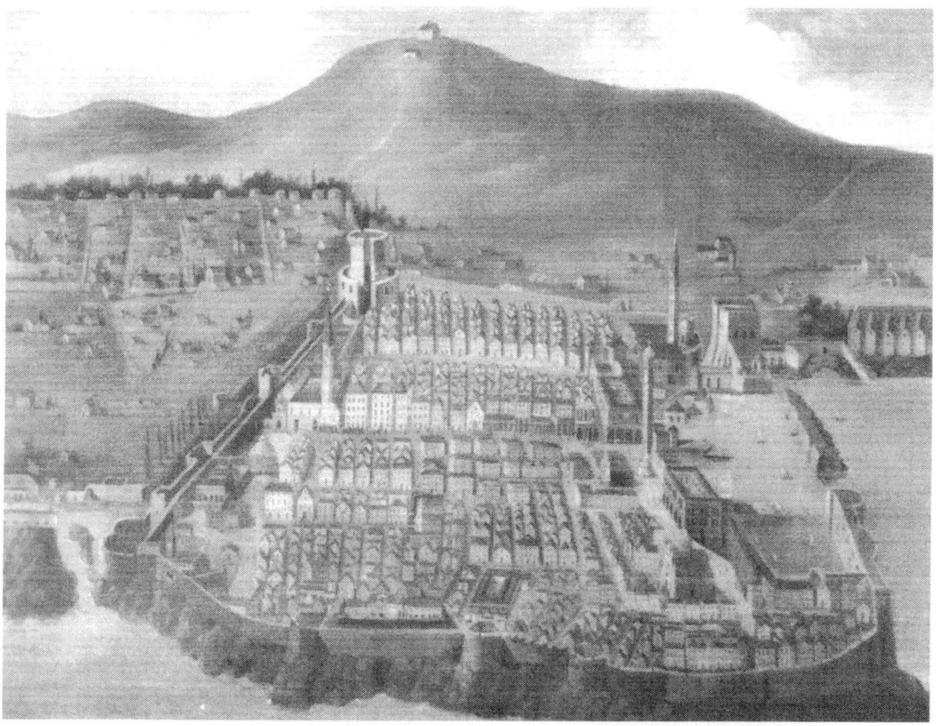

Figure 5.4 Zebedeo Picinni, *View of Ragusa (Dubrovnik) before the earthquake of 1667*, showing a procession entering the Franciscan church, nineteenth century. Dubrovnik, Franciscan monastery. Photo credit: Institute of Art History, Zagreb, Photoarchives collection, Inv. No. IPU-F-27924, photograph by Paolo Mofardin, 2013.

already being celebrated both universally and locally. In a list of public processions compiled by the eighteenth-century historiographer Seraphinus Maria Cerva, the processions not corresponding to universal feast days of the Catholic Church were only votive feasts instituted after the calamities that struck Ragusa during the seventeenth century – the fire of 1610, the earthquakes of 1639 and 1667, and the plague of 1691. The only other procession not paired with another feast day was the feast of the *translatio* of the hand of the Republic's patron saint, Blaise, brought from Venice in 1346.[40] The proclamation of a feast of a higher rank thus gave Fra Bonifacio's stone relics a hitherto unprecedented significance.

A parallel – at least regarding the significance assigned to relics of similar provenance – can be found in the well-known Florentine tradition of the *Scoppio del Carro*, created around three fragments of the Holy Sepulchre brought back to Florence by Pazzo (Pazzino) de' Pazzi after the First Crusade. According to the legend, he was the first to climb the walls of Jerusalem to place a troop banner as a sign of victory, and he was given three flint fragments from the Holy Sepulchre as a reward. Upon his return to Florence, Pazzo was greeted with great enthusiasm and taken through the city on a triumphal cart, while the ceremony of the *Scoppio del Carro* – in which the relics of the Holy Sepulchre were used to ignite the "holy fire" distributed among Florentine homes – was instituted to commemorate his heroic enterprise and the arrival of the relics in the city. The account of the Florentine fragments of the Holy Sepulchre is not supported by archival documentation dating from the late eleventh century (the assumed time of their arrival), but relies on a legend probably created in the first half of the fifteenth century.[41] The Florentine narrative of the origin of the *Scoppio del Carro* and the documented Ragusan episode reflect the same pattern: a starting point of a tradition centered on relics from the Holy Land that presupposed the participation of the whole city. In both cases the material evidence of the Holy Land procured by a fellow citizen laid the foundations for a civic ritual to be observed in the centuries to follow, a ritual that would celebrate not only the relic, a vivid representation of the physically distant Holy Land, but also the donor responsible for its translation.

The memory of the arrival of the relics of the Holy Sepulchre and the Column of the Flagellation, along with the relic of the wood of the Holy Cross donated by Fra Bonifacio to the Diocese of Ston (which was also carried in a procession through the city),[42] found a prominent place not only in the civic ritual of the Republic of Ragusa, but also in the works of later writers and chroniclers of Ragusan history. The Senate's decision that set the procession on December 15 was changed over time, probably before 1670,[43] so eighteenth-century sources list the procession of the fragment of the Holy Sepulchre on the Octave Day of Easter.[44] The altered date – more appropriate to the Easter cycle – entered the first ceremonial book compiled at the turn of the eighteenth century[45] and was kept until the fall of the Republic in 1808.[46] However, although the name of Fra Bonifacio is regularly associated with the story of their provenance, written accounts vary in important details, such as the date and manner of the donation of the relics. The notion of the exact date of their arrival had been obscured by the dates that marked Fra Bonifacio's ecclesiastical career: his return to his homeland in 1564, corresponding to the year of his appointment to the Diocese of Ston, was mistakenly identified with the year of the arrival of the relics by most of the manuscript sources (and adopted by subsequent authors). Cerva (1744) dates the donation of both fragments to 1564, but later within the same work notes that the relics of the Holy Sepulchre arrived in 1565.[47] In a manuscript compiled in the last quarter of the eighteenth century, the Jesuit Gian Maria Mattei recorded their presence among the cathedral treasury relics since 1564,[48] but his own transcription of an older source gives the year 1558 as the time of the donation.[49] Equally imprecise information is found in an anonymous chronicle most likely written during the course of the eighteenth century, which – without recording the year – notes that the two fragments were brought to Ragusa by Fra Bonifacio himself (and not sent from Jerusalem), which, given the details of his biography, would correspond to the year 1564.[50] In a passage dedicated to the fragment of the Holy Sepulchre, the anonymous author praises it as "the most unique and singular relic, not to be found in any other part of the world outside Jerusalem,"[51] evidently disregarding the presence of two similar items in the cathedral treasury.

The decision of the Ragusan Senate issued immediately after the arrival of the two relics prescribed their collocation in two separate containers; the relic of the Holy Sepulchre – the only one preserved of the two donated by Fra Bonifacio – was subsequently added to the two fragments of the Holy Sepulchre already in the cathedral treasury and placed in a silver casing covered by a reclining figure of the dead Christ (Figure 5.5).[52]

The relic of the Column of the Flagellation, last recorded in an inventory of the treasury compiled in 1784, has unfortunately been lost.[53] The Republic of Ragusa's enthusiastic reception and long-lasting reverence of the stone fragments sent by Fra Bonifacio de Stephanis were at odds with its attitude toward the donor. After the state's initial

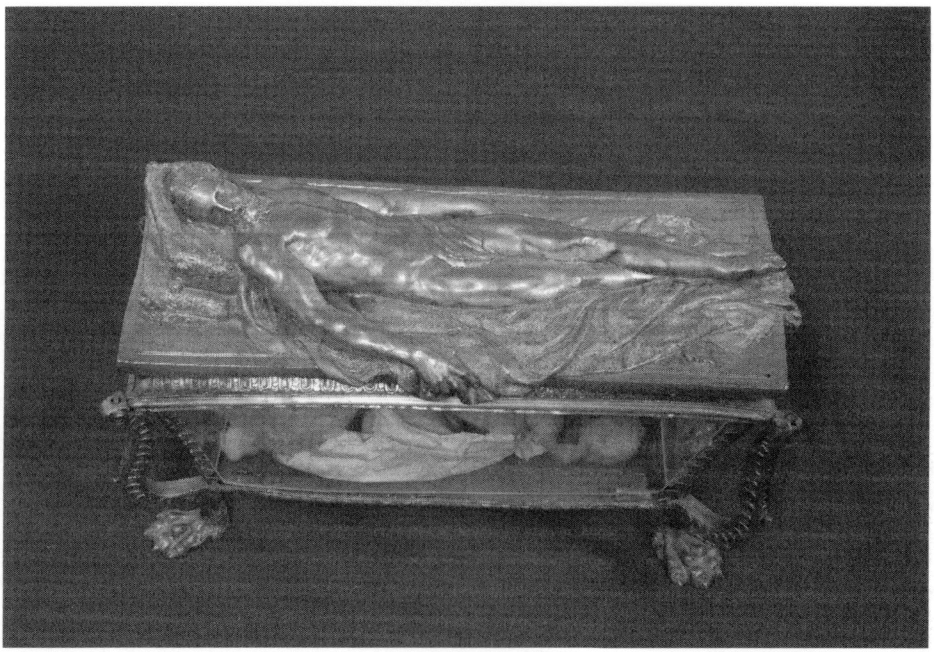

Figure 5.5 Reliquary of the fragment of the Holy Sepulchre, 17 × 32 × 21 cm, silver, late sixteenth century. Dubrovnik, cathedral treasury. Photograph by Božidar Gjukić, 2013.

praise, expressed in his appointment to the Diocese of Ston and the assignment of an important diplomatic mission to the Spanish King Philip II in 1566,[54] in 1581 the former Guardian of the Holy Land was officially banned from the Republic for refusing to deprive a priest accused of murder of the honor of clerical office without a proper trial; he spent his last days serving the Holy See as the apostolic visitor in the lands subject to Ottoman rule.[55] Notwithstanding the ill treatment of the donor, the reception of the two relics sent from Jerusalem in 1558 presents a singular case of the inclusion of spiritual and material symbols of the Holy Land in a civic ritual of the Republic of Ragusa that would last for the following two-and-a-half centuries.

Notes

1 For an overview of Ragusan history, see Robin Harris, *Dubrovnik: A History* (London: Saqi, 2003).

2 Lovro Kunčević, "The Rhetoric of the Frontier of Christendom in the Diplomacy of Renaissance Ragusa (Dubrovnik)," *Dubrovnik Annals* 17 (2013): 57.

3 Branislav M. Nedeljković, ed., *Liber viridis* (Belgrade: Srpska akademija nauka i umetnosti, 1984), 208–209.

4 Archbishop Ilija Saraka (archbishop 1342–1360) is reported to have advised the Ragusan government that another archbishop elected among the local nobility might misuse his powers and extend them to the secular sphere of government; his recommendation to elect only foreign archbishops was accepted and officially confirmed by the Republic of Ragusa and the Holy See in 1409. The succession of foreign – mainly Italian, although for political reasons never Venetian – archbishops was maintained until 1721, when the Republic itself, aware of the negative consequences of frequently nonresiding or negligent foreign archbishops, encouraged the election of Rajmund Jelić, a native of Ragusa. Serafino Razzi, "Narrazioni o vero storia degli arcivescovi di Raugia [1589]," in *Povijest dubrovačke metropolije i dubrovačkih nadbiskupa (X.–XVI. stoljeća)* ed. Stjepan Krasić (Dubrovnik and Split: Biskupski ordinarijat and Crkva u svijetu, 1999), 122; Kosta Vojnović, "Crkva i država u dubrovačkoj republici," *Rad Jugoslavenske akademije znanosti i umjetnosti* 41 (1894): 34–35.

5 Serafino Razzi, *La storia di Raugia* (Lucca: Per Vincentio Busdraghi, 1595), 177.

6 Razzi, *Storia di Raugia*, 136–137, 175.

7 Giacomo di Pietro Luccari, *Copioso ristretto degli annali di Rausa* (Venice: Ad instantia di Antonio Leonardi, 1605), 164; Stefano Skurla, *Ragusa. Cenni storici* (Zagreb: A spese dell'autore, 1876), 97.

8 Seraphinus Maria Cerva, *Prolegomena in sacram metropolim Ragusinam: editio princeps*, ed. Relja Seferović (Zagreb and Dubrovnik: Hrvatska akademija znanosti i umjetnosti, Zavod za povijesne znanosti u Dubrovniku, 2008), 452–458.

9 Ivan Šimić, ed., *Šematizam Dubrovačke biskupije* (Dubrovnik: Dubrovačka biskupija, 2011), 39. The most extensive study of the cathedral treasury of Ragusa was conducted by Vinicije Lupis, "Moćnik dubrovačke prvostolnice" (PhD diss., University of Zadar, 2003). On the cathedral treasury, see also Vinicije B. Lupis, "Moćnik katedrale," in *Katedrala Gospe Velike u Dubrovniku*, ed. Katarina Horvat-Levaj (Dubrovnik and Zagreb: Gradska župa Gospe Velike and Institut za povijest umjetnosti, 2014), 399–439.

10 Razzi, *Storia di Raugia*, 178.

11 Petar Marija Radelj, "Dubrovačke relikvije sv. Stjepana Ugarskog," *Dubrovnik* XI/3–4 (2000): 269–303.

12 Smiljka Malinar, "La descrizione di Ragusa nei resoconti dei viaggiatori italiani del Quattrocento e del Cinquecento," in *La Dalmazia nelle relazioni di viaggiatori e pellegrini da Venezia tra Quattro e Seicento*, ed. Sante Graciotti (Rome: Bardi Editore, 2009), 350–352.

13 Filip de Diversis, *Opis slavnoga grada Dubrovnika*, ed. and trans. Zdenka Janeković-Römer (Zagreb: Dom i svijet, 2004), 147. An English translation of the cited passage is in Harris, *Dubrovnik*, 223.

14 Cerva, *Prolegomena*, 458, 460, 462.

15 The variants of his surname are Drkolica, Drakulica, Drakolica, Drkolčić, Crassa, Stephani, Stjepović. For the biography of Fra Bonifacio de Stephanis, see especially Daniele Farlati and Jacopo Coleti, *Illyrici Sacri tomus sextus. Ecclesia Ragusina cum suffraganeis, et Ecclesia Rhiziniensis et Catharensis* (Venice: Apud Sebastianum Coleti, 1800), 353–354; Giovanni Evangelista Cusmich, *Cenni storici sui minori osservanti di Ragusa* (Trieste: Tipografia del Lloyd austriaco, 1864), 30–32; Antun Šimčik, *Hrvat obnovitelj Božjega groba: Fra Bonifacije Drkolica čuvar Svete zemlje* (Zagreb: Tiskara Narodne prosvjete, 1936); Seraphinus Maria Cerva, *Bibliotheca Ragusina*, ed. Stjepan Krasić, vol. 1 (Zagreb: Jugoslavenska akademija znanosti i umjetnosti, 1975), 206–207; Pijo M. Pejić, "Dubrovački franjevci u Svetoj zemlji," in *Samostan Male braće u Dubrovniku*, ed. Justin V. Velnić (Zagreb and Dubrovnik: Kršćanska sadašnjost and Samostan Male braće, 1985), 261–264; Dino Milinović, "'To neizrecivo mjesto, u kojemu je tri dana počivao Sin čovječji': Bonifacije Dubrovčanin i obnova Svetog Groba," in *Umjetnost i naručitelji: Zbornik Dana Cvita Fiskovića III*, ed. Jasenka Gudelj (Zagreb: Institut za povijest umjetnosti and Odsjek za povijest umjetnosti Filozofskog fakulteta Sveučilišta u Zagrebu, 2010), 73–80.

16 Farlati notes that Fra Bonifacio and Felice Peretti – also members of the Franciscan order – were "equals and colleagues" during their studies in Paris (Farlati and Coleti, *Illyrici Sacri*, 353); however, Felice Peretti was not a Parisian student; cf. Silvano Giordano, "Sisto V," in *Enciclopedia dei Papi* (2000), Digital facsimile: http://www.treccani.it/enciclopedia/sisto-v_(Enciclopedia-dei-Papi)/ (accessed November 7, 2013).

17 "Cuius quidem ligni crucem unam Jerosolymis reliquimus in phano divae Mariae de Apparitione juxta Domini Jesu Sepulcrum super altari eidem D. Cruci consecrato, partem aliam Romam nobiscum attulimus, inque particulas aliquot in cruces formam compositas eam divisimus; ex quibus unam Pontifici Summo, sacras tunc temporis totius Christi Eccclesiae habenas regenti Pio quarto obtulimus; duas Illustrissimis ac Reverendissimis Dominis Cardinalibus, de Carpo videlicet et de Ara Coeli nuncupatis, libenter donavimus, Crucem parvulam nobis servamus, qua, dum sacra peragimus, uti reverenter solemus." The letter was published as an appendix to the second edition of Fra Bonifacio's work *Liber de perenni cultu Terrae Sanctae et de fructuosa eius peregrinatione* (Venice: Ex Typ. L. merlo Ioh. Bapt. filii, 1875; first published in Venice in 1573), 278–284.

18 Upon his departure from the Republic in 1581, Fra Bonifacio left the relic of the wood of the Holy Cross to a member of the Poli family in Ston; it was eventually donated to the cathedral treasury of Ragusa in 1720. Farlati and Coleti, *Illyrici Sacri*, 353; Stjepan Skurla, *Moćnik stolne crkve dubrovačke* (Dubrovnik: Troškom i nakladom D. Pretnera, 1868), 91–92.

19 The donation of the relic to the Republic of Venice was mediated by former Venetian chancellor of Cyprus Vincenzo Fedeli, who traveled to the Holy Land in 1558. Renata Targhetta, "Fedeli, Vincenzo," *Dizionario biografico degli italiani* 45 (1995), Digital facsimile: http://www.treccani.it/enciclopedia/vincenzo-fedeli_(Dizionario-Biografico)/ (accessed October 25, 2013).

20 Henry Kamen, *Philip of Spain* (New Haven and London: Yale University Press, 1998), 189.

21 "Cum enim divina gratia huius Sacri montis Syon Guardianus et totius terrae sanctae apostolicus fuerim (licet imperibus meritis) commissarius constitutus, multa Sanctuaria multaeque pretiosae reliquiae ad manus meas devenerunt alia quidem Jure ordinario alia etiam ex nimia indagatione atque curiosa investigatione quibus iam deo dante plures ecclesias nostratum dictavi, maxime ecclesiam Beatorum apostolorum Petri et Pauli apud Romam, atque etiam ecclesiam Sancti Marci venetiarum cum alijs, eo semper zelo ductus quo Christianam decet Religionem et sine preiuditio horum sacrorum locorum." Dubrovnik State Archives (hereafter DAD), *Diplomata et acta*, XVI ct., n. 164. The letter was published by Skurla, *Moćnik*, 174–176, although with some errors in transcription: for instance, the passage regarding the donation of the relics to the Basilica of Saint Mark in Venice is transcribed as "Ecclesiam s. Mariae Venetiarum." The choice of the church dedicated to Saints Peter and Paul as the destination of the relics from the Holy Land is rather curious, as the only church dedicated to the two apostles extant in sixteenth-century Rome was a small one outside the Porta San Paolo used by the *Compagnia della Trinità di Ponte Sisto* (or the *Confraternita della Santissima Trinità dei pellegrini e dei convalescenti* founded by Saint Philip Neri in 1548), which gave the church an alternative denomination *La Trinità*. The church was

ceded to the *Compagnia della Trinità* only in 1562, after it was restored by four members of the confraternity. Cf. Ottavio Panciroli, *I tesori nascosti nell'alma città di Roma* (Rome: Appresso Luigi Zannetti, 1600), 693, 696; Pietro Martire Felini, *Trattato nuovo delle cose maravigliose dell'alma città di Roma ornato di molte figure, nel quale discorre di 300 e più chiese* (Rome: Per Bartolomeo Zanetti, 1610), 31; *Statuti della ven: Archiconfraternita della Santissima Trinità de' pellegrini, e convalescenti di Roma* (Rome: Dalle stampe di Crispino Puccinelli, 1821), 8. Fra Bonifacio's choice to send the relics from the Holy Land to the church affiliated with the confraternity might have been motivated by the fact that it provided assistance to the many pilgrims arriving in Rome every year, just as the Franciscans took care of the pilgrims who visited the Holy Land.

22 Denys Pringle, *The Churches of the Crusader Kingdom of Jerusalem: A Corpus*, vol. 3: *The City of Jerusalem* (Cambridge: Cambridge University Press, 2007), 93.

23 The Gospels do not make explicit mention of the (chronologically first) flagellation in the house of Caiphas. On the distinction between the Columns of the Flagellation in Jerusalem and Rome, see Prospero Lambertini (Benedetto XIV), *Delle feste di Gesù Cristo Signor Nostro, e della B. Vergine Maria* (Venice: Apresso Francesco Pitteri, 1767), 110; [Jacques] Mislin, *Les saints lieux, pèlerinage à Jérusalem* (Paris and Lyon: Lecoffre fils et c[ie], 1858), vol. 2, 314.

24 "Prima pars est de imponendo Dominis Procuratoribus ecclesiae sanctae Mariae. Debeant in duobus vasis honoratis, sicut eis melius videbitur, reponere reliquias ad nos messas a Reverendo Fratre Bonifacio Guardiano Montis Syon, quae Reliquiae sunt hae, videlicet, una pars lapidis abscissa ex illa vera rupe, in qua praecavata fuit sepultus CHRISTUS D. N. et Deus, et una pars Columnae ad quam Christus in pretorio Pilati fuit ligatus et flagellatus: ponendo quamlibet dictarum Reliquiarum in suo vase separato. Et de ordinando quod anno singulo in perpetuum Die XV[to] mensis Decembris debeat celebrari una processio generalis et in cathedrali Ecclesia decantari debeat una missa ad laudem et honorem Dei ac totius curiae celestis, et quod Pretorium nostrum tali die feriare debeat." DAD, *Acta Consilii Rogatorum* 54, fol. 211v; Skurla, *Moćnik*, 176–177.

25 On Lodovico Beccadelli see Razzi, "Narrazioni," 141–144; Giuseppe Alberigo, "Beccadelli, Ludovico," *Dizionario biografico degli italiani* 7 (1965): 407–413; Gigliola Fragnito, *In museo e in villa: saggi sul rinascimento perduto* (Venice: Arsenale, 1988); Guido Dall'Olio, "Beccadelli, Ludovico," *Dizionario storico dell'Inquisizione* 1 (2010): 165–166. Beccadelli resided in Ragusa from December 1555 until June 1560.

26 ". . . un Frate Bonifatio Raguseo dell'ordine di san Francesco Zoccolante si trova guardiano a Monte Sion in Hierusalem, et a giorni passati ha mandato a questi signori di Ragusa duo pezzi di pietra in una scattola ben serrati et involti et scrive che l'uno di quei pezzi è della propria pietra, ove fu sepulto nostro signore Iesu Christo, la qual sepoltura, dice ch'a questo tempo s'è scoperta per certo bisogno di fabrica essendo gia centenaia d'anni stata coperta da lavori di marmo, l'altra pietra scrive essere della propria colonna, ove Nostro signore fu battuto, et le manda a questi signori che le tengano per devotione fra molte altre reliquie che hanno. Li signori che sono devoti vorriano non solo tenerle in veneratione, ma che si instituisce una processione da farsi ogni anno con una messa solenne nel giorno che le furono presentate, che fu alli xv. di Decembre, ho detto loro che gli affetti pij tutti sono laudevoli pur che l'introdurre nove reliquie non sta a loro, ma alla sede Apostolica, et perche il detto fra Bonifacio Guardiano dice haver mandato a Roma di Hierusalem cose simili, prego V. S. che sia contenta avisarmi quello havrò da fare in questo caso, per non partirme dalla regola vera . . . " Lodovico Beccadelli to Giovanni Giacomo Barba, Ragusa, January 19, 1559. Parma, Biblioteca Palatina, MS. Pal. 1010, fols. 276v – 277.

27 Norman P. Tanner, ed., *Decrees of the Ecumenical Councils*, vol. 2: *Trent to Vatican II* (Washington: Sheed and Ward and Georgetown University Press, 1990), 774, 776.

28 *Annales ragusini anonymi item Nicolai de Ragnina*, ed. Speratus Nodilo, *Monumenta spectantia historiam Slavorum meridionalium*, vol. 14 (Zagreb: Academia scientarum et artium Slavorum meridionalium, 1883), 137.

29 Milorad Pavić, "Prilog poznavanju hodočasničkih putovanja od Venecije do Svete zemlje u XVI. stoljeću," *Croatica Christiana Periodica* 59 (2007): 37; Malinar, "Descrizione di Ragusa," 344.

30 Harris, *Dubrovnik*, 81.

31 Dragutin Kniewald, *Liturgika* (Zagreb: Tipografija, 1937), 263.
32 For example in the event of the translation of the relic of the arm of Saint Philip from the Holy Land to Florence, commissioned by the Florentine bishop Giovanni da Velletri at the beginning of the thirteenth century. Cf. Sally J. Cornelison, "Art Imitates Architecture: The Saint Philip Reliquary in Renaissance Florence," *The Art Bulletin* 86, no. 4 (2004): 642.
33 "De quelle havete costi mandate io dico alla S. V. R.ma che queste non sono reliquie nove, ne di sancto qual anchor non sia sollennizato da la s.ta chiesa, et dala sede apostolica, ma sono reliquie antiche, tanto osservate nel christianesimo quando s'hanno con verita." Giovanni Giacomo Barba to Lodovico Beccadelli, Rome, February 17, 1559. Parma, Biblioteca Palatina, MS. Pal. 1031/7, number 2.
34 "29. Le Processioni le comandano e fanno fare a loro voglia, et a quelle istituite dalla Chjesa dalla solennità in poi del Santissimo Sacramento, non intervengono. 30. Nelle reliquie non anno voluto che si registrino li nomi, opponendosi, come fecero al Visitatore; e le mandano a Casa de particolari d'autorità propria, e senza licenza." Gian Maria Mattei, *Zibaldone*, Dubrovnik, Library of the Franciscan monastery, MS. 433, vol. II, 297.
35 Edward Muir, *Civic Ritual in Renaissance Venice* (Princeton: Princeton University Press, 1981), 93–94.
36 Skurla, *Moćnik*, 171–172.
37 Ibid., 177.
38 Zdenka Janeković-Römer, "Public Rituals in the Political Discourse of Humanist Dubrovnik," *Dubrovnik Annals* 6 (2002): 14.
39 The 1583 Roman martyrology prescribed only the memory of less well-known martyred saints and the fifth-century bishop Saint Valerian. *Martyriologium Romanum: Ad novam Kalendarij rationem, & Ecclesiasticæ historiæ veritatem restitutum. Gregorij XIII. Pont. Max. iussu editum* (Venice: Apud Ioannem Baptistam Sessam, 1583), 228.
40 Cerva, *Prolegomena*, 442–445.
41 Sergio Raveggi, "Storia di una leggenda: Pazzo de' Pazzi e le pietre del Santo Sepolcro," in *I fiorentini alle crociate. Guerre, pellegrinaggi e immaginario "orientalistico" a Firenze tra Medioevo ed Età moderna*, eds. Silvia Agnoletti and Luca Mantelli (Florence: Edizioni della Meridiana, 2007), 22–44; Alexei M. Lidov, "Svyatoy Ogon. Ierotopicheskie i iskusstvovedcheskie aspekty sozdaniya 'Novykh Ierusalimov,'" in *Novye Ierusalimy. Ierotopiya i ikonografiya sakralnykh prostranstv / New Jerusalems. Hierotopy and Iconography of Sacred Spaces*, ed. Alexei M. Lidov (Moscow: Indrik, 2009), 286.
42 *Cronaca di Ragusa*, fols. 186v – 187.
43 Nella Lonza, *Kazalište vlasti: ceremonijal i državni blagdani Dubrovačke Republike u 17. i 18. stoljeću* (Zagreb and Dubrovnik: Hrvatska akademija znanosti i umjetnosti, Zavod za povijesne znanosti u Dubrovniku, 2009), 354.
44 *Cronaca di Ragusa*, fol. 187v; Cerva, *Prolegomena*, 445–446.
45 Lonza, *Kazalište vlasti*, 30–32.
46 Skurla, *Moćnik*, 171.
47 Cerva, *Prolegomena*, 445–446, 466–467.
48 Mattei, *Zibaldone*, I, 219: "Primo Catalogo delle SS. Reliquie che si conservano nella Chjesa Cattedrale di Ragusa . . . 5. Columna Christi particulo in pixide argentea F. Bonifazio de' Stephanis Min. Oss. della Isola di mezzo portò dà Gerusalemme questi due frammenti l'anno 1564. e ne fece dono alla Signoria. . . . In memoria di tal dono, e secondo il desiderio del donatore con pubblica Processione si portava per la Città nel mese Decembre. 6. Sepulcri Christi duo fragmenta in navicula argentea. Ancor queste reliquie si portavano in Processione dall'Arcivescovo colla Signoria nel mese di decembre perchè così bramo il suddetto Vescovo di Stagno, quando donò alla repubblica uno di questi pezzetti l'anno 1564. Ora si fà tal Processione nella Domenica in Albis. L'altro pezzetto si trova nell'Inventario del 1335. notato."
49 "Donò ancora a Paolo IIIJ ed all'Imperatore Ferdinando I, ed alla repub.ca di Venezia /l'anno 1558/ quando per i bisogni de luoghi santi probabilmente venne in Europa /ed ancor alla Signoria di Ragusa un pezzetto della Colonna a cui fù NS. G. C. flagellato. Anzi a questa ultima presentò ancora un notabile pezzo del S. Sepolcro, che si scheggiò quando fù fatta l'apertura, la quale ogni anno nell'Ottava di Pasqua si porta processionalmente per Ragusa e per memoria del donatore passa essa Processione per la Chjesa di S. Francesco." Ibid., II, 281–282.
50 *Cronaca di Ragusa*, fols. 187–187v.

51 ". . . unica, e singolare reliquia, la quale in fuori di Gerusaleme, non si trova in altra parte del mondo . . . " Ibid., fol. 187.
52 17 × 32 × 21 cm, tentatively dated to the late sixteenth century and ascribed to the "Venetian cultural circle." Vinicije B. Lupis, "Prilog poznavanju mletačkog zlatarstva u Dubrovniku i okolici," *Peristil: zbornik radova za povijest umjetnosti* 53 (2010): 34.
53 Skurla, *Moćnik*, 173.
54 *Dubrovačka akta i povelje=Acta et diplomata Ragusina*, ed. Jovan Radonić, vol. 2, no. 2 (Belgrade: Srpska kraljevska akademija, 1938), 136–137, 151–153.
55 The difficulties with the Republic and Fra Bonifacio's departure from Ragusa are recorded in detail in *Cronaca di Ragusa*, fols. 188–189.

6 The Stone of Grace in the Gareja Desert, Georgia

Zaza Skhirtladze

The monastic complex founded in the middle of the sixth century in the southwest part of Georgia, in historical Kakheti, by Saint David, one of the so-called Syrian Fathers, who came to the country with the blessings of Saint Simon the Stylite (521–592), is the largest Christian Orthodox spiritual center in the Caucasus (Figure 6.1 and Figure 6.2).[1]

According to tradition, Saint David lived in a natural cave on a rocky mountain slope near Tbilisi – the capital of the country – for a short time after he arrived. Eventually, a monastery was built there, which is now in the center of Tbilisi and is known as *Mamadaviti* (Father David) or *Mtatsmida* (the Holy Mountain).[2]

Soon after his arrival, Saint David left his initial dwelling and together with his disciples, Saint Dodo and Saint Lukiane, laid the foundation for the monastic association in the eastern part of the country. From there the complex gradually spread to the north of the Mtkvari River in the semidesert area of the Gareja Mountain ridge and the center of the Iori River Valley. Over the centuries, the monastic association of Gareja grew to be much larger than other such unions. Recent studies have indicated that there were more than two dozen complexes,[3] among them lavras,[4] monasteries, and hesychastirions.[5] The hagiographic treatises and historical sources, along with the artifacts, inscriptions, and donor images connected with the Gareja complex reflect the political, social, and cultural realities during different periods in the long history of the country. This large group of rock-cut monasteries became one of the most important spiritual and cultural centers of the Christian Caucasus at a very early stage, and its prime importance in the spiritual life of the country endured through the early twentieth century. The Gareja Desert monasteries are among those spiritual centers of the Eastern Christian world in which the activities of the local monks were tied in with their creative work – literary, architectural, and artistic endeavors. They were among the influential centers of culture and education, where kings, noblemen, and clergymen, as well as other renowned representatives of Georgian culture lived and worked.[6] Gareja was a significant literary center,[7] but it was also home to an original, local school of mural painting,[8] which was an outgrowth of the close ties forged with Byzantine and East Christian spiritual centers – all of which can be traced in historical sources and Georgian hagiography, as well as in epigraphy and art.[9]

According to the *Vita* of Saint David,[10] after monastic life was firmly established in the Gareja Desert, he, together with some of his disciples, left on a pilgrimage to Jerusalem. They approached at the place known as the Hill of Mercy (Mons Gaudii),[11] from where they were able to see the Holy City. Saint David expressed gratitude to the Lord and humbly announced to his companions: "No, brethren, I may not venture to

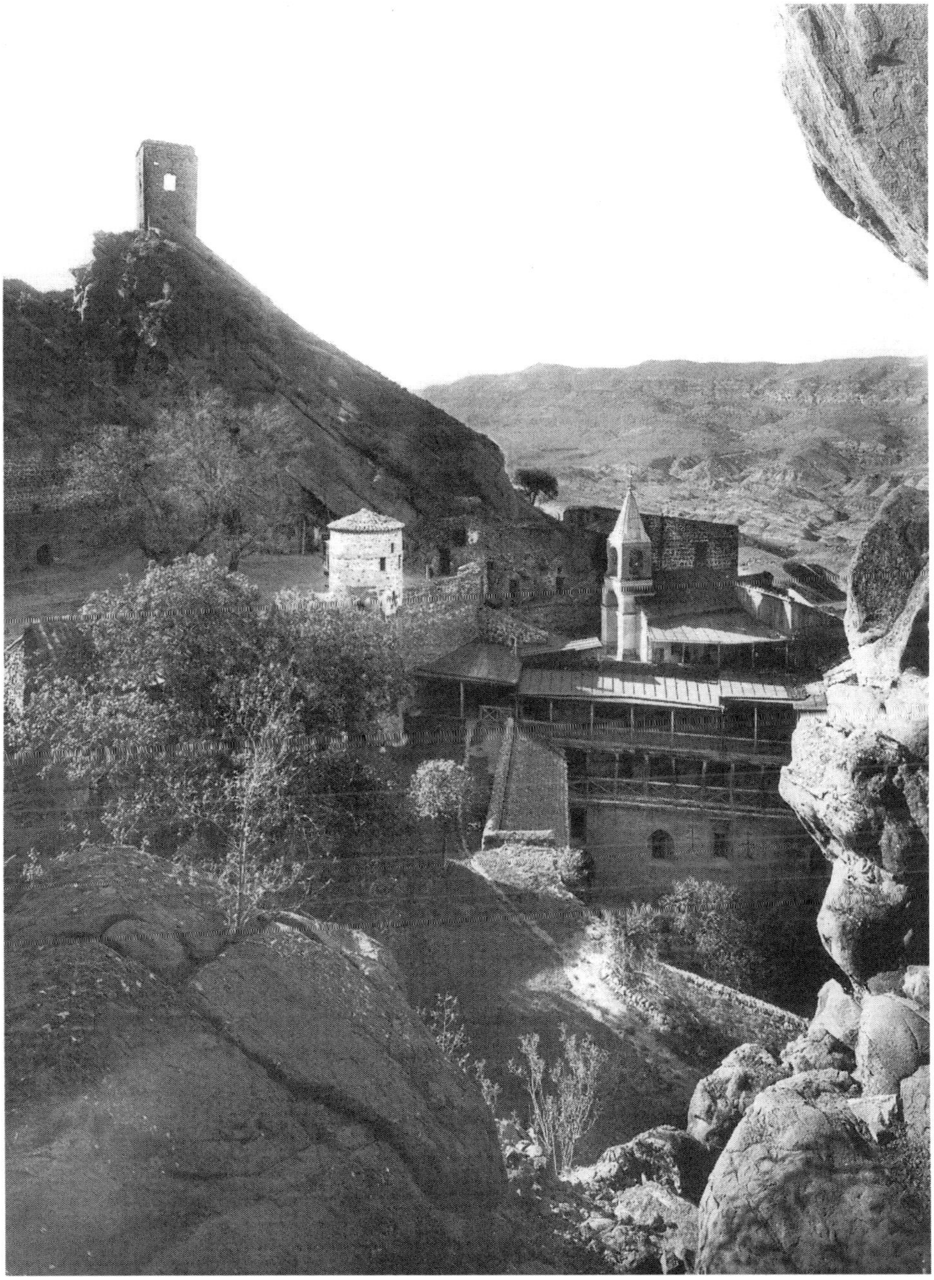

Figure 6.1 Gareja Desert, Georgia. Lavra of St. David. General view from south. Photo: Theodor Kühne (1913). Image source: Zaza Skhirtladze, *The Tomb of Saint David Garejeli* (Tbilisi, 2006), Figure 1.

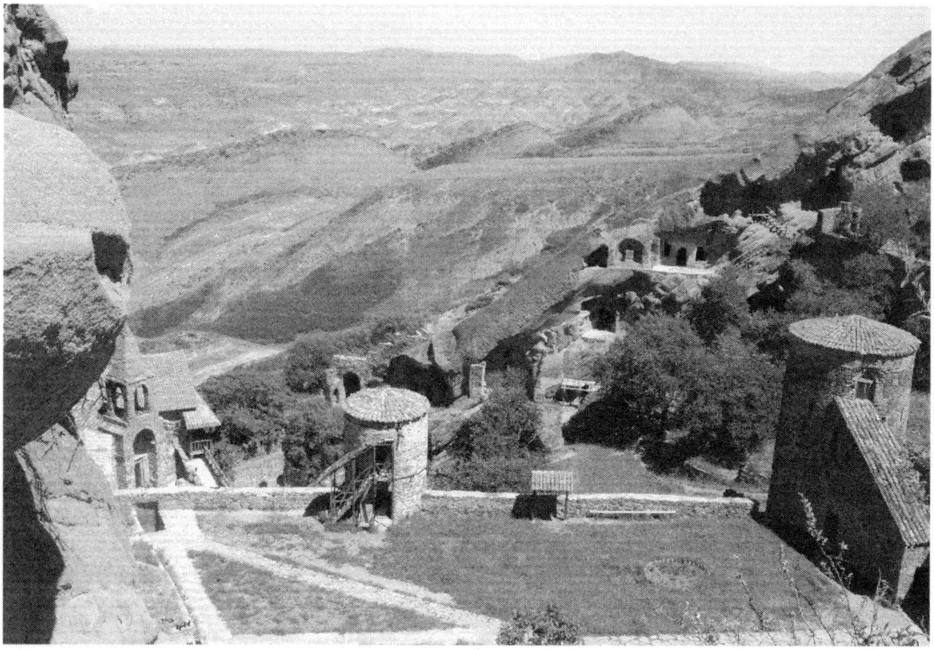

Figure 6.2 Lavra of St. David. General view from west. Photo: Zaza Skhirtladze.

advance further from this spot, for I judge myself unworthy even to approach those holy places. But you go and pray for me, a sinner I would not dare to step on the road of the God with my sinful feet."[12] He then sent his disciples on to Jerusalem while he himself looked at the city from afar. At the city gate,[13] he prayed fervently and took three stones for *eulogy* with the belief that "they were cut from Christ's grave."[14] When the pilgrims started back to Georgia, an angel appeared to Elija (459–516), patriarch of Jerusalem (according to another version the vision appeared to Makarios [r. 544 and 563–574]) and announced that: "With his belief he [Saint David] took the grace of Jerusalem."[15] The patriarch's envoys caught up with Saint David and his companions near Nablus (Neapolis) and told them of the angel's revelation. Following the patriarch's command, they took two stones back, "in order not to take away the full grace from it [Jerusalem], and one share do I give you for your desert; take this stone as the grace to your desert for memory and for declaration of your belief."[16] According to the *Vita* of the saint, "And even today that stone remains in the hermitage, effecting great miracles of healing right up to the present time."[17] This stone, the Stone of Grace, has been one of the most venerated relics of the Georgian Christian Orthodox Church since the sixth century.[18]

Textual accounts concerning the Stone of Grace can be found in the *Life* of Saint David,[19] as well as in various historical narratives[20] and documents. [21] The miraculous relic – a round, smooth stone somewhat bigger than an egg, of light yellowish-ruddy color with dark-red spots – was kept for centuries in the Church of the Transfiguration, the main sanctuary in the Lavra of Saint David, the central monastery of the Gareja monastic complex (Plate 3).[22]

According to tradition, ownership of the sacred stone endowed the Gareja Desert with a third part of Jerusalem's grace; accordingly, "praying three times there was equivalent to praying once in Jerusalem."[23] Thus, Gareja was considered to be a place of special sacredness and was highly venerated by pilgrims coming from different parts of Georgia and the Christian East.[24] Through the end of the nineteenth century, Georgian, Greek, Russian, and Armenian pilgrims and even Muslim visitors worshipped there.[25]

The way of pilgrimage to the burial place of Saint David (i.e., the Lavra of Saint David) for visitors from different parts of the Christian world included the monasteries of Saint John the Baptist and Saints Dodo, Udabno, and Bertubani. Especially sanctified sites for pilgrims in the Gareja Desert were the tombs of Saint David (the Lavra) and of his disciples Saint Lukiane (the Lavra) and Saint Dodo (the Saint Dodo Monastery) and the spring flowing in drops in the cave at the bottom of the rocky slope in the Lavra of Saint David – the miraculous healing water, which is still a holy *eulogy* for pilgrims coming to Gareja.[26]

The significance of the Stone of Grace is closely linked to local tradition, which influences all spheres of creativity, and, according to that tradition, the holy places in Georgia reflect the sanctity of Jerusalem. The most eloquent evidence of this notion is Mtskheta – an old capital of the country – which is the home of the most sacred relics venerated by Christians. Several places and sanctuaries in Mtskheta have copies of the topography of Jerusalem, thus pointing to the conveyance of the Jerusalem tradition to Georgia.[27]

The *Life* of Saint David tells of the miracles connected with the Stone of Grace. For centuries it healed "from diseases and pain all those, who believed in it." It is thought that to touch the Stone of Grace, as well as to pray in front of it "can heal patients from all diseases and pain."[28] The history of the Stone of Grace is represented in the murals of the diaconicon (north chamber) of the main church in the Udabno Monastery (Figure 6.3).

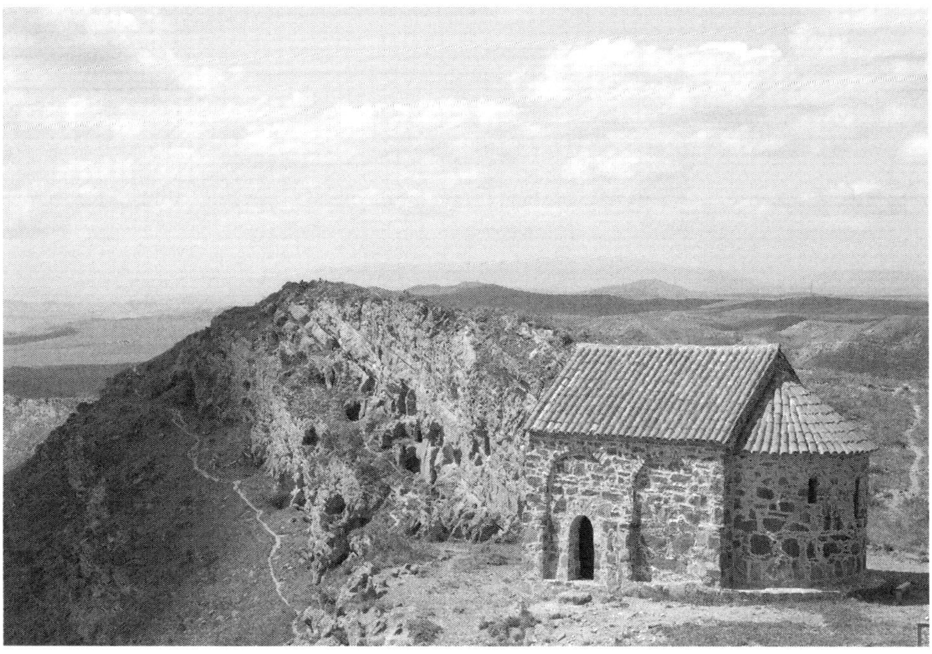

Figure 6.3 Gareja Desert. Udabno Monastery. General view from southeast. Photo: Zaza Skhirtladze.

The frescoes on the western and northern walls, which were executed in the first half of the tenth century, contain the pictorial cycle of the *Life* of Saint David, which consists of seven scenes: Saint David and his disciples settling in the Gareja Desert and the initial stage of monastic life there[29] and Saint David's pilgrimage to Jerusalem, the episode in which he did not dare to enter the Holy City and took the Stone of Grace (Figs. 6.4 and 6.5).[30]

The composition of this last scene consists of two parts: in the left section, against the background of a high wall, Saint David is shown with a canvas bag over his shoulder and a long stick in his right hand. In front of him in the center is the city of Jerusalem, shown as a single-nave edifice with a blue-tiled roof, a high arched gate, a coupled window on the side façade, and a comparatively small low annex at the side. On the other side of the building (i.e., in the right-hand section of the composition) we see two clergymen. The first, who is moving to the left, toward Saint David, is holding a scepter in his right hand, and his left hand, as a sign of address, is raised in front of his chest. He is wearing a long robe and a cloak, and his headdress is similar to Saint David's cockle. Along the head of the figure, in the upper part of the high wall passing near the right edge of the scene (along the upper section of the composition) is the fragment of single-line inscription in old Georgian *asomtavruli* (uncial) script. The preserved fragments of the letters suggest that it should read ელია (*Elija*). The presence of an image of Elija, patriarch of Jerusalem, is likely due to the episode of the stones. The clergyman accompanying the patriarch, with a cockle, with

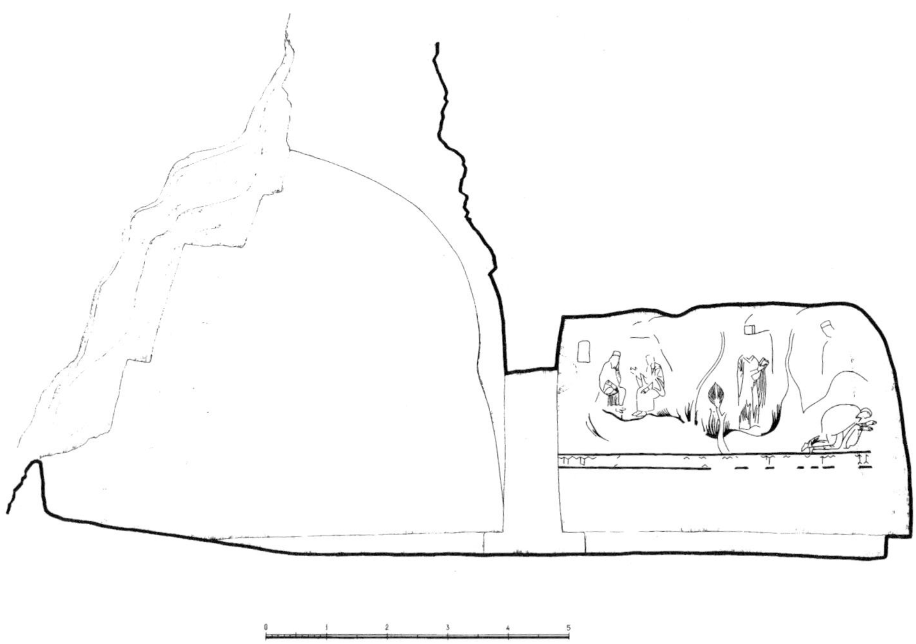

Figure 6.4 Udabno Monastery. North chamber of the main church. Scheme of the First Phase Painting on the west wall, with the life cycle of St. David Garejeli. Image source: A. Eastmond, Z. Skhirtladze, Udabno Monastery in Georgia: Innovation, Conservation and the Reinterpretation of Medieval Art, *Iconographica, Rivista di Iconografia Medievale e Moderna*, VII (2008), Figure 6.

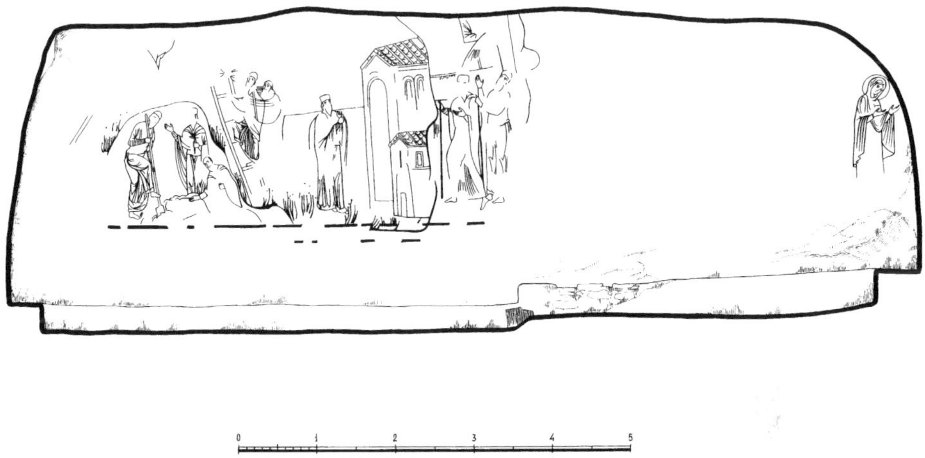

Figure 6.5 Udabno Monastery. North chamber of the main church. Scheme of the First Phase Painting on the north wall, with the life cycle of St. David Garejeli. Image source: A. Eastmond, Z. Skhirtladze, Udabno Monastery in Georgia: Innovation, Conservation and the Reinterpretation of Medieval Art, *Iconographica, Rivista di Iconografia Medievale e Moderna*, VII (2008), Figure 7.

a scepter in his left hand, is looking upward and pointing to the right with a raised right hand. Over his head, in the upper-right-hand corner of the composition, there are details of a half-figure, which probably represent the angel that appeared to the patriarch.[31]

The episode of Saint David's pilgrimage recorded in different recensions of his *Life* is thought to be a legend "in which the facts, names, and motives are mixed,"[32] but which is probably based on "the reflection of the real state of affairs."[33] In light of such a supposition, there were different opinions as to the chronological, confessional, and other issues connected with the saint's pilgrimage to Jerusalem. There are some discrepancies in the narrative of the hagiographer and the iconography of the life cycle of the saint in the monumental painting.[34] With regard to the debatable issues, future possible corrections do not relate to the main purpose of his pilgrimage and what is of principal importance – his acquiring the Stone of Grace.[35] This event is distinguished in the general picture of the tradition of worship of artifacts in Georgia because, according to centuries-old local tradition, the stone bears special sacral traits, as it holds a third share of Jerusalem's grace. Thus, the Stone of Grace turned out to be far more significant than the relics of the saints. A large number of pilgrim inscriptions across several monasteries located along the road leading to Saint David's grave can be explained by the age-old tradition according to which a pilgrimage to Gareja was rewarded by a third of the grace of traveling to Jerusalem. That is the reason that it was so highly venerated by pilgrims in Georgia and in the various parts of the Christian East. The tombs of Saint David and his disciples were distinguished sites of sanctity for pilgrimages to the Gareja Desert.[36]

Saint David placed the miraculous stone in the Church of the Transfiguration – the main sanctuary of the monastery he founded – which was cut into the rock during his lifetime, just opposite his cave (Figure 6.6).

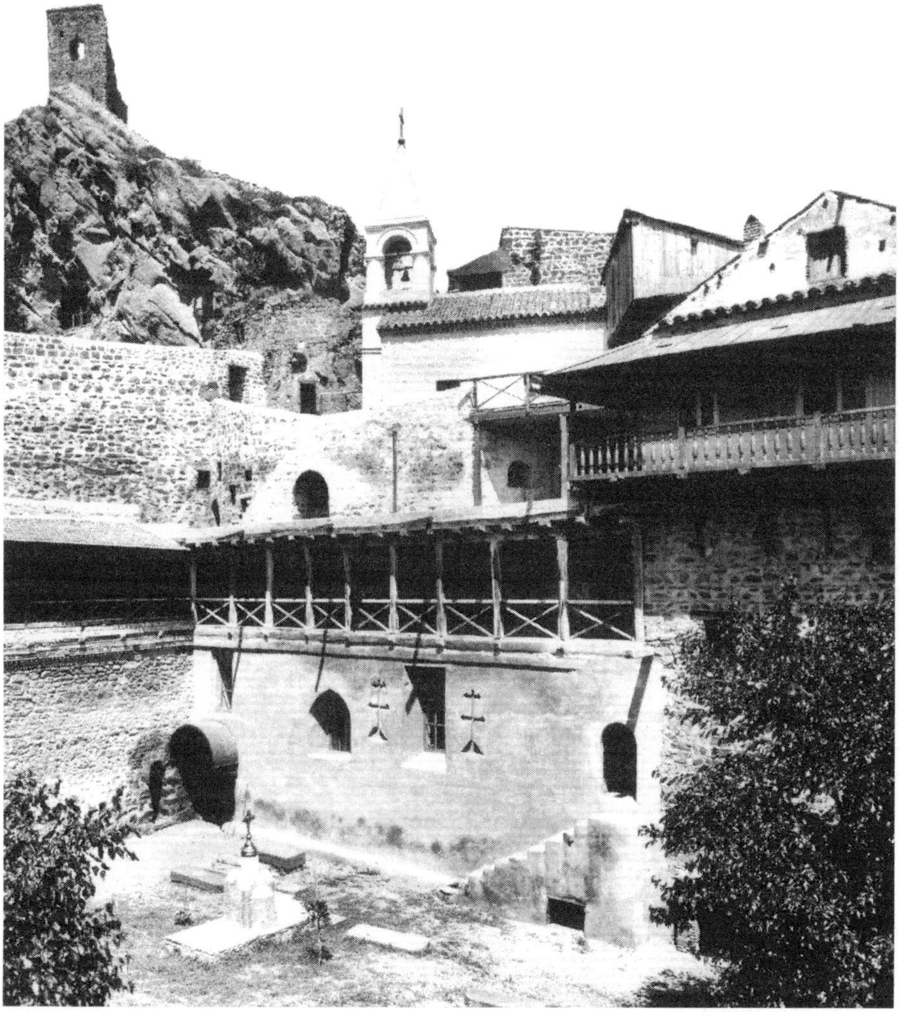

Figure 6.6 Lavra of St. David. View of the lower section of the complex with the Church of Transfiguration. Photo: Dimitri Ermakov (1880s). Image source: G. Chubinashvili, *Peshchernye Monastyri David-Garedzhi* (Tbilisi, 1948).

Bubakar, a ruler (*eristav*) of the District of Rustavi (a town on the edge of the Gareja Desert, some 20 kilometers to the northwest of the Lavra of Saint David), newly converted to Christianity, initiated the cutting out of the church and that became Saint David's burial site.[37] His tomb, which is carved into the rock and oriented along an east-west axis, is in a tall, arched bay near the chancel in the southeast section of the church (Figure 6.7).

For a long time, the Stone of Grace was on Saint David's grave, but it was only in the late medieval period that the sources indicated its exact position on the tombstone.[38]

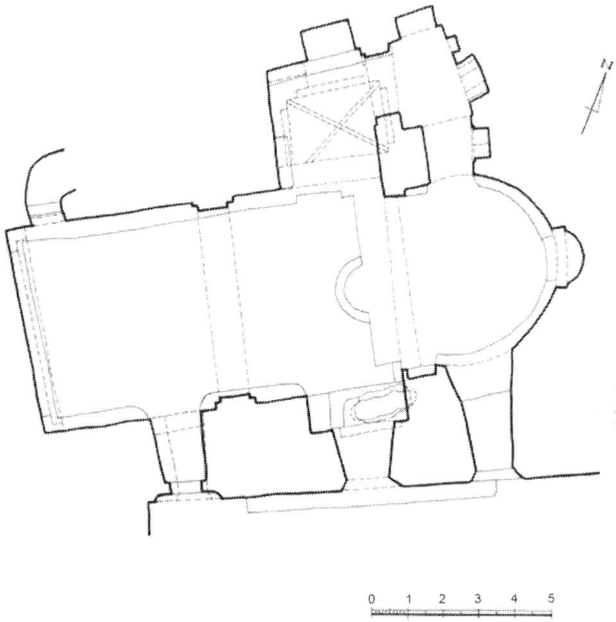

Figure 6.7 Church of Transfiguration. Groundplan. Image source: Zaza Skhirtladze, *The Tomb of Saint David Garejeli* (Tbilisi, 2006), Figure 5.

In June and July 2000, with the blessing of His Holiness Ilia II, Catholicos-Patriarch of All Georgia, Saint David's tomb was opened and cleaned. The archeological findings demonstrated the relevance and accuracy of the written sources and the local monastic traditions. The rock-cut portion of his resting place has survived nearly intact. At the same time, a comparison of written sources and the local monastic tradition enables to identify several stages in the construction of the burial site and its subsequent renovations and alterations.

Saint David's tomb was covered by two rows of ceiling slabs, each of which had a hole for removing holy dust. The hole for extending one's hand into the burial chamber to receive the blessing of the saint and to take a holy gift (ευλογία) and the Stone of Grace, whose presence was attested to in the written sources,[39] were the most important features of the shrine.

In the mid-ninth century, as a result of the renovation and extension of the Church of the Transfiguration on the initiative of Saint Ilarion the Georgian (ca. 822–875), Saint David's tomb was turned into a place of public worship.[40] The grave was then within the interior of the church, in its southern part. The vault was built up on the burial site with mortared rectangular upright slabs and an arched roof.

The time span between this alteration and the subsequent stages of the renovation of the resting place was considerable. There is no historical evidence regarding the burial site from the tenth century until the late medieval period. Thus, the periods to which the vestiges of the alterations discovered during the cleaning of the grave remain unclear.

By the end of the seventeenth century, after a long interlude, monastic life in the Gareja Desert began to flourish again. During the renovation and refurbishing of the spiritual center of the complex, the Father Superior of the Lavra of Saint David, Onopre Machutadze (1690–1733), would naturally have paid special attention to the saint's grave.[41] A document written in the name of a Father Superior known as the "new builder of Gareja" mentions several icons deposited on the tomb but makes no reference to the Stone of Grace.[42]

In the late Middle Ages, a period in which Georgia was often invaded, Gareja experienced a series of destructive raids.[43] During this difficult time, in order to protect this most sacred relic, the Stone of Grace was hidden in a special arched niche built into the lower section of the eastern wall of the southern bay of the church (immediately above the larnax of Saint David).[44] The niche most likely belongs to the period of the shrine's renovation around the turn of the seventeenth century. Toward the end of the nineteenth century, a large icon of the Savior hung on the lower part of that wall of the bay and it may be that the Stone of Grace was kept just at that place for a considerable period of time (Figure 6.8).[45]

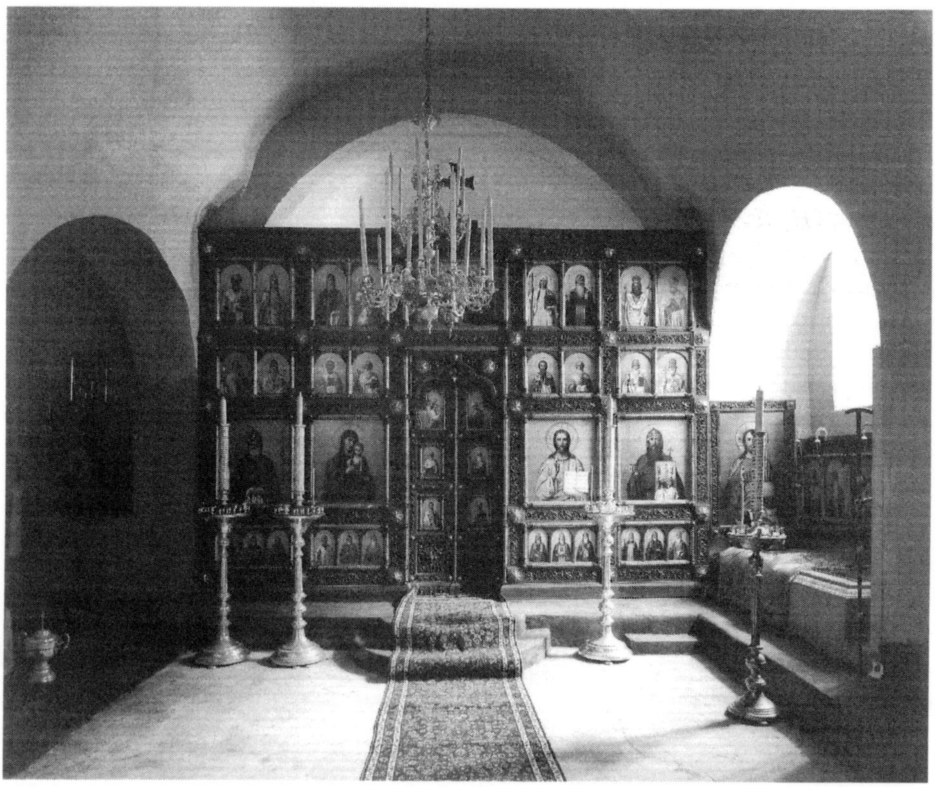

Figure 6.8 Lavra of St. David. Church of Transfiguration. Interior. View from west. Photo: Dimitri Ermakov (1880s). Image source: Zaza Skhirtladze, *The Tomb of Saint David Garejeli* (Tbilisi, 2006), Figure 3.

In 1811, Archimandrite Illarion (Beburishvili) began a major repair on the Church of the Transfiguration. When the plaster from the eastern wall of Saint David's tomb was removed, the small arched niche in which the Stone of Grace was preserved appeared. According to evidence revealed by Archimandrite Tarasi Aleksi-Meskhishvili (d.1874), the newly found relic was hidden again in another safe place. From that time on, until the early 1920s, the stone was taken out during liturgical celebrations and then returned to its hiding place.[46] It seems, however, that at some point it was returned to its previous place – on Saint David's tomb – for it was apparently there when A. Muraviev saw it in the mid-nineteenth century,[47] and when Bishop Kirion mentioned it at the end of that century.[48]

On July 18, 1851, Lezgians ravaged the Lavra of Saint David and destroyed nearly all of the monastery's treasures, but the invaders failed to find the Church of the Transfiguration and the grave of Saint David so they were left intact,[49] but they were subsequently despoiled in the Lezgins raid of October 1, 1857.[50] After dividing the booty, the Lezgins, unaware of its significance, hid the Stone of Grace in the ceiling of the house of one of the raiders. A captured hieromonk, Mitrophane (Natsvlishvili), saw where they had hidden the stone, and a year after he was set free, unnoticed by Lezgins, he removed the Stone of Grace and returned it to the Lavra of Saint David.[51]

In the twentieth century, major parts of the Gareja monastic complex, including the grave of its founder, were abandoned and damaged: in 1917–1918 thieves did not spare local holy places, and the vaults of Saint David and Saint Dodo were ransacked.[52] Early in the 1920s, shortly after the establishment of the Bolshevik regime, life finally came to an end in the still-active seats of Gareja – the Lavra of Saint David and the monastery of Saint John the Baptist – and the pillage and destruction of the principal sites in the desert began. In 1923, after the abolishment of monastic life in the Gareja Desert, the Stone of Grace and other relics were taken to Tbilisi, where they were preserved in the treasury of the State Museum of Georgian Art. In 1990, the stone was transferred to the Georgian Patriarchate and at present it is housed in the chancel of the Saint Trinity Cathedral in Tbilisi.

The history of the final resting place of Saint David Garejeli and the Stone of Grace is closely associated with the tradition of local Georgian monastic life, which spans many centuries. It enjoyed times of prosperity and renewal and endured a myriad of trials and tribulations. In the autumn of 2000, as soon as Saint David's tomb had been completely cleaned, the site was set in order: his larnax was faced with stone, a grave slab was adorned with images and an inscription, and ornaments were placed on the larnax (Figure 6.9).

Apart from this, the grave slab has a special, small-sized hole in which a copy of the Stone of Grace has been placed for the pilgrims to approach and touch for worship. The feast day of Saint David Garejeli is the Thursday after the Ascension. Many pilgrims visit the Lavra of Saint David during the course of the year, and they are permitted to touch and to kiss it.

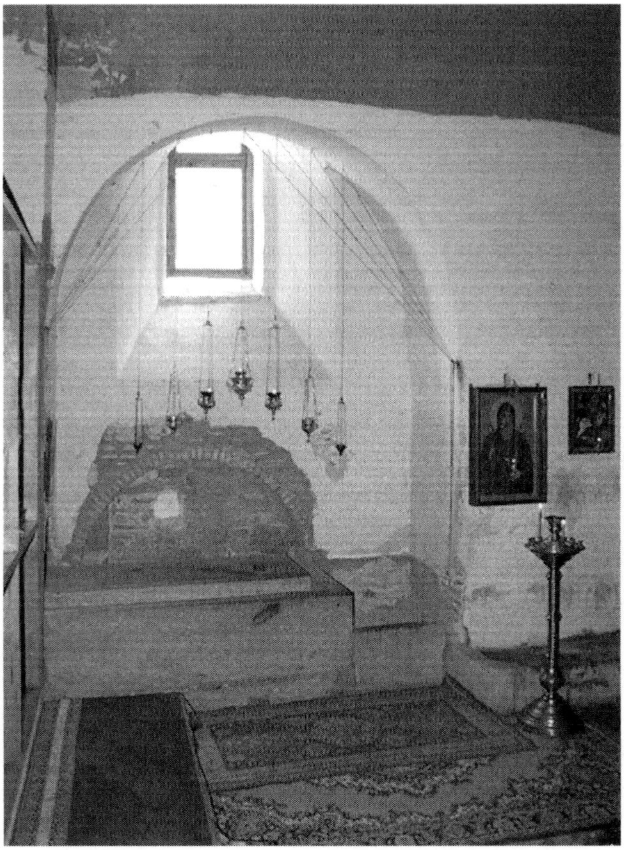

Figure 6.9 Lavra of St. David. Church of Transfiguration. Interior. Tomb of St. David Garejeli, view from north. Photo credit: Zaza Skhirtladze.

Notes

1 Giorgi Chubinashvili, *Peshchernye monastyri David-Garedzhi* [*The Rock-Cut Monasteries of David Gareja*] (Tbilisi, 1948); Levan Menabde, *Dzveli kartuli mtserlobis kerebi* [*Old Georgian Centres of Literary Activity*], vol. 1, no. 2 (Tbilisi, 1962), 273–319; Mikheil Kavtaria, *Davit Garejis literatutuli skola* [*The Literary School of Davit Gareja*] (Tbilisi, 1965); Guram Abramishvili, *Davit garejelis tsikli kartul kedlis mkhatvrobashi* [*The Cycle of Davit Garejeli in Georgian Mural Painting*] (Tbilisi, 1972); Aneli Vol'skaia, *Rospisi srednevekovykh trapeznykh Gruzii* [*Paintings in the Medieval Refectories of Georgia*] (Tbilisi, 1974), 30–87, 98–155; Zaza Skhirtladze, *Saberebis freskuli tsartserebi* [*Mural Inscriptions of Sabereebi*] (Tbilisi, 1985); Givi Gaprindashvili, *Gareji* (Tbilisi, 1987); Aneli Vol'skaia, "Rospisi peshchernykh monastyrei David-Garedji" ["Paintings in the Rock-Cut Monasteries of David Gareja"], in *Gareji: Trudy Kakhetskoi Arkheologicheskoi Ekspeditsii* [*Proceedings of Kakheti Archaeological Expedition*] 8 (Tbilisi, 1988), 130–155; Darejan Kldiashvili and Zaza Skhirtladze eds. *Garejis epigrafikuli dzeglebi* [*The Inscriptions of Gareja*], vol. 1, no. 1, Lavra of Saint David and Udabno Monastery (Tbilisi, 1999); Zaza Skhirtladze, *Istoriul pirta gamosakhulebebi garejis mravamltis kolagiris monastershi* [*Historical Figures at the Kolagiri Monastery in the Gareja Desert*] (Tbilisi, 2000); Zaza Skhirtladze, *The Tomb*

of Saint David Garejeli (Tbilisi, 2006); Dodo Lomidze, Manana Chumburidze, and Zaza Skhirtladze, eds., *Garejis istoriuli dokumentebi* [*Historical Documents of Gareja*], pt. 1 (Tbilisi, 2008), pt. 2 (Tbilisi, 2011). Note that throughout where no publisher is given the book was published by the Georgian Academy of Sciences.

2 Priest Nestor Macharashvili, *Aghtsera Mtatsmindis eklesiisa da tsminda mama Davit Garejeli* [*Description of the Mtatsminda Church and the Reverend Father David Garejeli*] (Tbilisi, 1895), 5–9.

3 Lavra of Saint David (sixth–twentieth centuries), Tetri Udabno (seventh–eighth centuries), Mravaltsqaro (ninth–eleventh centuries), Shavi Senakebi (ninth–eleventh centuries), Berta Mta (sixth–eleventh centuries), Tsamebuli (sixth–eleventh centuries), Natlismctsemeli (sixth–twentieth centuries), Chichkhituri (eleventh–seventeenth centuries), Dodos Rka (sixth–eighteenth centuries), Udabno (tenth–thirteenth centuries), Bertubani (sixth–thirteenth centuries), Verangareja (tenth–thirteenth centuries), Sabereebi (ninth–eleventh centuries), four complezes of Berebis Seri (ninth–eleventh centuries), two complexes of Pirukughmari (ninth–eleventh centuries), Patara Kvabebi (ninth–eleventh centuries), Didi Kvabebi (ninth–thirteenth centuries), Kolagiri (twelfth–thirteenth centuries), Kotsakhura (ninth–thirteenth centuries), and Satorge (eleventh–thirteenth centuries).

4 Lavra – a type of a large monastery, where hermits live with the community of monks. See: *Oxford Dictionary of Byzantium*, 2 (Washington, DC, 1991), 1190.

5 Hesychastirion – a small hermitage distinct from the large monastery.

6 For a brief overview of the history of Gareja in English, see Zaza Skhirtladze, "Newly Discovered Early Paintings in the Gareja Desert," in *Eastern Approaches to Byzantium* [*SPBS: 9*], ed. Antony Eastmond (Aldershot: Ashgate, 2001), 149–167, and the essays collected in Zaza Skhirtladze, ed., *Samonastro tskhovreba udabnoshi. Gareja da Kristianuli aghmosavleti* [*Desert Monasticism; Gareja and the Christian East*], *Proceedings of the Gareja Studies Centre:* 2 (Tbilisi, 2001).

7 Menabde, *Dzveli kartuli mtserlobis*, 317–319; Kavtaria, *Davit Garejis*.

8 Chubinashvili, *Peshchernye monastyri David-Garedzhi*. Sh. Amiranashvili, *Istoriia gruzinskoi monumental'noi zhivopisi* [*The History of Georgian Monumental Painting*], vol. 1 (Tbilisi, 1957), 41–76; G. Abramishvili, *Davit garejelis*; T. Virsaladze, "Osnovnye etapy razvitia gruzinskoi srednevekovoi monumental'noi zhivopisi" ["The Main Stages in the Development of Georgian Medieval Mural Painting"], in *IIe Symposium international sur l'art géorgien* (Tbilisi, 1977), offprint; Vol'skaia, *Rospisi srednevekovykh trapeznykh Gruzii*; idem, "Garedzhiiskaia zhivopisnaia shkola: Rospisi Bertubani" ["The Garejan School of Mural Painting: The Paintings of Bertubani"], in *Srednevekovoe iskusstvo: Rus' Gruziia* [*Medieval Art: Rus, Georgia*] (Moscow: Nauka, 1978), 92–105; A. Vol'skaia, "Peintures anciennes de Garedja," in *IVe Symposium international sur l'art géorgien* (Tbilisi, 1983), offprint; idem, "Rospisi peshchernykh monastyrei David-Garedji", 130–155; T. Sheviakova, *Monumental'naia zhivopis rannego srednevekov'ia Gruzii* [*Early Medieval Georgian Monumental Painting*] (Tbilisi, 1983), 9–16; Gaprindashvili, *Gareji*; Skhirtladze, "Newly Discovered Early Paintings", 149–167, and the essays collected in Z. Skhirtladze, ed., *Desert Monasticism: Gareja and the Christian East* (Tbilisi, 2001).

9 Z. Skhirtladze, *Sabereebis preskuli tsartserebi* [*Mural Inscriptions of Sabereebi*] (Tbilisi, 1985); D. Kldiashvili and Z. Skhirtladze, *Garejis epigrapiuli dzeglebi* [*The Inscriptions of Gareja*] vol. 1, pt. 1: *The Lavra of St David, Udabno Monastery (11th – 18th cc.)*, vol. 2 of *Garejis Sagandzuri* [*The Treasury of Gareja*] (Tbilisi, 1999).

10 "Tskhorebai da mokalaqobai tsmidisa mamisa chuenisa Davit Garejelisai" ["Life and Deeds of Our Saint Father David Garejeli"], in *Dzveli kartuli agiografiuli literaturis dzeglebi* [*Monuments of Old Georgian Hagiographic Literature*], vol. 1, eds. Ilia Abuladze et al. (Tbilisi, 1963), 237–239.

11 The *Vita* of Saint David provides early evidence for the Georgian name of the Hill of Mercy (მადლოვანების ქედი) near Jerusalem, known in Crusader literature as the Mount of Joy: Darejan Kldiashvili, "Gareja da piligrimoba adreuli shua saukuneebis kartul tsqaroebshi" ["Gareja and the Pilgrimage in Early Medieval Georgian Sources"], in *Desert Monasticism: Gareja and the Christian East*, ed. Skhirtladze (Tbilisi, 2001), 92–93, 101.

12 "Tskhorebai da mokalaqobai tsmidisa mamisa chuenisa Davit Garejelisai", 237.

13 The name of the gate is not indicated in the *Vita* of the saint.

14 "Tskhorebai da mokalaqobai tsmidisa da ghmertsahemosilisa mamisa chuenisa Davit Garesjelisai" 200; "Tskhorebai tsmidisa ghmertshemosilisa mamisa chuenisa Davit Garesjelisa meudabnoesi" ["Life of Our Saint and God-Blessed Father David Garejeli the Hermit"], in *Dzveli kartuli agiografiuli literaturis dzeglebi* [*Monuments of Old Georgian Hagiographic Literature*] (Tbilisi, 1963), 3, eds. Abuladze et al., 306.

15 "Tskhorebai da mokalaqobai tsmidisa mamisa chuenisa Davit Garejelisai," 238; "Tskhorebai da mokalaqobai tsmidisa da ghmertsahemosilisa mamisa chuenisa Davit Garesjelisai," in *Dzveli kartuli agiografiuli literaturis dzeglebi*, vol. 3, eds. Abuladze et al., 200–201; "Tskhorebai tsmidisa ghmertshemosilisa mamisa chuenisa Davit Garesjelisa meudabnoesi," ibid., 306. For the English translation cf. *Lives and Legends of the Georgian Saints*, ed. and trans. D.M. Lang (London and Oxford: A.R. Mowbray, 1976), 91–92.

16 "Tskhorebai da mokalaqobai tsmidisa mamisa chuenisa Davit Garejelisai," in *Dzveli kartuli agiografiuli literaturis dzeglebi*, vol.1, eds. Abuladze et al. (Tbilisi, 1963), 238; "Tskhorebai da mokalaqobai tsmidisa da ghmertsahemosilisa mamisa chuenisa Davit Garesjelisai," ibid., 201; "Tskhorebai tsmidisa ghmertshemosilisa mamisa chuenisa Davit Garesjelisa meudabnoesi," ibid., 307.

17 *Lives and Legends of the Georgian Saints*, 92.

18 The sources do not contain any information concerning the further history of the other two stones.

19 In the keimenic recension, presumably written by the Catholicos of Kartli (East Georgia) Arseni in the tenth century: "Tskhorebai da mokalaqobai tsmidisa mamisa chuenisa Davit Garejelisai" ["Life and Deeds of Our Saint Father David Garejeli"], in *Dzveli kartuli agiografiuli literaturis dzeglebi* [*Monuments of Old Georgian Hagiographic Literature*], vol. 1, eds. Abuladze et al. (Tbilisi, 1963), 198–199; in two metaphrastic recensions compiled in the twelfth and eighteenth centuries: "Tskhorebai da mokalaqobai tsmidisa da ghmertsahemosilisa mamisa chuenisa Davit Garesjelisai" ["Life and Deeds of Our Saint and God-Blessed Father David Garejeli"], ibid., 3, 197–203, and "Tskhorebai tsmidisa ghmertshemosilisa mamisa chuenisa Davit Garesjelisa meudabnoesi" ["Life of Our Saint and God-Blessed Father David Garejeli the Hermit"], ibid., 304–306; the synaxarion recension of the first half of the eighteenth century: "Tskhorebai tsmidisa mamisa chuenisa Davit Gareelisa" ["Life of Our Saint Father David Garejeli"], ibid., vol. 4, ed. Enriko Gabidzashvili (Tbilisi, 1968), 412–414.

20 Simon Qaukhchishvili, ed., *Kartlis Tskhovreba* [*Life of Kartli/Georgia*], vol. 1 (Tbilisi, 1955), 210.

21 Lomidze, Chumburidze, and Skhirtladze, eds., *Garejis istoriuli dokumentebi, XIII–XVIII ss.*, vols. 1 and 2 (Tbilisi, 2008, 2011).

22 Kldiashvili, "Gareja da piligrimoba adreuli shua saukuneebis kartul tsqaroebshi," 77–78.

23 "Tskhorebai da mokalaqobai tsmidisa mamisa chuenisa Davit Garejelisai," in *Dzveli kartuli agiografiuli literaturis dzeglebi*, vol. 1, 238; "Tskhorebai tsmidisa ghmertshemosilisa mamisa chuenisa Davit Garesjelisa meudabnoesi," ibid., 3, 306.

24 Kldiashvili, "Gareja da piligrimoba adreuli shua saukuneebis kartul tsqaroebshi," 85–88.

25 Cf. the statement of the hieromonk Kalistrate that "Georgians, Russians, and Armenians have a great piety and respect for this stone and the tomb of Saint David" (Hieromonk Kalistrate, *Tsminda Davit Garesjelis Udabno* [*The Desert of Saint David Garejeli*] (Tbilisi, 1884), 16.

26 On the pilgrimage of Saint David Garejeli to Jerusalem see: Kldiashvili, "Gareja da piligrimoba adreuli shua saukuneebis kartul tsqaroebshi," 92–93.

27 Korneli Kekelidze, "K voprosy ob Ierusalimskom Proiskhojdenii Gruzinskoi Tserkvi" ["Toward the Jerusalem Origin of the Georgian Church"], in *Etiudebi Dzveli Kartuli Literaturis Istoriidan* [*Studies in Old Georgian Literature*], vol. 4, ed. Korneli Kekelidze (Tbilisi, 1957), 358–363; Zurab Kiknadze, *Kartli Gakristianebis Gzaze: Adamianebi da Sitsmindeebi* [*Georgia on the Way of Christianity: People and Sanctities*] (Tbilisi, 2009), 172–177. The connection between Jerusalem and Georgia has always been exceptionally strong. It was expressed in several ways: in the topography of the old capital of the country – Mtskheta, in the tradition of the veneration of the Holy Cross beginning from the conversion of the country, in the church architecture, etc.; cf. Zaza Skhirtladze, "The Mother of All the Churches: Remarks on the Iconographic Programme of the Apse Decoration of Dort Kilise," *Cahiers Archeologiques* 43 (1995): 101–116.

28 Cf. "Tskhorebai da mokalaqobai tsmidisa da ghmertsahemosilisa mamisa chuenisa Davit Garesjelisai," in *Dzveli kartuli agiografiuli literaturis dzeglebi*, 3, eds. Abuladze et al., 197, 203, 206–207; "Tskhorebai tsmidisa mamisa chuenisa Davit Gareelisa," ibid., vol. 4, ed. Gabidzashvili (Tbilisi, 1963), 414.

29 Zaza Skhirtladze and Antony Eastmond, "Tsminda davit garejelis tskhovrebis tsikli udabnos monastris mtavari eklesiis sadiakvnes mokhatulobashi: akhali monatsemebi da dakvirvebebi" ["The Life of Saint David Garejeli in the Wall Paintings of the Diakonikon of the Main Church at Udabno: New Data and Observations"], *Sakartvelos Sidzveleni* [*Georgian Antiquities*] 2 (2002): 28–49; Antony Eastmond and Zaza Skhirtladze, "Udabno Monastery in Georgia: Innovation, Conservation and Reinterpretation of Medieval Art," *Iconographica* 7 (2008): 23–43.

30 "Tskhorebai da mokalaqobai tsmidisa mamisa chuenisa Davit Garejelisai," in *X Dzveli kartuli agiografiuli literaturis dzeglebi*, vol. 1, 238.

31 Skhirtladze and Eastmond, "Tsminda davit garejelis tskhovrebis tsikli," 45.

32 Ilia Abuladze, *Asurel moghvatseta tskhovrebis tsignta dzveli redakciebi* [*Old Versions of the Lives of the Syrian Fathers*] (Tbilisi, 1955), 179–180.

33 Korneli Kekelidze, "Sakitkhi asurel moghvatseta kartlshi mosvlis shesakheb" ["On the Arrival of the Syrian Fathers in Kartli"], in *Etiudebi dzveli kartuli literaturis istoriidan* [*Studies on Old Georgian Literature*], vol. 1, ed. Korneli Kekelidze (Tbilisi, 1956), 36.

34 Abramishvili, *Davit Garejelis tsikli*, 56–57.

35 Skhirtladze and Eastmond, "Tsminda davit garejelis tskhovrebis tsikli," 46.

36 Kldiashvili, "Gareja da piligrimoba adreuli shua saukuneebis kartul tsqaroebshi," 77–79.

37 "And as soon as Bubakar converted to Christianity, the baptized son came over to Saint David and brought with him some tools and workmen and built as a place of worship for his brothers a church, which was later expanded and blessed by the Holy and Blessed-by-God Father Ilarion, who built a beautiful larnax at the south edge of the church where the remains of our Holy Father David are buried. And until now, the spirit of wonderful fragrances has emanated from his grave and everyone who breathes in his faith the dust of that holy grave enjoys the healing of all illnesses and weaknesses of men."; "Tskhorebai da mokalaqobai tsmidisa mamisa chuenisa Davit Garejelisai," 236; cf. Z. Skhirtladze, *Tsm: Davit Garejelis gansasvenebeli* [*The Tomb of Saint David Garejeli*] (Tbilisi, 2006), 5, 16–17, 50–51, 106–107.

38 Bishop Kirion, *Ambavi Davit Garesjis udabnos sakvirvelt-mokmedis madlis kvisa* [*The History of the Miraculous Stone of Grace of the David Gareja Desert*] (Tbilisi: M. Sharadze, 1899), 8–9.

39 The presence of holes for blessing in the upper and lower covers of Saint David's vault are mentioned in a passage of Revered Gabriel Mtsire's *Spiritual Stories*, which relates the removal of a holy relic from the burial site and a miracle connected with it.

40 Skhirtladze and Eastmond, *Tsm. Davit Garejelis gansasvenebeli*, 123.

41 Ibid., 125–126.

42 "For laid on the grave of Saint David are: a chased icon of the Saviour, one; reveted icon of Being, one; a reveted icon of the All Holy Virgin, one; an icon of the All Holy Virgin with wings, one; an icon of the Descent from the Cross, one; a reveted icon of the Trinity, one; an icon of Saint Nicholas, one." See *Kartuli samartlis dzeglebi* [*Monuments of Georgian Law*], vol. 3, *Monuments of Ecclesiastical Canon (11th – 19th Centuries)*, ed. I. Dolidze (Tbilisi, 1970), 681.

43 At the turn of the of the fifteenth century, Tamerlane raided several times, among which raids the fifth (1394) and the sixth (1400) were the most brutal; at the beginning of the seventeenth century, the invasion of Persian Shah Abbas I; from the late seventeenth to the early eighteenth century the Lezgins raided continuously: see Anton I Catholikos, "Martirika," in *Dzveli kartuli agiografiuli literaturis dzeglebi* [*Monuments of Old Georgian Hagiographic Literature*], vol. 4, ed. E. Gabidzashvili (Tbilisi, 1980), 189–208; Lomidze, Chumburidze, and Skhirtladze, eds., *Garejis istoriuli dokumentebi*, vol. 1, 31ff; cf. also, Chubinashvili, *Peshchernye monastyri David-Garedzhi*, 84–118; Menabde, *Dzveli kartuli mtserlobis kerebi*, vol. 1, 284–297.

44 Archimandrite Tarasi, "Mokle aghtsera sakartvelos monasterta" ["Brief Description of the Monasteries in Georgia"], *Sakartuelos sasuliero makharebeli* [*Spiritual Annunciation of Georgia*] (August, 1868): 42–43; cf.: Kirion, *Ambavi Davit Garesjis*, 10–11; Bishop Kirion,

Prepodobnyi David Garejeli i ego Lavra [*Reverend David Garejeli and His Lavra*] (Tbilisi, 1901), 23–24; Hieromonk Kalistrate, *Tsminda Davit Garesjelis Udabno*, 15–16.

45 Kirion, *Prepodobnyi David Garejeli i ego Lavra*, 26–27.

46 "Having started to build it from the foundation [upward], it became necessary to straighten the old quadrangular pillar of the sanctuary near the chancel screen, which is at Saint David's foot, and to extend its vault, under which is the larnax of the reverend; and upon the demolition of half of the pillar on the side of the saint's feet, a wall plastered with burnt gypsum came to light, with the icon of Saint David painted on it, and of the Reverend Dodo, in the position of kissing each other's hands, and their pleased faces expressing gladness at their fresh meeting, as was made clear by the words written above his head: 'The icon of Saint David's arrival from Jerusalem.' Complete renovation of the church at the time demanded the demolition of the wall with the icon in order to straighten the chancel, and when it was demolished, another wall appeared and in it a small shelf plastered with burnt gypsum, and in the shelf a small hollow in which lay the precious stone brought by Saint David from Jerusalem, the shelf bearing signs of candles having been lit on it and the molten [wax] discolored with age. This stone is slightly less than a goose egg in size and longish, like it; on top it has a mark of the color of dried blood, and on one side a streak of the same color running round it; the color of the stone is between red and white; it healed those who touched it with faith. As the stone had been buried in olden times in the mentioned wall by Holy Fathers, the present brethren who had entered the monastery following the example of the Fathers of old, after several years of finding the stone, buried it in his church in a secret place, not known to everyone but the old brethren, and on the feast day of the saint, they took it out to kiss it and then put it away again." See Archimandrite Tarasi, "Mokle aghtsera sakartvelos monasterta," 42–43.

47 Andrei Muraviov, *Gruzija i Armenija* [*Georgia and Armenia*], vol. 1 (Saint Petersburg, 1848), 106.

48 Bishop Kirion, *Prepodobnyi David Garejeli i ego Lavra*, 26–27.

49 Novice Spiridon Grdzelishvili, "Aokhreba tsmidis Davit Garesjis udabnosa" ["Ravage of the Holy Desert of David Gareja"], *Sakartuelos sasuliero makharebeli* [*Spiritual Annunciation of Georgia*] (March 1868): 81.

50 "They found the door of the church, where Saint David was buried and broke it and took away all the treasures remaining there, pearled phelonion, the chalices, the crosses, the Gospels and the icon lamps of silver, the gold-treated cover of the grave, miters, and all the valuable things; and they took away the Stone of Grace, which was brought by Saint David from Jerusalem": ibid.

51 Kirion, *Ambavi Davit Garesjis udabnos sakvirvelt-mokmedis madlis kvisa*, 12.

52 Skhirtladze, *Tsm. Davit Garejelis gansasvenebeli*, 138–139.

7 Earth from elsewhere

Burial in *terra sancta* beyond the Holy Land*

Lucy Donkin

In 1508, Pope Julius II gave Florian and Barbara von Waldenstein a twofold gift. He not only granted the burial chapel they had founded in Hall in Austria privileges equivalent to those of the Campo Santo Teutonico in Rome, but also gave the couple the right "to strew its surface with the dust or holy earth of the cemetery of the Campo Santo."[1] In its turn, the Campo Santo Teutonico had a material connection with the Holy Land, as it was understood to contain earth from the burial ground of Akeldama in Jerusalem. This chapter explores these and other contexts in which Christian burial places were equated through the movement of earth in the later Middle Ages, allowing people buried in one location to enjoy the benefits of another.[2] As the papal grant suggests, this phenomenon was associated with Jerusalem-based piety, but it was far from restricted to interest in the Holy Land. In what follows, I therefore juxtapose Jerusalem and Rome as sources of earth for networks of recipient sites, encompassing a wide range of devotional geographics.[3] My intention here is not to suggest dependent relationships among the examples or to compare the local circumstances that gave rise to each, but rather to clarify the spatial relationships created between burial places and the part that matter played in this process. This focus offers a perspective on the materiality of place and the sacrality of matter by asking how the movement of earth relates to less tangible means of equating places, such as the bestowal of equivalent indulgences, and how the sanctity of soil fits within a spectrum of other properties, including the power to decompose and discriminate among the bodies buried in it. Although it is connected to practices of taking dust from tombs and stones from holy sites, moving earth between cemeteries had a particular logic and role within the sphere of transportable topography.[4]

Transporting Jerusalem

The Campo Santo Teutonico, with which Julius II's bull was concerned, is central to this chapter because it was understood as both a destination and a source of earth. In order to clarify the relationship between these roles, I begin by considering the former alongside comparable cases in which soil was said to be taken from Jerusalem. I draw here particularly on the work of Albrecht Weiland on the Campo Santo and Nine Miedema on the indulgence literature, but by focusing solely on the movement of soil, aim to illuminate the expressive potential of this material. The Campo Santo Teutonico lies immediately to the south of the Saint Peter's and served as a burial place for German-speaking residents of the city as well as for pilgrims to Rome. Already in the ninth century, the church associated with the *schola Francorum* seems to have been connected with the burial of pilgrims.[5] Only in the later Middle Ages, however, was the cemetery thought to contain soil from Jerusalem.[6] The traditions surrounding the site seem to have developed and been elaborated on particularly in German and Dutch texts that listed the indulgences unofficially associated with the churches of Rome.[7] Indulgences had been available for visiting individual churches in the city from the twelfth century, but the number said to be obtainable at each increased dramatically in the fourteenth century.[8] A German guide to Rome, dated to 1377 and probably the work of Leopold of Vienna, though based on the *Mirabilia Romae* and the *Indulgentiae ecclesiarum urbis Romæ*, noted that whoever went to the "*Goczakcher*" reverently received 1,500 years pardon.[9] *Libri indulgentiarum* and related texts promoted Rome vis-à-vis other pilgrimage centers; indeed the anonymous late fourteenth-century *Stacions of Rome* specifically noted that so much pardon was available there that there was no need to travel to Jerusalem.[10] It may be in this spirit of competition that the Campo Santo Teutonico was claimed to contain soil from the Holy Land.

Specifically, the soil was associated with Akeldama. In Saint Matthew's Gospel, this site is described as having been bought by the chief priests as a burial ground for strangers, and it was used in that way for Christian pilgrims from at least the sixth century, when the Piacenza Pilgrim noted the practice.[11] Weiland convincingly suggests that the association of the Campo Santo Teutonico with Akeldama was due to the long-standing use of the former as a burial place for pilgrims and strangers in Rome.[12] It is hard to establish the point at which this association was first made. What has been seen as the earliest reference is found in a Dutch-language version of the *Indulgentiae ecclesiarum urbis Romae* in The Hague, Koninklijke Bibliotheek, MS 76 E 5, a manuscript generally dated to the end of the fourteenth century. Here the soil is said to have been brought from the field acquired with Judas Iscariot's thirty pieces of silver. The cemetery is described as a place of burial for poor pilgrims, and 15,000 years remission are said to be granted to those who recited the Lord's Prayer there.[13] This miscellany begins with a computus table with the date 1374, and the same date has therefore been ascribed to the text "*De aflaten der zeven kerken van Rome*" on folios 57v – 61r.[14] Although the manuscript is described as having been written in the same hand throughout, the computus text is found on a single leaf at the very beginning, and it has been suggested that the Rome text is far later, even by a century.[15] This argument is based partly on the large number of indulgences awarded and the arrangement of the text according to the seven churches of Rome, although a similar organization is found in Leopold's work. A connection with Akeldama is made in several other versions of the *Indulgentiae ecclesiarum* dating from the fifteenth century.[16] The first document associated with the Campo Santo Teutonico to mention the site in Jerusalem seems to be a notarial act of 1454, which records that one of the founders of the Confraternity of the Campo Santo buried the poor for free, since the "*ager sanctus*" was bought with the money for which Christ was sold and could not itself be sold.[17] The earth is referred to explicitly in a petition of 1476 addressed to Pope Sixtus IV, in which the confraternity asked for permission to send two pardoners to Germany to collect funds for the restoration of the church.[18] At a later date, probably during the seventeenth century, the associations of the earth shifted to Calvary.[19]

The extent to which the rationale for the indulgences at the Campo Santo Teutonico derived from its association with Jerusalem is not clear. One pilgrim to both Rome and Jerusalem mentioned celebrating Palm Sunday at the cemetery in 1470, although the stational church for this feast day was Saint John Lateran, which might reflect its Jerusalem associations.[20] Most frequently, however, references in the indulgence literature simply note relatively small numbers of indulgences granted for visiting the site.[21] At the same time, in a number of cases they do seem to reflect its status as a cemetery. The Campo Santo is the only place described in the manuscript in The Hague in which indulgences are connected with saying the Lord's Prayer, and it is possible that this is related to the practice of churchyard indulgences. In the fifteenth century, a prayer ascribed to Pope John XXII and sometimes said to be "written behind the altar of St Peter" in Rome offered indulgences to anyone passing a cemetery who said a Pater Noster and a Hail Mary, and the prayer "*Avete omnes Christi fidelibus animae.*"[22] The Lord's Prayer and the Ave Maria are mentioned in several other versions of the *Indulgentiae ecclesiarum*, with the text occasionally specifying that the prayers were said for the souls of those buried in the Campo Santo.[23] All this is compatible with the treatment of Akeldama itself. One of the earliest accounts of Holy Land pilgrimage to enumerate indulgences for the *loca sancta*, Niccolò da Poggibonsi's *Libro d'Oltramare*, of ca. 1345, notes the "*perdonanza grandissima*" available at Akeldama, and describes how pilgrims walked around it, reciting psalms, paternosters, and prayers for the souls of those buried there.[24] However, there does not seem to have been any attempt to link the Campo Santo indulgences explicitly to those available in Jerusalem.

Only a few of the versions of the *Indulgentiae ecclesiarum* listed by Miedema in her comprehensive study of these texts mention that indulgences were available for burial in the Campo Santo, even if the effect of the indulgenced prayers was to enhance the site for those buried there.[25] In these cases, which date from the second half of the fifteenth century onward, burial in the cemetery was said to obtain particular indulgences or even the forgiveness of all sins. Presumably, the relative silence on this matter reflects the fact that pilgrims did not anticipate being buried in the cemetery themselves, although those who died in Rome were likely to be interred there. It was claimed that one of the founders of the Confraternity of the Campo Santo, Frédéric de Meydenburg, buried 3,500 strangers at the Campo Santo in the Jubilee year of 1450.[26] Although it is likely that this figure was inflated, it is possible that a high mortality rate among pilgrims that year contributed to the renown of the Campo Santo as a burial place. Equally, the founding of the confraternity is likely to have influenced the inclusion of the benefits of burial in the indulgence literature. Although the confraternity issued disconnected indulgences in order to restore the chapel, it is possible to identify a shift in the significance of the site, with consequences for the later history of its soil.

Before tracing these subsequent developments, however, the material presence of earth from Jerusalem demands further scrutiny regarding the particular qualities it possessed, the way in which it translated place, and the extent and nature of its sanctity. One characteristic of Akeldama noted by travelers was its distinctive capacity to consume the flesh of corpses within a matter of days, and this property was also commonly attributed to the Campo Santo Teutonico.[27] The claim is found in fifteenth-century versions of the *Indulgentiae*, the earliest of which may be a manuscript from Augsburg, dated to 1448.[28] Here the capacity of the shared soil of Akeldama and the Campo Santo is expressed as "*gnaud*" or "grace," more fully as "*die gnaud in im selbe*"; in another manuscript, dated to 1500, this is termed its "*natúre*."[29] This property of the Campo Santo earth authenticated its origin in the Holy Land and demonstrated the material equivalence of the two places. However, it also contributes to a wider distinction that can be drawn between earth from Akeldama and that from elsewhere in the region. During the Middle Ages, earth and stones from *loca sancta* throughout the Holy Land were treated as relics, not only commonly included in relic collections and placed in reliquaries, but even used in the consecration of churches.[30] Although sites such as Calvary and the Mount of Olives were particularly popular sources, such relics came from a wide range of locations associated with the life of Christ and other biblical figures. Yet, earth from Akeldama was not central to these practices; as far as I am aware, it is rarely described in relic lists and church consecration accounts, although there is some evidence that it was venerated and taken away by pilgrims: in 1431, Mariano da Siena mentioned removing earth from Akeldama for his "*devozione*."[31] There is similar ambiguity of status regarding the earth in the Campo Santo in the Roman indulgence literature. In contrast with its later designation in papal bulls, it is not expressly described as *terra sancta*, although this may be because the term was used to describe the Holy Land as a whole. At the same time, the soil was associated with Holy Land relic collecting. From the mid-fifteenth century, there is evidence of interest in the manner in which the earth reached Rome, and some versions of the *Indulgentiae* credit the Empress Helena – known for her discovery of the True Cross – with bringing the soil from Jerusalem.[32] There was also a tradition that saw Akeldama soil as having been contained in the four bronze columns in Saint John Lateran, themselves often said to have come from Jerusalem.[33] While many texts simply described the columns as filled with "*terra sancta*" from Jerusalem, in the mid-fifteenth century Nikolaus Muffel also included reference to the "*gotzacker*," probably referring to Akeldama.[34] Nevertheless, in sources that focus on the Campo Santo Teutonico and its relationship to Jerusalem, the soil is not unambiguously described as holy.

The place of the Campo Santo soil within the indulgence tradition and its position vis-à-vis other material fragments of the Holy Land raise questions regarding the relationship between properties intrinsic to a substance and those bestowed upon it. Although often connected with visiting a holy site or seeing or kissing a holy object, indulgences were conferred on these activities by ecclesiastical authorities rather than emanating from that place or object. At the same time, Diana Webb has suggested that the draw for pilgrims might still have been the less quantifiable holy qualities of the place or the object itself.[35] In the case of the substance of the Holy Land, further distinctions can perhaps be drawn between the ways in which sites assumed such qualities in the first place. The Holy Land as a whole drew its sanctity primarily from the presence of Christ and other biblical figures and was venerated as the location of salvific events.[36] Thus, earth and stones from the Holy Land were often valued as the substance of the ground that the Lord had trodden; even relics from locations of key events are often described in terms of physical contact, emphasizing their status as contact relics.[37]

In contrast, Akeldama drew its significance not from the physical presence of Christ, but from its association with his Passion.[38] Differentiating Akeldama soil from earthen contact relics that transmitted the sanctity of a holy figure invites us to give more weight to the qualities of the material itself and its relationship with place. In aiding decomposition, the property of the soil – although striking – was not wholly divorced from the potential properties of earth more generally. Indeed, the language that could be used to describe it – "*die gnaud in im selbe*" – suggests that it was understood to be inherent in the material. It is perhaps significant, then, that during the Middle Ages there are other indications of the belief that active properties of certain places resided in their soil and could be moved with it. For example, stones and soil from Thanet were credited with the capacity to kill or ward off snakes, one of a number of similar claims made for various islands since antiquity.[39] More unusually, in a passage found in some manuscripts of Marco Polo's *Description of the World*, the warlike nature of the people of Kerman is ascribed to the local soil, for after this was spread under carpets in Shiraz, the usually peaceable citizens who walked on it became quarrelsome.[40] As transportable topography has been couched primarily in terms of holy places, these examples are useful in demonstrating how properties of place could transcend sanctity. In the case of the Campo Santo Teutonico, the earth was clearly bound up with the story of salvation, and its qualities should be seen in this context. However, the wider capacity of earth to render one location present in another suggests that this material essentially effected a translation of a holy burial place, rather than transmitting sanctity in the manner of a contact relic.

The importance of the identity of the linked sites as burial places is confirmed by a second active property of the Campo Santo Teutonico. This derives more from the designation of the site than from its substance, as it was connected to a wider characterization of consecrated burial space, as well as to aspects of the Roman indulgence system. The Augsburg manuscript mentioned above notes that the cemetery was exclusively for pilgrims and other foreigners and that the ground would not hold Romans.[41] The idea of the ability of the earth to discriminate among bodies buried in it does not appear to stem from traditions regarding Akeldama, but a similar capacity is found in accounts of burials from late antiquity onward. Most concern unsuitable candidates for burial in holy or consecrated ground, such as the excommunicated knight described by Adhemar of Chabannes in his record of the Council of Limoges in 1031, whose body was forcibly ejected several times by the ground of the cemetery in which it had been placed.[42] The ground here is defined by the authority of the Church, and it is possible that this example reflects the development of consecration rites for cemeteries in continental Europe, as well as episcopal attempts to control burial at the time of the Peace of God.[43]

In the case of the Campo Santo Teutonico, however, the exclusion of Romans is not condemnatory, even when reference is later made to the earth expelling bodies.[44] It probably derived more immediately from the practice of differentiating among pilgrims according to their geographical origins and the length and type of their journeys when determining the indulgences available to them or the time they had to spend in Rome. In 1291, Pope Nicholas IV had divided pilgrims into three groups – inhabitants of Rome, Campagna, and Le Marche; pilgrims from Tuscany and Lombardy; and those from across the Alps or arriving by sea – when assigning indulgences to those visiting Saint John Lateran; shortly afterward, Boniface VIII simply distinguished between Romans and other "*peregrini . . . aut forenses*."[45] In the Campo Santo Teutonico, that distinction was policed by the ground itself. The properties of the soil in the cemetery as described in the indulgence literature were thus framed in different ways – one intrinsic to the substance and the place from which it came, the other probably derived from ecclesiastical designations of its present site. However, in both cases they involve not so much the sanctity of matter as the materiality of sacred place.

As noted above, the beliefs surrounding the Campo Santo were part of a wider phenomenon in which places claimed possession of earth from Jerusalem. Within the still broader context of Jerusalem as the center of a network of relic distribution and replication elsewhere, this particular process of translation is distinguished by involving quantities of earth destined to be returned to the ground. In Rome itself, the Church of Santa Croce in Gerusalemme also claimed soil from the Holy Land from at least the fifteenth century, and the interpretative traditions regarding the two Roman sites likely reflect a measure of competitive cross-fertilization within the city. The difficulties of dating the first reference to Akeldama earth in the Campo Santo Teutonico mean that it is hard to establish which tradition had precedence. More significant in this context, however, is the different characterization of the earth at the two sites. Santa Croce had a long association with the Holy Land through its relics of the passion, and understandings of the soil built on this.[46] Pero Tafur, who visited Rome in the 1430s, stated that the whole church was constructed with earth from Jerusalem employed as ballast in the ships used to bring back the relics.[47] Within the church, the earth was associated with Calvary and clearly defined as holy, in contrast to the vocabulary used in the Roman indulgence literature to describe the soil from Akeldama. In 1520/1521, an inscription was set up by Cardinal Bernardino Carvajal in the entrance to the Jerusalem chapel that stated this to have been filled with "holy earth of Mount Calvary," which the Empress Helena had had shipped to Rome, making it a second Jerusalem.[48] A marble slab probably set in the pavement at around the same time recorded a similar claim.[49] The first of the four narrative scenes in the chapel's vault mosaic, of ca. 1500, which show the story of the True Cross, has also been seen by some scholars to depict the bringing of the earth.[50] In Carvajal's inscription, the earth was holy because it was soaked with the blood of Christ, again distinguishing it from the claims made for the Campo Santo earth. Indeed, even without explicit reference to Christ's blood, stones and soil from Calvary were among the most popular of earthen relics, and the bronze columns at the Lateran too came to be seen as filled with earth from that site.[51]

Instead, the traditions surrounding the Campo Santo Teutonico bear a closer relationship to those associated with cemeteries elsewhere in Italy and the Mediterranean region. Famously, the Camposanto in Pisa was understood to contain soil from the Holy Land from at least the mid-fourteenth century.[52] The earliest reference is found in a chronicle preceding that of Raniero Sardo in Florence, Biblioteca Nazionale, MS Magl. XXV-491, which narrates the history of Pisa up to 1354. This states that the cemetery was founded by Archbishop Ubaldo in 1200 and explains its name as deriving from the presence of the holy earth: "*è decto chanposanto perché vi fu messa della terra sancta d'oltra mare.*" The Pisans are said to have brought it back from an expedition in the 1180s and scattered it all over the cemetery: "*sparsolla per tucto nel dicto luogho a onore di Dio.*"[53] An anonymous description of Pisa dating to around 1430 attributes the name to the bringing of "*terra sancta asai quantità*" when the Pisans took Jerusalem.[54] Although these sources do not give a specific place of origin for the soil, in the 1430s Pero Tafur understood the earth in the Camposanto to derive from Akeldama and to have the power to strip corpses in thirty days.[55] Within Italy, Pisa and Rome appear to have been the only places to claim possession of soil from Jerusalem during the Middle Ages. There are, however, fifteenth- and sixteenth-century accounts of earth from Jerusalem being spread on cemetery sites in Cyprus.

In an account of his travels in the Holy Land and other parts of the Mediterranean in the 1470s, the Swiss pilgrim and merchant Ulrich Leman described Akeldama and its properties, noting that earth had been taken from there to both Rome and Cyprus.[56] His subsequent discussion of Nicosia includes a passage on the *camposanto* there that contained this earth, again mentioning Rome as another recipient as well as referring to the capacity of the earth to reduce bodies to ashes in three days. His description of the "*gotzaker*" in Rome is shorter and simply characterizes it as a burial place for pilgrims.[57] Given the direct links between Cyprus and the Latin Kingdom of Jerusalem, and the existence of the Crusader Kingdom of Cyprus until 1489, it is noteworthy that Cyprus and Rome are associated in this way. A few decades after Leman, an anonymous compilation of Iberian pilgrimage accounts, of ca. 1520, described the burial ground in Nicosia with no reference to Jerusalem, noting that the Dominican convent possessed a *camposanto* that stripped the flesh off the bodies buried there in three days just like the Campo Santo in Rome behind the church of Saint Peter.[58] Although the ultimate implicit reference here is to the characteristics of Akeldama itself, this is evidence of the independent fame of the Campo Santo Teutonico in a period in which its soil was being distributed. A Jewish commentator, however, was more interested in the connection with the Land of Israel: Rabbi Basola of Pesaro visited Famagusta between 1521 and 1523, during a trip to Syria and Palestine, and wrote in his travel account that "there is a place where many ships deposited dust of the Land of Israel and where they were accustomed to bury prominent people of old, and they call it Campo Santo, this being an indication that everyone recognizes the sanctity of the Land of Israel."[59] In general, however, the examples of Pisa and Cyprus support the evidence concerning the Campo Santo Teutonico in showing Jerusalem and Akeldama in particular to have been points of reference for a number of burial sites in the fourteenth to sixteenth centuries. Interest was expressed in terms of material translation, and the striking qualities of the shared soil created a certain equivalence of place. Although locations that claimed earth from Jerusalem might present theirs as an exclusive relationship, travelers who visited the sites might see them as one of a number of recipients and recognize a likeness between them.

Transporting Rome

From the early sixteenth century, earth from the Campo Santo Teutonico was itself taken to other cemeteries, including sites in Austria, Germany, Poland, and Spain as well as elsewhere in Italy.[60] The movement of earth from a place that claimed soil from elsewhere does not simply present a new set of spatial relations for analysis, but also raises questions regarding how far these were informed by the original transaction. The remainder of this chapter addresses the granting of soil from the Campo Santo in this light, before comparing it with accounts of transporting soil from other sites in Rome found in hagiographical literature. I draw particularly on the work of Nikolaus Grass on the papal bulls and Anja Tietz on early modern cemeteries, but again focus on the role of the earth itself, as well as adding new comparative material. In its provision of soil, the Campo Santo is unusual among sites claiming Akeldama earth – neither Pisa nor Cyprus seems to have been used in this way – and this reflects the particular nature of the Roman cemetery. Specifically, it seems likely to derive from an increased emphasis on the burial of pilgrims there from the mid-fifteenth century onward, the involvement of particular foreign communities with the site, and promotion by the papacy. The practice of granting the soil, along with the indulgences associated with the Campo Santo, is first found in a series of grants made by Popes Julius II (1503–1513) and Leo X (1513–1521) to sites mainly located in Austria and southern Germany in the early sixteenth century, often to individuals connected with the imperial court. As noted above, in 1508 Pope Julius II granted Florian and Barbara von Waldenstein the right to establish their burial chapel in Hall "with the apostolic authority in the Campo Santo" and to "strew its surface with the dust or holy earth of the cemetery."[61] This grant was followed shortly afterward by others for Kolsaß in Unterinntal and the extramural cemetery in Innsbruck, and in 1513 the new cemetery of Saint Stephen in Vienna was established on the model of that in Innsbruck.[62] That same year, Leo X granted Campo Santo earth and privileges to the cemetery of Freiburg im Breisgau (Baden-Württemberg) at the behest of Emperor Maximilian I.[63] Tietz discusses a further grant made to Annaberg (Saxony) in 1517.[64] The geographical outlier in this first era of donations is the grant of earth and the corresponding privileges and pardons made to Gniezno Cathedral in Poland in 1515, two years after its Archbishop Jan Łaski the Elder had attended the Fifth Lateran Council in Rome.[65] From the later sixteenth century and into the seventeenth, earth from the Campo Santo was taken to a number of sites within the Italian peninsula, Spain, and Poland.[66]

The appeal of these grants seems to have been focused primarily – even exclusively – on Rome. At no point in any of the documents is reference made to the Campo Santo's association with Jerusalem or Akeldama, or indeed to the reputed properties of the earth. Instead, the grants simply state that those buried in the chapels or cemeteries concerned were to have exactly the same rights and pardons as those in the Roman Campo Santo itself. This suggests that the Campo Santo Teutonico had an independent status as a burial place, reinforcing the impression created by the comparison drawn between Nicosia and Rome in the anonymous Iberian pilgrim's account of ca. 1512. Nevertheless, from the point of view of the recipient site the relationship achieved with the Campo Santo was not necessarily an exclusive or dominant one along the lines of the relationship with Jerusalem that characterized the cemeteries discussed above. Rather it could form part of a wider portfolio of benefits and attractions, at least in the case of the burial chapel if not in that of the cemeteries. Arguably the defining feature of the chapel at Hall was the vast collection of relics assembled by Florian von Waldenstein, transferred to the chapel in 1501, and exhibited annually until 1524.[67] These were described by von Waldenstein himself in the *Haller Heiltumbuch*, of ca. 1508–1509, along with the various privileges and indulgences possessed by the chapel.

At the time of Julius II's bull, the chapel had already received a grant of Roman indulgences from Pope Alexander VI, which were available to those visiting it on particular feast days.[68] In comparison, the Campo Santo bull was directed more specifically at the chapel as a family burial place, although it did offer those visiting the chapel the same indulgences as were available to those visiting the Campo Santo.[69] Interestingly, the account of the grant of earth and indulgences given in the Heiltumbuch also contains no reference to Jerusalem. This absence in the bulls themselves might be seen to reflect the priorities of the papal donors. However, that the same is also true of the *Heiltumbuch*, with its interest in the accumulation of relics – including examples from the Holy Land – confirms that, for the recipients too, the appeal of the Campo Santo grant lay with Rome. At the same time, it may speak of more fundamental limitations to the capacity of earth to link places; once removed from Rome, the earth was essentially expressive of its immediate point of origin and could not easily sustain its previous associations with Jerusalem.

Even if the point of the grants was to make certain places as effective for the salvation of the souls of those buried there as the Campo Santo Teutonico, it is still notable that this was achieved through a combination of the papal grant of indulgences and the material transfer of the earth. Since this was an era in which indulgences equivalent to those available in Rome were regularly granted to churches elsewhere, it remains to be asked why these particular grants were effected through the transfer of soil as well.[70] A partial answer can be found in the designation of the soil and its association with earth from the graves of saints. First, the soil is now explicitly referred to as holy, as was not the case in the *Indulgentiae* or the petition of 1476. It is described in the papal bulls for Hall, Kolsaß, and Innsbruck as "*terra sancta*," while a letter from the imperial councilor Jacob Villinger similarly refers to the presence of "*heilig erdtreich von dem gotsackher zu Rom*" in the cemetery of Freiburg im Breisgau.[71] Second, earth from saintly graves could be described similarly and spread on the same spot. The *Haller Heiltumbuch* records that the chapel at Hall was already spread with the dust and ashes of the virgins of Cologne and with holy earth (*heiligen erdrich*) from the grave of Saint Ursula.[72] Together these factors suggest that the Campo Santo soil was understood in the context of the practice of taking dust from the tombs of the saints, that is to say in terms of contact relics. Indeed, Tietz has suggested that in cases where new extramural cemeteries distanced graves from the *communio sanctorum* enjoyed by churchyard burials, the connection with the Campo Santo in Rome served as compensation.[73] While part of the appeal laid in the indulgences, in its materiality the *terra sancta* was certainly closer to these traditional sources of sanctity and salvation.

This is not to suggest that the Roman earth was necessarily understood to relate to saintly burials there. The soil from the Campo Santo may have been associated with Saint Peter through the proximity of the cemetery to the basilica, or even have evoked Rome's wider reputation as the resting place of the martyrs and a site of martyrdom; introducing a description of the city's cemeteries, a German version of the *Mirabilia urbis Romae* noted that the earth of Rome was sacred on account of the blood of the martyrs shed there.[74] Yet this is not made explicit, nor is such a relationship clear when the Campo Santo soil was linked with that from elsewhere in the city. The papal bull for Gniezno mentions earth from the cemeteries of both the Campo Santo and San Gregorio de Urbe, two places often associated in contemporary wills of Germans resident in Rome.[75]Although there seems to have been a chapel dedicated to Saint Gregory at the Campo Santo and the churches of San Gregorio de Palatio and San Gregorio de Cortina were both close to Saint Peter's, where Gregory the Great was buried, it is probable that these references are to the church of San Gregorio on the Celian Hill.[76] In the entry for this church in the German language *Historia et descriptio urbis Romae*, Saint Gregory is said to have been buried there for fifteen years and to have ensured that anyone who joined the Confraternity of Saint Gregory or who was buried in the cemetery there would be protected from eternal damnation.[77] While the idea that Gregory had in fact been buried at the church corresponds to the value placed on saintly remains and resting places, it is arguably the protection secured by the saint in another manner that was of most relevance to those buried there and that the soil from the site would have been instrumental in translating. The Roman soil thus provided a material connection to a privileged burial place in a manner reminiscent of, but not directly replicating, dust from the tombs of saints.

At the same time, the way in which the earth was treated suggests that it played a distinctive part in rendering this connection visible and tangible at the time of the establishment of the cemetery and in defining the extent of the sacred space. It also went beyond the indulgences it accompanied in effecting a translation rather than just an equivalence of place. The bulls grant the recipients sufficient earth "to cover or strew the whole or part of the surface" of the spaces concerned.[78] Indeed, the Austrian Vice Chancellor Lorenz Saurer is said to have requested more earth for the Vienna cemetery than had been provided for Innsbruck, on the grounds that it was far larger than the latter.[79] The formula in the bulls is reminiscent of the description of the earth from Jerusalem "*sparsolla per tucto*" in the Pisan Camposanto.[80] Moreover, the dust of the Cologne virgins and the holy earth from Saint Ursula's grave, said to have been brought from Cologne in a large chest, was also used to mark a particular space within the chapel at Hall, being spread beneath the paving slabs of red marble.[81] This formed a precedent for the use of the Campo Santo earth and a parallel with the paving of the Santa Croce in Gerusalemme chapel. The spreading of the earth may have taken place in public. It was probably not until 1635, toward the beginning of another series of grants of earth, that the confraternity instituted a specific set of procedures for those who wanted a portion of the soil, including a ceremonial handover.[82] However, when the arrival of the soil coincided with the consecration of the cemetery, it is likely that it was integrated into the established ceremony. The rite for the blessing of a cemetery as given in the late thirteenth-century Pontifical of William Durandus, and in the *editio princeps* of the Roman Pontifical of 1485 that was based on it, included sprinkling the ground with holy water while circumambulating the site and then processing between five crosses placed in the center and at the cardinal directions.[83] An eighteenth-century account of the consecration of the cemetery at Annaberg, which draws upon earlier sources and presents the ceremony as based on that at Innsbruck, gives a sense of the part the soil might have played in this ritual.[84] The earth was taken to the cemetery in a chest and then divided into five basins, four of which were taken to the corners of the space by priests and the fifth kept for the consecrating bishop. After the cemetery had been sprinkled with holy water, the priests scattered the earth over the ground in turn, followed by the bishop, in such a way that the whole space was strewn five times. Finally, the papal bull was read out to the assembled congregation. In common with the holy water, the scattered earth was not only sanctifying, but also had a performative value in rendering visible the spoken words of the ceremony and their implications for the definition of space. At the same time, in all these cases scattering earth goes beyond defining the extent of the site to create a stratigraphic relationship between places; as the recipient site is overlaid with earth from elsewhere, it is as if one place is carpeted with another.

If the removal of earth from the Campo Santo Teutonico still corresponds closely to the reception of soil from Akeldama in its capacity to translate place, it was also not unprecedented in Rome. Earth is said to be taken from Rome to cemeteries in Ireland and Scotland in saints' *Lives* dating from the twelfth to the sixteenth centuries, including those of Colmán son of Lúachán, Colmán Élo, Molaise, and Lolan.[85] Although the two phenomena are unlikely to be directly connected, they share some important features. Notably, in the hagiographical literature, the soil spread over the cemetery achieves a similar layering and elision of place: burial in earth from Rome is as good as burial in Rome, and allows those interred there to obtain salvation. However, here the soil is explicitly associated with the cults of Roman saints and brings about a long-distance burial *ad sanctos*. For example, in an account of the life of Colmán son of Lúachán, we read that soil from the graves of the Roman saints was brought to the cemetery of Lann Mic Lúacháin and "scattered in every direction . . . so that it is a burial in the soil of Rome for each one who has been buried there from that [point] onward."[86] Similarly, it is recorded in the *Life* of Colmán Élo that the saint received a gift of "seven sacks full of the soil of Rome" and was instructed to "shake it over the length and breadth of thy cemetery, and anyone who is buried in it shall not see hell."[87] In an early-sixteenth-century *Life* of Saint Molaise, the saint visits Rome and brings back to Devenish "a load of Rome's soil" as well as relics of Roman saints.[88]

These *Lives* illuminate the way in which the soil of the Campo Santo Teutonico could function despite shedding its associations with Jerusalem. On the one hand, they show the role of earth in transportable topography with no connection to the Holy Land; on the other, they reinforce the Central European evidence by providing a comparison for the reception and framing of the Campo Santo soil within a devotional setting that also focused on the graves of saints. Perhaps the closest parallel, however, is found in the *Breviary of Aberdeen*, printed in the early sixteenth century; here, Saint Lolan has four ass-loads of the dust of Saint Peter's cemetery (*pulveris cimiterii beati petri*) sent for the consecration of the cemetery in which he is to be buried, and prays that anyone buried there will receive the same indulgences as if he had been buried in Saint Peter's cemetery.[89] Two separate traditions reach a remarkable convergence here. In illustrating how indulgences were incorporated into a long-standing interest in Roman soil, the *Breviary* shows the double gift of Julius II to be a product of its time.

A final feature of the hagiographies brings us back to the dual nature of the Campo Santo Teutonico as both a destination and a source of earth, as well as to the expressive potential of the material itself. The *Lives* of Colmán son of Lúachán and Colmán Élo describe attempts to steal the soil, confirming its prestige: in the former, Colmán's mother takes a bag of the earth to her brothers' kin, only to have the saint deny its efficacy away from Lann; in the latter, the family of Durrow only succeeds in stealing some of the soil from outside the stone enclosure of the cemetery, but Colmán Élo prays that it might have "the virtue of the earth of Rome from henceforth."[90] This fits within narrative traditions in which theft or the desire for theft could help to construct the sanctity or enhance the status of holy remains.[91] At the same time, it also speaks of the particular relationship between soil and place. In neither of the Irish examples is earth taken from the consecrated cemetery itself. However, both the denial that the Roman earth would be effective beyond its immediate destination and the need to bestow Roman properties on soil taken from outside the cemetery support the hypothesis that the movement of soil between two places did not easily express a relationship with a third.

In conclusion, the phenomenon of moving earth between burial places is found in a number of different sources, including travel narratives, indulgence literature, relic catalogs, hagiographies, and papal bulls. These reflect the perspectives of both donors and recipients of earth, as well as of those who visited the sites concerned. It is important to make the distinction that some of them record the moving of earth, whereas others simply reflect claims that this had taken place. However, their collective significance transcends the actual movement of soil to demonstrate the importance of this material for envisaging and constructing relationships between places more generally in the later medieval West. There has been no intention in this chapter to present the instances discussed as part of a single coherent tradition or to provide a developmental account of the phenomenon. Nevertheless, as seen here in sources dating from the twelfth or thirteenth century to the early sixteenth century, accounts of moving soil are both rooted in a common religious culture and testify to its diversity and changing emphases, expressing interest in different religious centers and reflecting the broad move from a devotional currency focused on saintly remains to one articulated by indulgences.

Ultimately, however, bringing together two networks of material translation focused on Jerusalem and Rome, intersecting at the Campo Santo Teutonico, has emphasized the substance of soil itself rather than that of the Holy Land or any other sacred place. On the one hand, earth could both retain and transmit the properties of particular places, and also embody them more fundamentally. Spread on the ground, it created a stratigraphic elision of place in which one location overlaid the other, and rendered visible and tangible transformations that were also brought about verbally. At the same time, soil was also subject to some expressive limitations. Moving earth from one location to another created a powerful connection between the two places and potentially also an element of likeness between recipient sites, but it could not easily express a reference to a more distant point of origin.

Notes

* This chapter presents initial results of a research project on Earth and the Portability of Place. An earlier version was delivered at a workshop in the AHRC Network, Remembered Places and Invented Traditions: Thinking about the Holy Land in the Late Medieval West. I would like to thank the AHRC and Professor Anthony Bale for the opportunity to participate in the network, and the editors of this volume for their helpful comments on a draft of the chapter. I am also grateful to Professor Stefan Heid and Marjan Rebernik of the Görres-Gesellschaft in Rome for their kind assistance.

1 "eandem capellam sacram auctoritate apostolica in Campum Sanctum et ad instar capelle cemiterii Campi Sancti de Urbe erigi, creari et institui et illud ex pulveribus seu terra sancta cimiterii ipsius Campi Sancti de Urbe accipiendis vel accipienda et ad illud deferendos seu deferendam in superficiem eius aspergi facere": *Cartularium Vetus Campi Sancti Teutonicorum de Urbe / Urkunden zur Geschichte des deutschen Gottesackers bei Sankt Peter in Rom*, ed. Paul M. Baumgarten, Römische Quartalschrift, Supplementheft 16 (Rome: Forzani, 1908), XXXIII, 69–72, esp. 70.

2 For reasons of space, this chapter cannot engage with the place of earth in Jewish burial practice; on this, see Isaiah M. Gafni, *Land, Center and Diaspora: Jewish Constructs in Late Antiquity* (Sheffield: Sheffield Academic Press, 1997), ch. 4.

3 This focus means omitting discussion of other places, such as Glastonbury, from which earth was said to be taken to burial sites elsewhere: James P. Carley, ed., and David Townsend, tr., *The Chronicle of Glastonbury Abbey* (Woodbridge: Boydell, 1985), lviii, 32–35.

4 For this term and attendant ideas, see Robert Ousterhout, "Flexible Geography and Transportable Topography," in *The Real and the Ideal in Jewish, Christian and Islamic Art*, ed. Bianca Kühnel (Jerusalem: Centre for Jewish Art, Hebrew University of Jerusalem, 1998), 393–404.

5 On the early history of the church and cemetery, see Albrecht Weiland, *Der Campo Santo Teutonico in Rom*, vol. 1, *Der Campo Santo Teutonico in Rom und seine Grabdenkmäler*, ed. Erwin Gatz, Römische Quartalschrift, Supplementheft 43 (Rome: Herder, 1988), 37–54; Rudolf Schieffer, "Karl der Große, die *schola Francorum* und die Kirchen der Fremden in Rom," *Römische Quartalschrift für christliche Altertumskunde und Kirchengeschichte* 93, nos. 1/2 (1998): 20–37.

6 Marjatta Wis, "Gottesacker und Campo Santo: Spuren mittelalterlicher Glaubensvorstellungen im deutschen und italiensichen Wortschatz," *Neuphilologische Mitteilungen* 58 (1957): 71–108, esp. 89–91; Weiland, *Der Campo Santo Teutonico*, 38–39.

7 Nine Robijntje Miedema, *Die römischen Kirchen im Spätmittelalter nach den "Indulgentiae ecclesiarum urbis Romae"* (Tübingen: M. Niemeyer, 2001), 365–367, for the Campo Santo Teutonico; Miedema, *Rompilgerführer in Spätmittelalter und Früher Neuzeit: Die "Indulgentiae ecclesiarum urbis Romae" (deutsch/niederländisch). Edition und Kommentar* (Tübingen: M. Niemeyer, 2003), 18–37.

8 Nikolaus Paulus, *Geschichte des Ablasses im Mittelalter*, 2nd ed. (Darmstadt: Wissenschaftliche Buchgesellschaft, 2000), vol. 2, 228–239; Debra Birch, *Pilgrimage to Rome in the Middle Ages: Continuity and Change* (Woodbridge: Boydell, 1998), 179–180; Diana Webb, "Pardons and Pilgrims," in *Promissory Notes on the Treasury of Merits: Indulgences in Late Medieval Europe*, ed. R.N. Swanson (Leiden: Brill, 2006), 241–275.

9 "wie offt ein mensch get in der andakcht (sic) auf den Goczakcher bey sand peters chirichen der hat fumfczehen hundert iar antlas": Leopold of Vienna, *Von der Stat ze Rom*, in Josef Haupt, "Philippi Liber de terra sancta in der deutschen Übersetzung des Augustiner Lesemeister Leupold vom Jahre 1377," *Österreichische Vierteljahrschrift für katholische Theologie* 10 (1871): 511–540; 517–526, esp. 520, for the text on Rome.

10 "ȝif men wuste grete and smale / þe pardoun þat is at grete Rome /þei wolde tellen in heore dome / Hit were no neod to mon in cristiante / To passe in to þe holy lond over þe see / To Jerusalem ne to kateryne": F.J. Furnivall, ed., *The Stacions of Rome* (London: N. Trübner for the Early English Text Society, 1867), 10; cited in Webb, "Pardons and Pilgrims," 261.

11 Matt. 27: 8; "Alcheldemach, hoc est ager sanguinis, in quo sepeliuntur omnes peregrini": Piacenza Pilgrim, *Itinerarium*, 26, ed. P. Geyer in *Itineraria et Alia Geographica* (Turnhout: Brepols, 1965), vol. 1, 143.

12 Weiland, *Campo Santo Teutonico*, 41.

13 The Hague, Koninklijke Bibliotheek, MS 76 E 5, fols 57v–61r; "Aen sente Peters kerke daer steet een kerchof dat heet gods acker d'op greef men die arme pelgrime dat es dat lant daer dir XXX penige om ghegheuen waren daer god mede vercocht was also dicke alse een mensche daer op gheet soe heeft hi xvm iaer aflaets. Die enen pater noster. spreect": N.C. Kist, "De aflaten der zeven kerken van Rome," *Archief voor Kerkelijke Geschiedenis inzonderheid van Nederland* 6 (1835): 303–318, esp. 310–311.

14 Kist, "De aflaten," 306, followed by Miedema, *Die römischen Kirchen*, 108; Miedema, *Rompilgerführer*, 25.

15 For a description of the manuscript, see G.I. Lieftinck, "Beschrijving van het Handschrift," in *Beatrijs: Eerste Integrale Reproductie van het Handschrift*, ed. A.L. Verhofstede (Antwerp: De Vlijt, 1949), 20–31. For the later dating, see Paulus, *Geschichte des Ablasses*, vol. 2, 237–238; Jan van Herwaarden, *Between Saint James and Erasmus: Studies in Late Medieval Religious Life: Devotion and Pilgrimage in the Netherlands* (Leiden: Brill, 2003), 102–103.

16 Miedema, *Die römischen Kirchen*, 367–397; two manuscripts that associate the Campo Santo Teutonico with Akeldama are dated broadly to the fifteenth century; the others date from the mid-fifteenth century onward.

17 "Tunc dictus Fridericus considerans quod ager sanctus et pretio quo Christus venditus idem ager emptus existit et eundem vendere non licet, et cum cetere nationes officium huiusmodi tumulatoris pro pauperibus defunctis gratis exercere voluerunt, extunc supradictus Fridericus pauperes defunctos gratis sepeliri desideravit et fecit": Karl August Fink, "Die Anfänge der Bruderschaft am deutschen Campo Santo in Rom," *Römische Quartalschrift* 44 (1936): 221–230, esp. 225.

18 "compatientes eidem loco in quo, sicut per omnes eiusdem nationis Christifideles pie creditur prout etiam esse dicitur, terra agri illius Alchedemach (sic) existere": Baumgarten, *Cartularium*, 37, doc. 18.

19 Anton de Waal, *Der Campo Santo der Deutschen zu Rom: Geschichte der nationalen Stiftung zum elfhundertjährigen Jubiläum ihrer Gründung durch Karl den Grossen* (Freiburg im Breisgau: Herder, 1896), 18–19, mentions references to Calvary in seventeenth-century sources.

20 "Dominica palmarum celebravi uff dem Gotzacker": Reinhold Röhricht, "Die Jerusalemfahrt des Kanonikus Ulrich Brunner vom Haugstift in Würzburg (1470)," *Zeitschrift des deutschen Palästina-Vereins* 29 (1906): 1–50, esp. 10.

21 Miedema, *Die römischen Kirchen*, 367–397: D3, D15, D20, D27, D36, D44, D45a, D48, D49, D51, D56, D57, D61, D62, D63, D70, D71, D72, D74, D75, D76, N14a/2; Leopold of Vienna, *Von der Stat ze Rom*, 520.

22 R.N. Swanson, "Praying for Pardon: Devotional Indulgences in Late Medieval England," in *Promissory Notes on the Treasury of Merits. Indulgences in Late Medieval Europe*, ed. R. N. Swanson (Leiden: Brill, 2006), 215–240, esp. 223–224; R.N. Swanson, *Indulgences*

in Late Medieval England: Passports to Paradise? (Cambridge: Cambridge University Press, 2007), 251–252; for the reference to Saint Peter's, see Virginia Reinburg, *French Books of Hours: Making an Archive of Prayer, c.1400–1600* (Cambridge: Cambridge University Press, 2012), 204.

23 Miedema, *Die römischen Kirchen*, 367–397: D5, D23, D39, D52, N2/2, N5, N13, N14 (for Pater Noster); MSS D26a, D67/2, N7/2 (for Ave Maria); German *Historia et Descriptio* (for Ave Maria and burial); D46 (for Pater Noster and Ave Maria). The souls are mentioned, for example, in D46, Munich, Staatsbibliothek, MS Cgm 6342, fols 107r-v; "wer den gotsacker amsten umbget und ainen pater noster und ein ave maria sprechet den selbigen seln zu hilf und zu trost der verdiennt taussenndt jare ablass."

24 "Tutti i peregrini che vanno da quella santo campo, si lo circuiscono tre volte, dicendo salmi e paternostri, e orazioni per l'anime di tutti i cristiani, che vi sono sepelliti, e questo si chiama Campo Santo. Ecci perdonanza grandissima": Niccolò da Poggibonsi, *Libro d'Oltramare*, ed. A. Bacchi della Lega (Bologna: G. Romagnoli, 1881–1882), vol. 1, 152; on Holy Land indulgences, see Paulus, *Geschichte des Ablasses*, vol. 2, 241–242.

25 Miedema, *Die römischen Kirchen*, 367–397: D46 (late fifteenth–early sixteenth centuries), D67 (1500), D76 (1448), and German *Historia et descriptio*. For example, "Auch habent vil bäbst darzw geben, wer sich beichtet vnd rw hat vber sein sund vnd die goczrecht enpfacht, vnd stirbt er vnd lat sich graben jn den Gotzacker, dem jst vergeben all sein sund": D76, Miedema, *Rompilgerführer*, 133; "der mensch der darein begraben wirt hat grosse und besundre genad von got": D46, Munich, Bayerische Staatsbibliothek, MS Cgm 6342, fol. 107r.

26 Fink, "Die Anfänge der Bruderschaft," 225.

27 An early testimony is Ludolph of Sudheim in the fourteenth century: *De itinere terre sancte*, ed. G.A. Neumann, in *Archives de l'Orient latin*, ed. Paul Riant, 2 vols (Paris: Ernest Leroux, 1881–1884), vol. 2: pt. 2, 305–377, esp. 355: "Ibi est spelunca profunda testudinata, ubi desuper per foramina proiciuntur cadavera, que mox in triduo sunt consumpta."

28 "vnd hat die gnaud jn jm selbe, daz ain jetlich mensch, daz man dareijn vergrebt, daz es nit mer leit wen biß an den driten tag, so jst es verwesen": Miedema, *Rompilgerführer*, 132–133; for the manuscript (D76) and the tradition more generally see Miedema, *Die römischen Kirchen*, 91–92, 367.

29 "und der acker hait de natúre, so wer daryn begraben wirt, der licham verzert bynnent III dagen bis uff de beyne" (D67/2); Miedema, *Rompilgerführer*, 148–149; Miedema, *Die römischen Kirchen*, 85–87.

30 For examples of earthen relics used in consecrations, see Otfried Ellger, *Die Michaelskirche zu Fulda als Zeugnis der Totensorge: Zur Konzeption einer Friedhofs- und Grabkirche im karolingischen Kloster Fulda* (Fulda: Verlag Parzeller, 1989), 82–86, 238; Matthias Thiel and Odilo Engels, eds., *Die Traditionen, Urkunden und Urbare des Klosters Münchsmünster*, Quellen und Erörterungen zur Bayerischen Geschichte n.s. 20 (Munich: Beck, 1961), 52; Otto Holder-Egger, ed., *Dedicationes Monasterii Lacensis*, Monumenta Germaniae Historica, Scriptores, 15/2 (Hannover: Hahn, 1888), 970.

31 One late example is the relic of the "*gotszacker*" in the *Hallesches Heiltumsbuch* (1520), fol. 26v. "Io recai per mia devozione uno taschetto di quella terra": Mariano da Siena, *Viaggio in Terra Sancta*, ed. Domenico Moreni (Florence: Stamperia Magheri, 1822), 57–58. Further evidence for the removal of soil from Akeldama is reviewed by Neta Bodner in "Earth from Jerusalem in the Pisan Camposanto", in *Between Jerusalem and Europe: Essays in Honour of Bianca Kühnel*, eds. Renana Bartal and Hanna Vorholt (Leiden: Brill, 2015), 74–93. I am most grateful to Neta Bodner for sending me her work prior to publication.

32 For example, D67/2, "Desen acker dede sent Helena ußgraben und dede den van Iherusalem zu Rome foeren": Miedema, *Rompilgerführer*, 148.

33 Miedema, *Die römischen Kirchen*, 194.

34 "Et hec omnia apportaverunt Vespasianus et Titus de Hierusalem cum quatuor columnis ereis plenis de terra sancta di Hierusalem": *Mirabilia Romae vel potius Historia et descriptio urbis Romae* (Rome: Silber, 1491); "Quatro colonne ve press all'altare / di bronzo pene della terra santa / le quale vespasiano vso portare / a Roma": Giovanni Dati, *Tractato de Sancto Ioanni laterano* (Rome: Besicken, 1499). "die mit dem heiligen ertrich und stetten zu Jerusalem und auß dem gotzacker ausgefult sein worden": Nikolaus Muffel, *Der ablas und die heiligen stet zu Rom*, in *Descrizione della città di Roma nel 1452*, ed. and tr. Gerhard

Wiedmann (Bologna: Pàtron Editore, 1999), 36. For the description of Akeldama as the Gottesacker, see Wis, "Gottesacker und Campo Santo," 77–80.

35 Webb, "Pardons and Pilgrims," 269–275.

36 For the development of ideas of the Holy Land, see Robert L. Wilken, *The Land Called Holy: Palestine in Christian History and Thought* (New Haven: Yale University Press, 1992); for relics of places associated with events see Bruno Reudenbach, "Reliquien von Orten: Ein frühchristliches Reliquiar als Gedächtnisort," in *Reliquiare im Mittelalter*, eds. Bruno Reudenbach and Gia Toussaint (Berlin: Akademie Verlag, 2005), 21–41.

37 Some examples are discussed in Lucy Donkin, "Stones of St Michael: Venerating Fragments of Holy Ground in Medieval France and Italy," in *Matter of Faith: An Interdisciplinary Study of Relics and Relic Veneration in the Medieval Period*, eds. James Robinson, Lloyd de Beer, and Anna Harnden (London: British Museum, 2014), 23–31.

38 For further discussion, see Bodner, "Earth from Jerusalem."

39 "Tanatos, cujus terra, quovis gentium portata, serpentes perimit": Honorius Augustodunensis, *De imagine mundi*, 1.31, Migne, *PL*, 172.130b; Philip Freeman, *Ireland and the Classical World* (Austin: University of Texas Press, 2001), 87–88.

40 Marco Polo, *The Description of the World*, 35, eds. and trans. A. C. Moule and Paul Pelliot, 2 vols (London: Routledge, 1938), vol. 1, 119–120, 515; vol. 2: fol. xii; the manuscripts that contain the passage date from the fifteenth century onward.

41 "Auch mag kain Riemer darjn ligen, wen daz ertirch wil kain behalten vnd wil auch nit anders denn pilgrin oder sinst fremd leit": Miedema, *Rompilgerführer*, 133.

42 Concilium Lemovicense II, *Sacrorum Conciliorum Nova et Amplissima Collectio*, 19, ed. Giovanni Mansi (Florence: A. Zatta, 1774), cols. 507–548, esp. 541. For the episode, see Daniel F. Callahan, "The Cult of the Saints in Aquitaine," in *The Peace of God: Social Violence and Religious Response in France around the Year 1000*, eds. Thomas Head and Richard Landes (Ithaca and London: Cornell University Press, 1992), 165–183.

43 On cemetery consecration in the tenth and eleventh centuries, see Barbara H. Rosenwein, *Negotiating Space: Power, Restraint, and Privileges of Immunity in Early Medieval Europe* (Manchester: Manchester University Press, 1999), 178–182; Élizabeth Zadora-Rio, "Lieux d'inhumation et espaces consacrés: Le voyage du Pape Urbain II en France (Août 1095 – Août 1096)," in *Lieux sacrés, lieux de culte, sanctuaires: Approches terminologiques, méthodologiques, historiques et monographiques*, ed. André Vauchez (Rome: École française de Rome, 2000), 197–213; Cécile Treffort, "Consécration de cimetière et contrôle épiscopal des lieux d'inhumation au Xe Siècle," in *Le sacré et son inscription dans l'espace à Byzance et en Occident*, ed. Michel Kaplan (Paris: Publications de la Sorbonne, 2001), 285–299.

44 "Romanos enim respuens, sola peregrinorum corpora ad sepulturam admittit": Christiaan van Adrichem, *Theatrum Terrae Sanctae et Biblicarum Historiarum* (Cologne: Birkmann, 1590), 173.

45 Ernest Langlois, ed., *Les registres de Nicolas IV* (Paris: Thorin, 1886–1893), 677, no. 4712; Georges Digard, Maurice Faucon, Antoine Thomas, and Robert Fawtier, eds., *Les registres de Boniface VIII* (Paris: Boccard, 1884–1935), cols. 922–923, no. 3875; Webb, "Pardons and Pilgrims," 243, 248.

46 On the development of the association, see Sible de Blaauw, "Jerusalem and the Cult of the Cross," in *Pratum Romanum: Richard Krautheimer zum 100: Geburtstag*, ed. Renate L. Colella (Wiesbaden: L. Reichert, 1997), 55–73; on the soil see Michael Friedrich, "*Tradition – Imagination – Legitimation: Untersuchungen zur Visualisierung lokaler Sonderformen allgemeiner Heiligentraditionen am Beispiel der Hl: Helena: Rom Trier Köln Bonn Xanten*" (PhD dissertation, University of Trier, 2000), 21–22, 41–53.

47 "é toda esta yglesia, ansí el suelo como las paredes é toda la obra, fué fecha de la tierra de Ierusalem trayda por laste en los navíos, quando Santa Elena envió las santas reliquias de Ierusalem á Roma": Pero Tafur, *Andanças e Viajes de Pero Tafur por Diversas Partes del Mundo Avidos*, ed. Marcos Jiménez de la Espada (Madrid: M. Ginesta, 1874), 32. I owe this reference to John Lansdowne.

48 "terraque sancti montis Calvariae navi inde advecta supra quam Christi sanguis effusus fuit redemptionis humanae praecium [cuiusque vigore in celestem Hierusalem mortalibus aditus patuit] ad primum usque inferiorem fornicem repleverit ex quo sacellum ipsum et tota

basilica ac universa urbs secunda Hierusalem meruit appelari": Ilaria Toesca, "A Majolica Inscription in S. Croce in Gerusalemme," in *Essays in the History of Art Presented to Rudolf Wittkower on his Fifty Sixth Birthday*, eds. Douglas Fraser, Howard Hibbard, and Milton J. Lewine (London: Phaidon, 1967), 102–105, missing text in brackets.

49 "Hic tellus samctae [sic] Calvarie solime ab beata Helena in inferiorem fornicem demissa servata est atq[ue] inde nomen Hierusalem capelle indita": Friedrich, "Tradition," 41, citing Raimondo Besozzi, *La Storia della Basilica di Santa Croce in Gerusalemme* (Rome: G. Salomoni, 1750), 85–86, for the location in the pavement.

50 Cynthia Payne, "'*In the Fullness of Time': The Vault Mosaic in the Cappella Sant'Elena, Santa Croce in Gerusalemme*" (PhD dissertation, University of Georgia, 2003), 10, note 30; Friedrich, "Tradition," 40–41.

51 See, e.g., James P. Carley and Martin Howley, "Relics at Glastonbury in the Fourteenth Century: An Annotated Edition of British Library, Cotton Titus D.vii, fols 2r–13v," *Arthurian Literature* 16 (1998): 83–129, esp. 94–95, 107, where relics of Calvary are mentioned at least four times. For the belief that the Lateran columns contained soil from Calvary, see Jack Freiburg, *The Lateran in 1600: Christian Concord in Counter-Reformation Rome* (Cambridge: Cambridge University Press, 1995), 134.

52 Mauro Ronzani, "Dal 'cimitero della chiesa maggiore di Santa Maria' al Camposanto: aspetti giuridici e istituzionali," in *Il Camposanto di Pisa*, eds. Clara Baracchini and Enrico Castelnuovo (Turin: Einaudi, 1996), 49–56, esp. 52; Roberto Paolo Ciardi, "'Quest'insigne dormentorio de' morti': chiesa, cimitero, museo," *Il Camposanto*, 57–68, esp. 57; Bodner "Earth from Jerusalem."

53 Raniero Sardo, *Cronaca di Pisa*, ed. Ottavio Banti (Rome: Istituto Storico Italiano per il Medio Evo, 1963), 37.

54 Mauro Ronzani, *Un'idea trecentesca di cimitero: La costruzione e l'uso del Camposanto nella Pisa del secolo XIV* (Pisa: PLUS, 2005), 91–93.

55 "é una claustra quel suelo della es de la tierra de aquel Campo Santo que fué comprado por los treynta dineros, que es en Ierusalem, é ansí le llaman aquí en Pisa el Campo Santo, é dizen que persona que entierren allí non tura más de treynta dias, que la tierra lo gasta": Tafur, *Andanças e Viajes*, 295.

56 "Dasselb etrich hat man ain tail gen Rom gefürt, der noch ze Rom der gotzaker haist, ain tail in Tschipre": Monika Reininger, ed., *Ulrich Lemans Reisen: Erfahungen eines Kaufmanns aus St. Gallen vom Ende des 15. Jahrhunderts im Mittelmeer und in der Provence* (Würzburg: Königshausen & Neumann, 2007), 99.

57 Reininger, *Ulrich Lemans Reisen*, 101–102 for Nicosia "Vnder demselben kilchhoff oder schopff ist das halig ertrich, das koft ward vmb die XXX pfening, durch die Jhesus Cristus durch Judas verkoft ward, das hat man von Jerusalem dahin gefürt vnd gen Rom, haist campo santo, hatt die kraft, daz die tötten körpell in drÿ tagen verwesen sind vnd zů äschen worden, die man darin vergrebt, wie ich vor von gesait hân"; 157 for Rome "vff dem gotzaker, da man die bilgri vergrapt, da sind all tag XV hundart jar aplass".

58 "Yten ay en esta iglesia de Nicoxía un campo sancto que a tercero día come & gasta la tierra de los cuerpos de los diffunctos que allí son enterrados ansí como lo haze el campo sancto que está en Roma detrás de la iglesia de Sant Pedro": Joseph R. Jones, ed., *Viajeros Españoles a Tierra Santa (siglos XVI y XVII)* (Madrid: Miraguano, 1998), 141. I owe this reference and the following one to Dr. Michalis Olympios.

59 "Hay allí un lugar donde muchas embarcaciones depositaron polvo de la Tierra de Israel y allí acostumbraran a enterrar a los príncipes en tiempos antiguos, y le llaman Campo Santo, siendo eso indicio de que todos reconocen la santidad de la Tierra de Israel": José R. M. Nom de Déu, tr., *Relatos de viajes y epistolas de peregrinos judíos a Jerusalén (1481–1523)* (Sabadell: Editorial AUSA, 1987), 174.

60 Weiland, *Campo Santo Teutonico*, 43–45; for the Austrian examples, see Nikolaus Grass, "Camposanto-Teutonico-Privilegien für Österreich: Ein Beitrag zur Sakralkultur im Zeitalter Kaiser Maximillians I", in *Domus Austriae. Festgabe H. Wiesflecker zum 70: Geburtstag*, eds. W. Höflechner, H. J. Mezler, and O. Pickl (Graz: Akademische Druck- und Verlagsanstalt, 1983), 137–158. Austrian and German examples are discussed in Anja A. Tietz, *Der frühneuzeitliche Gottesacker: Entstehung und Entwicklung unter besonderer Berücksichtigung des Architekturtypus Camposanto in Mitteldeutschland* (Halle: Landesamt für

Denkmalpflege und Archäologie Sachsen-Anhalt, Landesmuseum für Vorgeschichte, 2012), esp. 31, 34–38.

61 Baumgarten, *Chartularium*, XXXIII, 69–72.

62 Baumgarten, *Chartularium*, XXXIV, 72–74 for Kolsaß; XXXV, 74–77 for Innsbruck. On the cemetery in Vienna, see Grass, "Camposanto-Teutonico-Privilegien," 146–147.

63 Tietz, *Der frühneuzeitliche Gottesacker*, 35; part of the text of the bull is reproduced there from Joseph A. S. von Riegger, *Analecta academiae Friburgensis ad historiam et jurisprudentiam praecipue ecclesiasticam illustrandam* (Ulm: apud Aug. Lebrecht. Stettinium, 1774), 91–95.

64 Tietz, *Der frühneuzeitliche Gottesacker*, 35–36.

65 Baumgarten, *Chartularium*, XXXVII, 79–81.

66 Aloys Schmidt, *Das Archiv des Campo Santo Teutonico nebst geschichtlicher Einleitung*, Römische Quartalschrift Supplementheft 31 (Rome: Herder, 1967), 118–125.

67 Josef Garber, "Das Haller Heiltumbuch mit den Unika-Holzschnitten Hans Burkmairs des Älteren," *Jahrbuch der kunsthistorischen Sammlungen des Kaiserhauses* 32, no. 6 (1915): XX, XXIV.

68 Garber, "Das Haller Heiltumbuch," XCV – XCVII.

69 Baumgarten, *Cartularium*, XXXIII, 71; Garber, "Das Haller Heiltumbuch," C.

70 For the granting of indulgences equivalent to places in Rome, see Webb, "Pardons and Pilgrims," 256.

71 Tietz, *Der frühneuzeitliche Gottesacker*, 31.

72 "allenthalben unter den roten merblein phlasterstainen ganz uber und uber bestreut, bedeckt, geeeret, gewirdigt und geheiligt mit vil gepainen von den lieben heiligen auch staub und aschen von der geselschaft der aindliftausent maide und von andern vil lieben heiligen und darzu auch mit dem heiligen erdrich von sand Vrsulen grab, des der stifter ain grosse truhen voll aus der heiligen stat Cöln gen Hall gebracht hat": Garber, "Das Haller Heiltumbuch," LXXV, CLXXII; cited in Grass, "Camposanto-Teutonico-Privilegien," 149, n. 17.

73 Tietz, *Der frühneuzeitliche Gottesacker*, 30–31.

74 "Auch is die erde zu Rom heylig, wenn sie mit der heiligen plute verspenget vnd pegossen ist uberall": Nine Miedema, *Die "Mirabilia Romae": Untersuchungen zu ihrer Überlieferung mit Edition der deutschen und niederländischen Texte*, Münchener Texte und Untersuchungen zur deutschen Literatur des Mittelalters 108 (Tübingen: M. Niemeyer, 1996), 345.

75 "terra seu superficie terre cimiteriorum Campi Sancti et Sancti Gregorii de Urbe recipere": Baumgarten, *Cartularium*, XXXVII, 79–81, esp. 80–81. Knut Schulz, "Die Anfänge der Bruderschaft des Campo Santo Teutonico," *Römische Quartalschrift* 93 (1998): 38–61, esp. 52; Christiane Schuchard, "Vier Testamente für die römische Anima-Bruderschaft (1524/1527)," in *Kurie und Region: Festschrift für Brigide Schwarz zum 65: Geburtstag*, eds. Brigitte Flug, Michael Matheus, and Andreas Rehberg (Stuttgart: Steiner, 2005), 307–324, esp. 323.

76 Schulz, "Die Anfänge der Bruderschaft," 49–53. On San Gregorio in Palatio and San Gregorio in Cortina, see C. Huelsen, *Le chiese di Roma nel medio evo: cataloghi e appunti* (Florence: Olschki, 1927), 257, 259. I am grateful to the anonymous reviewer of this chapter for the reference to San Gregorio in Palatio. Generally but not universally thought to have been located to the south of Saint Peter's, it was identified by Anton de Waal with the church of the same dedication at the Campo Santo: de Waal, "San Gregorio in Palatio," *Römische Quartalschrift* 18 (1904): 35–39.

77 "Daselbst hat sanctus Gregorius fierzehen iar gepuest in der grufft . . . Sanctus Gregorius hat erworben vmb Got, das er alle die behuet vor ewiger verdampnis, die sich zu sant Gregorio in die brudershaft schriben ader dohin begraben lassen": Miedema, *Rompilgerführer*, 266.

78 Baumgarten, *Chartularium*, XXXIII, 71, "ex terra seu superficie terre sancte cimiterii ipsius Campi Sancti de Urbe tantum recipere . . . quod ex huiusmodi terra deferenda superficies ipsius sacre capelle tota vel in parte cooperiri seu aspergi possit"; XXXV, 76, "ut ex terra seu superficie terre sancte cimiterii ipsius Campi Sancti de Urbe tantum recipere . . . quod ex huiusmodi terra deferenda superficies ipsius cimiterii tota vel in parte cooperiri seu spargi possit."

79 Hermann Wiesflecker, *Kaiser Maximilian I: Das Reich, Österreich und Europa an der Wende zur Neuzeit*, 5, *Der Kaiser und seine Umwelt: Hof, Staat, Wirtschaft, Gesellschaft und Kultur* (Vienna: Verlag für Geschichte und Politik, 1986), 268.

80 Raniero Sardo, *Cronaca di Pisa*, ed. Banti, 37.
81 Garber, "Das Haller Heiltumbuch," LXXV, CLXXII.
82 Schmidt, *Das Archiv*, 120.
83 Michel Andrieu, *Le Pontifical Romain au Moyen-Age* (Vatican City: Biblioteca Apostolica Vaticana, 1960), vol. 3, 506–508; *Pontificalis Liber*, eds. Agostino Patrizi Piccolomini and Johann Burchard (Rome: Planck, 1485), consulted as Rome, Biblioteca Apostolica Vaticana, Inc.II.252, fols 194v–202r: Digital facsimile: http://digi.vatlib.it/view/Inc.II.252.
84 Adam Richter, *Umständliche aus zuverläßigen Nachrichten zusammengetragene Chronica der im Meissnischen Ober-Ertz-Gebürge gelegenen Königl: Churfl: Sächßischen freyen Berg-Stadt St. Annaberg*, 2 vols (St. Annaberg: Friese, 1746–1748), vol. 1, 239–247; Moritz Spiess, *Der Gottesacker zu Annaberg* (Annaberg, 1860), vol. 1, 130–131.
85 Recently, some of these examples have been discussed in Tomás Ó Carragáin, *Churches in Early Medieval Ireland: Architecture, Ritual and Memory* (New Haven and London: Yale University Press, 2010), 83, 190.
86 Kuno Meyer, ed., *Betha Colmáin maic Lúacháin: Life of Colmán Son of Lúachan*, 76–82 (Dublin and London: Hodges, Figgis; Williams & Norgate, 1911), 79–85, esp. 82, 85. For the dating of the life to the early twelfth century, see Richard Sharpe, *Medieval Irish Saints' Lives: An Introduction to Vitae Sanctorum Hiberniae* (Oxford: Clarendon Press, 1991), 26; James F. Kenney, *The Sources for the Early History of Ireland: Ecclesiastical* (Dublin: Pádraic Ó Táilliúir, 1979), 454–455. For a later date, see Pádraig Ó Riain, *A Dictionary of Irish Saints* (Dublin: Four Courts Press, 2011), 197.
87 Charles Plummer, ed., *Bethada Náem nÉrenn: Lives of Irish Saints*, 2nd ed. (Oxford: Clarendon Press, 1968), vol. 1, 173; vol. 2, 166–167; for the life as a later medieval compilation of legends, see Kenney, *Sources*, 399–400.
88 Standish H. O'Grady, ed., *Silva Gadelica: A Collection of Tales in Irish with Extracts Illustrating Persons and Places* (London: Williams and Norgate, 1892), vol. 1, 32 (Irish text), vol. 2, 29–30 (translation); for the dating, see Raymond Gillespie, "Saints and Manuscripts in Sixteenth-Century Breifne," *Breifne* 11, no. 44 (2008): 533–558, esp. 539–541. I am most grateful to Professor Gillespie for this reference and for sending me a copy of the article.
89 "tantas et quantas indulgentias a deo consequeretur ac si in cimiterio beatus petri sepultus fuisset": William Blew, ed., *Breviarium Aberdonense* (London: J. Toovey, 1854), pars estiva, pt. 3, fol. cxiiiir. Examining the *lectiones* of saints from the Strathclyde area, Alan MacQuarrie suggests that these draw on *vitae* from the eleventh and twelfth centuries: Alan MacQuarrie, "Lives of Scottish Saints in the Aberdeen Breviary: Some Problems of Sources for Strathclyde Saints," *Scottish Church History Society* 26 (1996): 31–54.
90 Plummer, *Bethada Náem nÉrenn*, vol. 2, 167.
91 On this tradition, see Patrick J. Geary, *Furta Sacra: Thefts of Relics in the Central Middle Ages*, rev. ed. (Princeton: Princeton University Press, 1990).

8 Materiality and liminality

Nonmimetic evocations of Jerusalem along the Venetian sea routes to the Holy Land*

Michele Bacci

Imitating buildings vs. evoking sites

The debate among art historians over Jerusalem's impact on medieval Christianity has often emphasized the role played by the celebrated buildings marking the most worshipped loca sancta as architectural models. In Richard Krautheimer's pioneering study, the different (and often ambiguous) ways in which the Anastasis Rotunda was replicated in Romanesque or Gothic buildings were viewed as exemplary case studies enabling iconographic and iconological analyses of medieval architecture.[1] Since then, many new data have been collected and several new studies have dealt with the monumental replicas of the Jerusalem holy sites between the ninth and the fifteenth centuries. Scholars have become more and more aware that the motivations underlying the construction of *ad instar* churches might have been multifarious and dependent on many different factors, such as an association with particular institutions connected with the Holy Land and the Crusaders, the varying conditions of pilgrimage, the use of a building as a burial site for individual donors or as visual or spatial support for the performance of collective or private devotions, or the wish to efficaciously evoke the heavenly Jerusalem by hinting at its terrestrial double.[2] Less clearly discernible before the late Middle Ages and the invention of *sacri monti* is the purpose of a *translatio Hierusalem*, that is, creation of a surrogate goal for pilgrims, not only replicating the forms of buildings and their settings in the urban or natural landscape, but also reproducing the sanctity attributed to the holy sites and their memorial qualities.[3]

The more-or-less accurate imitation of the outward appearance of the loca sancta proved efficacious inasmuch as buildings could be treated as icons, thus enabling their proliferation and association with various kinds of cultic phenomena.[4] However, it must be stressed that structures were not themselves regarded as holy and their mimetic reproduction did not always represent the best way to evoke and "translocate" the holiness attributed to Jerusalem, as I argue in this chapter. Indeed, architecture, decorations, and furnishings contributed in large measure toward shaping the auratic power attributed to Palestinian shrines. The latter stood out for their specific nature: they were mostly committed to fostering worship for portions of hallowed ground associated with the commemoration of relevant moments in biblical history. The cultic focus corresponded to the site itself: unlike relics or miraculous images, it was deemed to be grafted directly onto the soil. Buildings worked essentially as monumental frames providing the sites with a spatial arrangement, but their presence was not perceived as a decisive factor in conveying sanctity. Holiness was thought to be indissolubly bound to Palestine's ground, irrespective of its being included within a building.

Places vs. spaces: the architectural framing of site-bound holiness in Palestine

This notion was well known even among the opponents of Christianity. When Saladin conquered Jerusalem in 1197, some of his advisers tried to convince him that it was necessary to completely destroy the Holy Sepulchre, as that would discourage the Franks from laying claims to the Holy City. However, Saladin was not persuaded, and other wise men contended that razing the church would be ineffective, as "what they [the Christians] adore is the sanctity of Jerusalem of which the Refuse [i.e., the Holy Sepulchre] is only the noblest place."[5]

In many cases, the architectural mise-en-scène contributed toward overemphasizing this tension between monumental frame and holy site. The rock of Golgotha within the Holy Sepulchre was hidden in the Chapel of Calvary, a monumental parallelepiped that served as a surrogate mountain and a simulacrum of verticality, symbolically visualizing its sacred content.[6] Other contexts were characterized by an axial arrangement of ritual spaces and sites of worship, where the former's luxuriousness starkly contrasted with the latter's deliberately unpretentious or even bare appearance. This feature was especially remarkable in the Nativity Church in Bethlehem, where pilgrims tended to consider its wonderful decor as negligible as compared with the spirituality engendered by the tiny underground grotto where Christ was born. For example, the Italian friar Anthony of Cremona, who visited Bethlehem in 1327, wrote:

> In Bethlehem there is a church, known as St. Mary, in the very place where Christ was born. It is so beautiful that I cannot say I ever saw any other so good-looking, rich and big church, and so much decorated with columns and images as this one in Bethlehem, which is worship-worthy for the whole world. It would take too long to describe seriously and in detail its size, width, and length, its decorum made of different marbles and stones, the sequence of wonderful and multiple columns, its multifarious paintings, its shape and various embellishments, its floor made out of marvelous marble incrustations, and its roof covered with lead. Therefore let us forget such worldly matters (*temporalia*), and let us report only on the spiritual things (*spiritualia*) that are in the church.[7]

The distinction is explicit. Architecture and monumental ornaments belonged to the category of *temporalia*: they could contribute toward enhancing the spiritual meaning of the site, but worship was due only to the sections of ground (the place of Christ's birth, the manger, the star's well, and so on) that were deemed to be imbued not only with memorial qualities, but also with supernatural power. Indeed, the ruined appearance of many Palestinian shrines in the Mamluke period must have contributed toward making pilgrims growingly aware that sites associated with biblical events should be considered holy even if their architectural frame proved to be unappealing, decaying, or even entirely absent. The contrast between the poor state of then-contemporary Jerusalem and its spiritual worthiness was remarked, for example, by the Italian pilgrim Jacopo da Verona in 1335 in clear terms:

> Concerning the appearance of Jerusalem, that is, its ancient buildings, walls, fortifications, and gates, you have to know that its description would take too long and would be of no interest. The town has been altered and mostly destroyed, so

> I went around several times and could hardly see any gate or old building anywhere. Therefore I will limit myself to mentioning those sites that are remarkable and worship-worthy because of their connection with both the Old and New Testament.[8]

The somewhat disappointing appearance of the town and its buildings did not prevent pilgrims from acknowledging its unparalleled site-specific holiness. On the contrary, with the settling of the Minor friars from 1334 onward and the slightly later creation of the Custodia di Terra Santa, there were many efforts to reinvigorate pilgrimage and to foster worship for a number of new holy places: instrumental in this context were the reconstruction of Christ's stations along the Way of the Cross[9] and the attribution of precise indulgences to each site both inside and outside the walls of the Holy City – even if the latter proved to be simply bare rocks on the city's sacred soil or an otherwise anonymous stone embedded in a street wall, as can still be observed today (Figure 8.1).[10]

The friars insisted that the religious importance of such scanty places was appreciably deeper than any other cult object, given that the earth itself, which had been so frequently touched by Christ's incarnated body, had been sanctified. This concept was explained in the following terms by Francesco Suriano, a Franciscan writing in the second half of the fifteenth century, to the addressee of his *Treatise on the Holy Land*, an Italian nun:

> Those things which touched Christ only for a while were strongly imbued with His virtue and grace. Indeed the more one approached Him, the more he was sanctified. And whoever got very near to Him, he received much more grace. Yet this blessed earth was touched by Him at the highest degree, and consequently it is completely filled with divine virtues and turned into a most holy abode. . . . Because of that the [Holy Land's] trees, woods, vegetables, grass, bread, water, stones, and everything else there is holy, and filled with virtue.[11]

In other words, by a kind of transitive act, everything being produced by that Holy Land was in itself imbued with extraordinary qualities and divine virtue.

A self-evident witness to this principle was the fruit of the so-called *Musi* tree, our banana, which was considered to be the paradisiac fruit of the Tree of Knowledge of Good and Evil. Pilgrims were astonished at seeing that its pulp bore the imprint of Christ's crucified body,[12] as though it were a "natural" acheiropoieton. All this was enough to boast that no other pilgrimage goal could be compared to Palestine, given that this geographical entity itself was to be considered a kind of Christological contact relic, which even enabled a direct experience of the incarnated one's salvific action. Its holiness was so firmly grafted in the ground that there was almost no need to enhance it by means of resplendent architectural frames or to manifest it by way of a synecdoche through the mise-en-scène of an especially attractive holy object. It was almost automatically perceived as the cultic and visual focus for worshippers and could eventually be turned into a movable relic – when one was so impudent as those pilgrims who were clever enough to bring a chisel in their bags and, having no scruples regarding the friars' strictures, managed to remove some fragments out of the rock of Calvary or the walls of the old churches.[13] This practice was so common that metal grates, such as those placed around the Rock of the *Templum Domini* (the Dome of

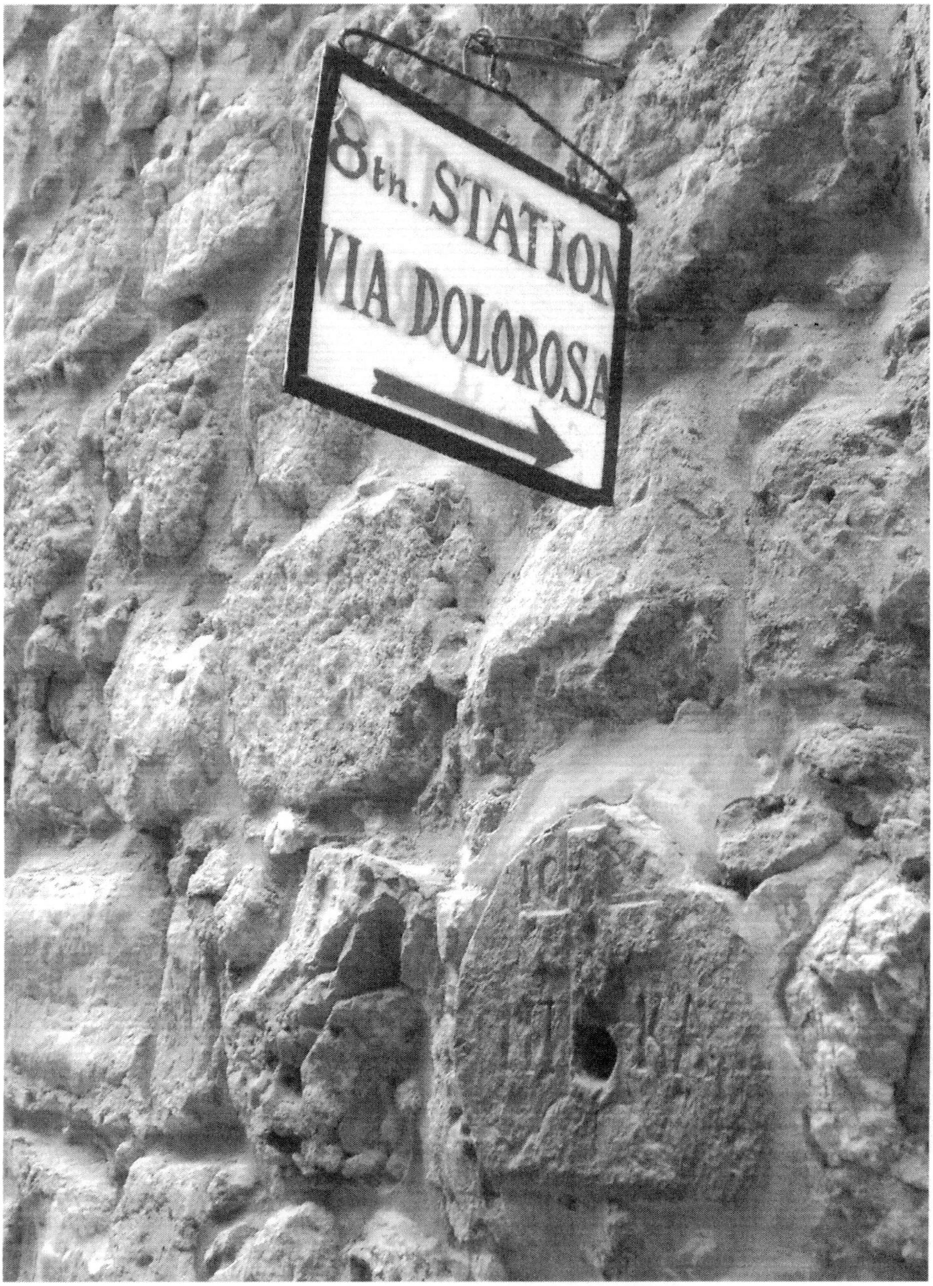

Figure 8.1 Jerusalem, wall of the Greek Monastery of Saint Charalampos, identified in the nineteenth century as the eighth station (*The Women of Jerusalem*) on the Way of the Cross. Photo credit: Michele Bacci.

the Rock) in Crusader times, were installed so as to prevent visitors from chiseling Jerusalem's holy rocks away.[14]

The absence of relics in the Holy Land and their spread along the maritime routes of pilgrimage

Nonetheless, some pilgrims to Mamluke Palestine were not thoroughly at ease when observing that the most celebrated places in Christendom were completely deprived of the relics, images, and other sacred "attractions" that in Western Europe they were accustomed to associate with major shrines or with the treasures of town cathedrals. In 1458, the Englishman William Wey, in his effort to define a kind of "geography" of Christian relics, found it difficult to include the Holy Land and was forced to just mention the objects that were known to have been transferred from there to Constantinople or to Rome and hinted only, in a somewhat contradictory way, at the stones imbued with the Virgin's milk in the homonymous grotto in Bethlehem, which were also segments of the sanctified earth of Palestine.[15]

In order to find authentic relics it proved necessary to leave the Holy Land and cross the sea from Jaffa, Beirut, or Alexandria toward Venice that, from the fourteenth through the sixteenth century, held a sort of monopoly over the navigation routes between Western Europe and the holy sites. According to Wey, elements of an articulated network of shrines that were absolutely worthy of worship by pious pilgrims corresponded to stops along these routes:

> First, twenty miles from the port of Salines [Larnaka, Cyprus] there is the cross of the Good Thief, which is said to hang within a chapel although it is not fixed to any support. Then in Rhodes there is a thorn of Christ's crown, which flourishes for one hour on Holy Fridays, at the very moment when the Passion of Christ is being read: there is also the virgin and martyr Catherine of Alexandria's left arm, as well as a cross made out of the basin which Christ used to wash his disciples' feet. There is also one of the thirty coins, by which Christ was sold by the betrayer Judas. Further on in Candia is preserved the head of Saint Titus, disciple of Saint Paul. Then in Casopoli [Kassiopi, Corfu] is a lamp hanging in front of [the image] of Our Lady, which is filled with oil only once in a year and nonetheless keeps burning with that same oil during the whole year. Then in Ragusa is preserved the arm of Saint Blasius. Then in Zadar in Dalmatia is the whole body of the fair Symeon, who held Christ in his arms; then the bodies of Saint Servolus the Confessor and Saint Anastasia the martyr.[16]

Such sacred attractions enriched the major ports of the Eastern Mediterranean and together constituted a sacred topography that was perceived as an important religious experience, being preliminary and in some way introductory to that of the Palestinian holy sites. Extraordinary mementoes marked all the Latin, mostly Venetian-ruled outposts in the Aegean and the Levant, and they formed a kind of symbolic front against the Turkish enemy. The sites, being invested with such new religious meaning, corresponded to the locations of safe landing places along the basic sea route connecting Venice with the Mamluke Empire via Corfu, Modon, Crete, Rhodes, and Cyprus.[17]

In each of the ports at which they happened to stop, their wish to appropriate the holiness of their final goal stimulated them to search for holy signs that proved to anticipate, even if only partially, the extraordinary experience that they expected to undergo in Jerusalem. In order to accommodate such needs, existing cult phenomena were invested with thoroughly new meanings, and new shrines were set up and provided with a specific legendary pedigree.

Alternative Holy Lands and extensions of the loca sancta network

Admittedly, creating holy places that would retain the flavor of the Holy Land outside of the Holy Land itself proved to be a stressful undertaking, and there were various strategies for achieving this outcome. The following were the main approaches. The most common (and most traditional) was the construction of an alternative Holy Land, boasting a sacred meaning analogous to or even more important than the Palestinian holy sites. Such were the major Christian sites of Rome, Santiago de Compostela, or, on the opposite side of the Christian world, the holy city of Echmiadzin, the site of the first evangelization of Armenia and the resting place of the most important Armenian saints, including Hripsime and Gayane. The Armenians consider the town so holy that it is constantly evoked and visualized even within their own holy places in Jerusalem, as if the holiness of Echmiadzin was transplanted into the holy city itself.[18]

Some of the countries neighboring Palestine could aspire to being considered "extensions" of the Holy Land's holiness, predominantly Egypt, whose soil had admittedly been trod on by Christ's feet. Consequently, the stopping places of the Holy Family along the Nile Valley, such as the garden in Matariya or the crypt of Abu Sargha in Old Cairo, were and are regarded as important mementoes of Jesus' life. The Matariya garden was used by Mary to wash the baby and his garments and it was there that a supernatural balsam started pouring from a nearby tree, which is still worshipped as a cult object. In Abu Sargha, the main altar was erected over an underground cave that was said to have been the first dwelling place of the Holy Family in Egypt (Figure 8.2).[19]

However, if the holiness of the land resided in its memorial meaning, one could also allege that the places visited by the apostles during their frequent voyages around the Mediterranean constituted an "extension" of the Holy Land. The memorial sites of Paul's landing on Crete – a grotto close to Kaloí Liménes and a small church in the deserted village of Agia Roumeli – were "hot spots" within this sacred topography, but their location on the island's southern coast, far from the major navigation routes, probably precluded their becoming prominent pilgrimage sites.[20] On the other hand, from the late fifteenth century onward, when Venetian ships started landing at the small port of Paphos, in the southwestern part of Cyprus, pilgrims found interest in the ruins of some ancient baths that were thought to be the prison where the apostle and his pupil Barnabas had been forced to stay for a while.[21]

Analogous memorial sites were established as far afield as Italy: by the thirteenth century at the very latest, the Church of San Piero a Grado near Pisa was said to mark the very place where Saint Peter had disembarked onto Italian soil and where he had erected the first Christian altar. In this case, the cult phenomenon was designed to rival Rome's claims to primacy and to emphasize Pisa's leading role as an important pilgrimage site. The church was conceived as a splendid architectural frame around a small column, marking the spot where Saint Peter had built his altar (Figure 8.3).

The latter's marble table was worshipped as a relic, as believers saw drops of blood that fell from Pope Clement I's nose at the moment that he performed the Mass there.

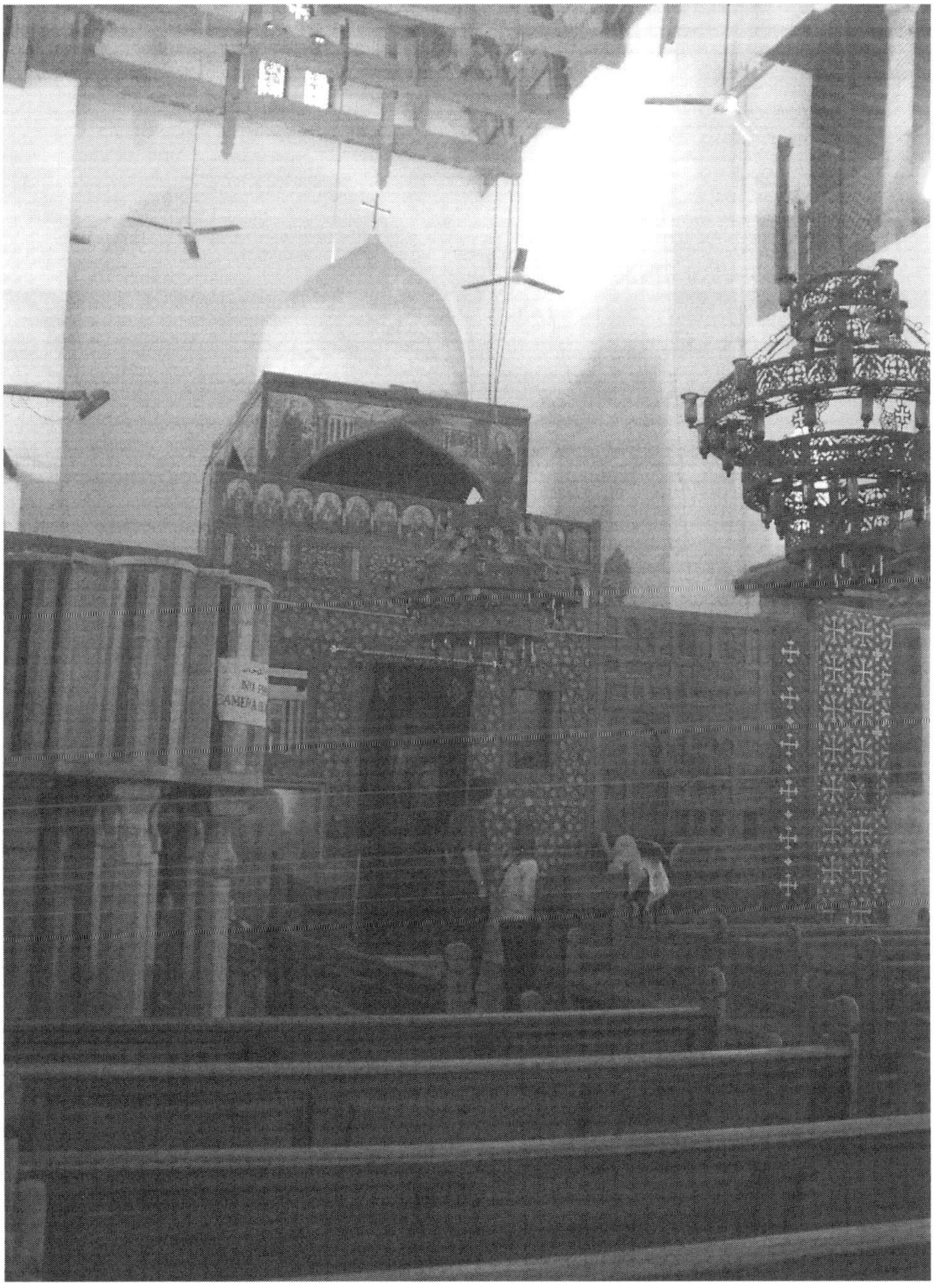

Figure 8.2 Old Cairo, Abu Sargha, view of the altar marking the first dwelling place of the Holy Family in Egypt. Photo credit: Michele Bacci.

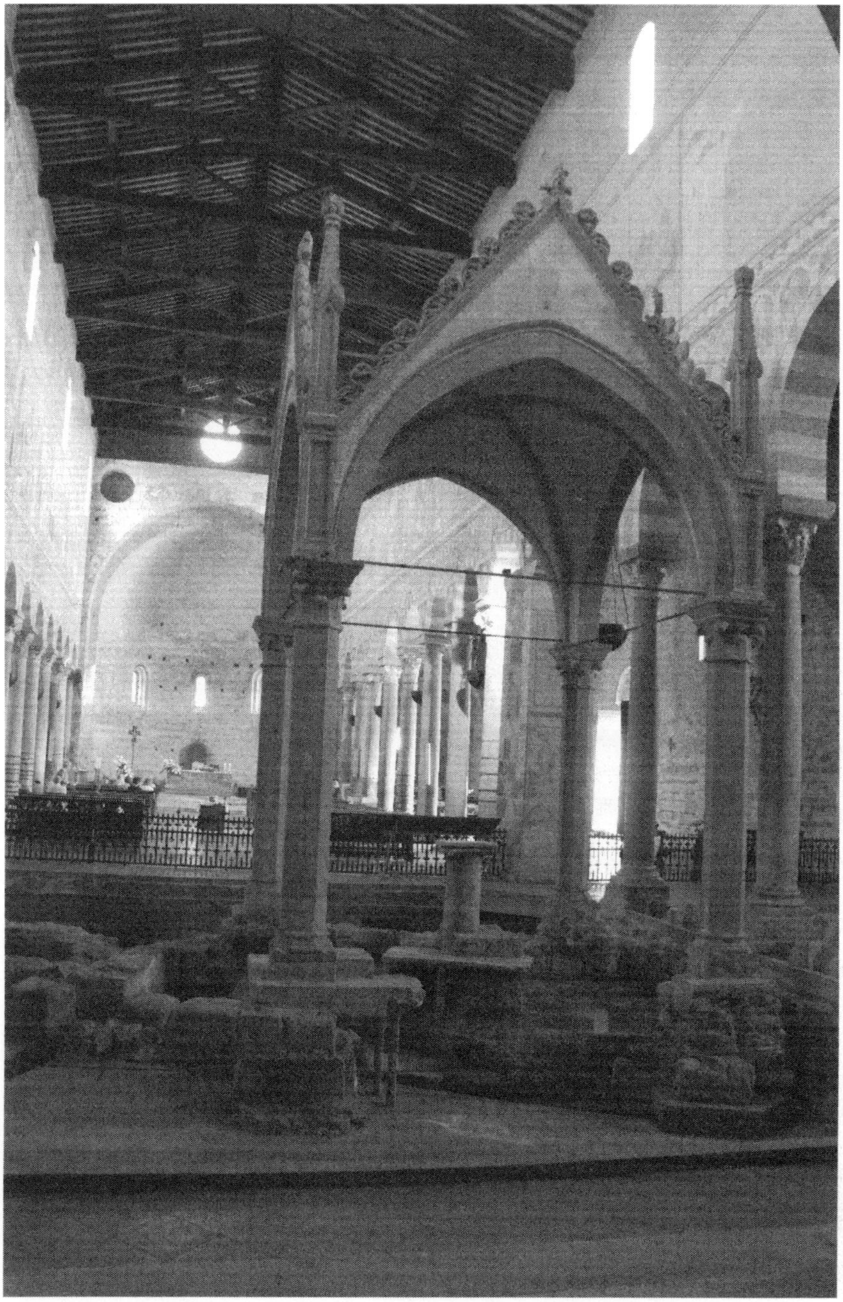

Figure 8.3 Pisa (neighborhoods), view of the church interior with the column marking the spot of the first altar erected by Saint Peter. Photo credit: Michele Bacci.

In the late thirteenth century, the church was adorned with a huge cycle of scenes representing the story of the apostles and partially reproducing the program that once decorated the atrium of Saint Peter's in Rome; a more elaborate mise-en-scène of this holy site was obtained in the fourteenth century when a Gothic ciborium was erected over it as a visual indicator of its site-bound holiness.[22] By then, the place had become so famous that Emperor Charles IV of Bohemia obtained a fragment of the altar table, which he deposited in the Church of Saint Peter of the Vyšegrad in Prague. Strikingly enough, the Pisan relic was used to invest the latter institution with Roman symbolism and thus to serve as a sort of *translatio Romae*.[23]

Evocations by synecdoche

More developed strategies for Holy Land evocation were worked out during the late Middle Ages. A particularly popular one functioned by synecdoche, by installing some of the sanctified soil of Palestine as a visual focus of a new shrine. A case in point is the Pisan Camposanto, which was erected around a field of holy earth transferred to Tuscany from the Jerusalem Akeldama, that is, the "field of blood" bought by Judas with his thirty coins, which was said to possess the miraculous power to carry out the decomposition of corpses in just three days. In this case, the Holy Land was clearly treated as a movable relic.[24] Similar strategies had already been evolved in earlier centuries. On Mount Gerizim, which was identified by the Samaritans as the biblical Moriah, the Christians included a fragment of the rock of Golgotha in the walls of the basilica erected there in the sixth century. In this way, the Samaritan Temple Mount was symbolically superseded by its Christian counterpart on the Mount of Golgotha.[25] In Jerusalem itself, the holiness associated with other holy mountains was often evoked by synecdoche, for example, worship of the stones of Tabor, Gerizim, and Sinai included in the altar of the holy Echmiadzin in the Armenian Cathedral of Saint James. On that very spot it was as if the sacred dynamis of five holy sites were being manifested at the same time. An analogous, yet less complex, case is that of the Greek Orthodox Chapel of Saint George (Figure 8.4) in the annex of the Holy Sepulchre, where the worship of a stone from Mount Sinai enables pilgrims to understand the principle by which the Mosaic law was just a shadow, an anticipation, of the Age of Grace starting with Christ's sacrifice on Calvary.[26]

Typologically similar to this *pars pro toto* strategy was evocation by means of a collection of Christological and especially Passion relics. This was a very common phenomenon in the Middle Ages, and almost every town or political entity wanting to assert its own prominence made all possible efforts to provide itself with prestigious remnants from the crucial events of Christ's sacrifice on the cross. The absence of Christ and the Virgin's earthly remains precluded Christians from using bodily relics as foci of Christological or Marian cult phenomena, as was the rule in martyria and cult sites associated with saints. For centuries, theologians denied that detachable parts of Christ's body, such as hair, nails, or blood might have been preserved on Earth; anything associated with his body was basically thought to have been involved in his resurrection, even if some exceptions did exist. Indeed, contact relics, *appendicia exteriora*, or material extensions of a holy body were assigned a prominent role in this context. These included not only the instruments of the passion, but more generally all kinds of objects sanctified by contact with the holy persons.[27]

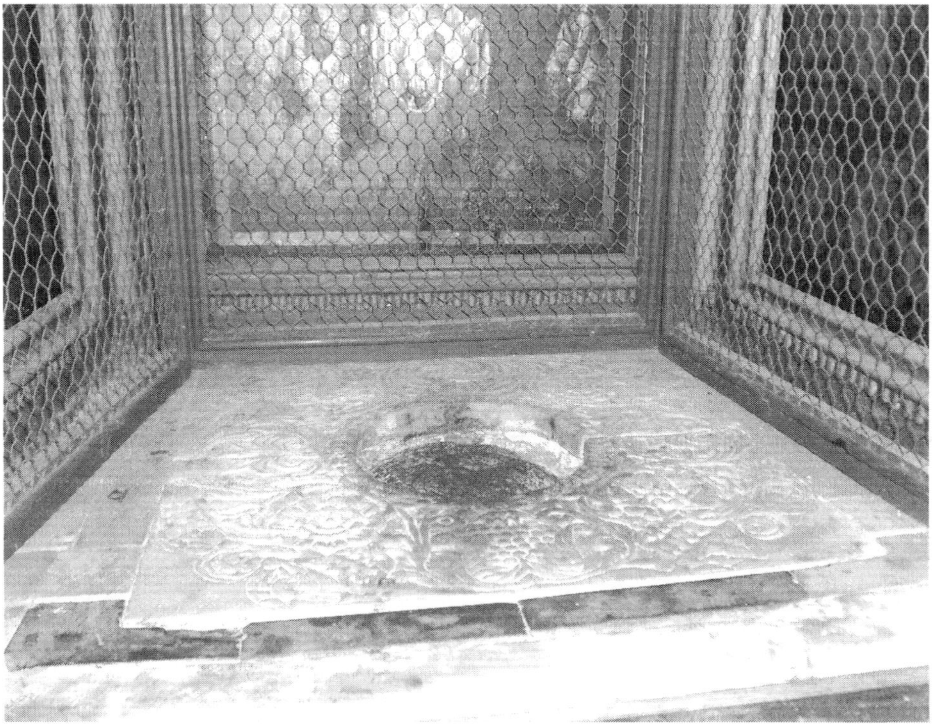

Figure 8.4 Jerusalem, complex of the Holy Sepulchre, Greek Orthodox Chapel of Saint George, altar with a stone from Mount Sinai. Photo credit: Michele Bacci.

In any case, such objects as the holy wood of the cross itself, the nails, the crown of thorns, the veils and funerary lintels, and so on were regarded not only as contact relics, but also as elements of the Jerusalem cityscape at the very moment of the passion, so the title of "New Jerusalem" could be attributed to any other place where such mementos were preserved. The presence of one relic was probably not enough to turn a church into a pilgrimage site, but this was certainly the case if a systematic collection of Christological and Marian mementos was offered for public worship. A complete set of bits of all the noncorporal objects associated with the Son of God, his temporally limited passage on Earth, and the hallowed landscape where his major deeds had taken place enabled believers to thoroughly experience and take spiritual benefit from the most important moments in the sacred history. Around the midpoint of the Middle Ages the most important treasures of this kind were in the Pharos Chapel within the Imperial Palace of Constantinople, the Lateran Church, and the nearby private chapel of the Popes – the *Sancta Sanctorum* – in Rome, and, from the mid-thirteenth century onward, the *Sainte Chapelle* of the French kings in Paris.[28] Whereas the core of such collections was represented by the instruments of Christ's suffering on the cross, they also included stones and earth from Jerusalem, suggesting that the latter were also perceived as Christological relics. A number of passion relics could have served simultaneously as topographical mementos: notable cases were the fragments of the rock of the Ascension on the Mount of Olives worshipped in the Lateran[29] and the pillow stone of the Holy Sepulchre kept in the Pharos Chapel.[30]

Owing to their association with the most eminent courts of the Christian world, such objects invested their owners with political legitimacy. In the fourteenth century, when the kings of Cyprus designated the port of Famagusta as the capital in exile of the lost Kingdom of Jerusalem, they made all possible efforts to provide the Latin Cathedral of Saint Nicholas with at least one important Christological relic. An alleged jar of Cana was exposed in a small church dedicated to Sancta Maria de Hydria. It was an astonishing piece of Islamic pottery, a sumptuous vase from Andalusia, whose exotic appearance stimulated its perception as an important memento of Christian origin, and it was accordingly exhibited within a special domed chapel. After the Cypriot sack of Alexandria in 1367, King Peter I of Lusignan even managed to obtain a piece of the flagellation column from the Mamluke sultan – a memento that could serve as a mediator of a partial, synecdochical *translatio Hierusalem*. The column could be interpreted as a miniscule part of the Jerusalem landscape that could be transferred elsewhere; this had long since taken place in Jerusalem itself, where the column was no longer worshipped on its original site (i.e., in the hardly localized Pretorium of Pilate), but in the Holy Sepulchre in at least two or even three forms, corresponding to the rival cult phenomena promoted by the Latins, the Armenians, and the Greek Orthodox. Not surprisingly, other fragments of the column – each of a different color – were also venerated in Constantinople and in Rome.[31]

Analogical, mimetic, and topomimetic evocations of Jerusalem

Other countries were, for the most part, forced to limit themselves to more or less complex analogical evocations, that is, attributing the cultic meaning associated with the holy sites of Jerusalem to some focal elements of a natural or urban landscape. Thus, for example, a high hill could be designated as a "Golgotha" or a "Calvary." In this respect, the inhabitants of Impruneta, near Florence, were naturally inclined to attribute the title of "Sancta Hierusalem" to a small nearby village located on the top of a hill. Such a designation could be combined with a mimetic imitation of the actual buildings framing the holy sites themselves. Finally, as is especially evident in the case of the later *Sacri Monti* and *Kalvarienberge*, there could be more precise efforts to arrange a topographical transposition, something that occurred only when a system of spatial relations between different holy places was applied to a different geographic context outside the Holy Land.[32]

Art historical research tends to place much more emphasis on analogical, mimetic, and topomimetic approaches. However, these were far from being the most common strategies by which site-bound holiness was evoked and materialized in medieval religious experience. The alternative approaches summarily sketched above – evocations by synecdoche and the creation of various holy lands and memorial sites associated with the deeds of the apostles that could be perceived as extensions of Palestine's network of loca sancta – played a much more prominent role. In the following, I describe the extent to which the nonmimetic strategies of evocation of Jerusalem's site-bound holiness helped shape the cultic landscape of the maritime towns located along the Venetian sea routes to the Eastern Mediterranean. In the fourteenth, fifteenth, and sixteenth centuries, Holy Land pilgrims regularly visited the most important ports-of-call between Venice and Jaffa or Alexandria. Thus, they came in some way to be integrated into the religious experience of those places, and the churches and relics of these ports were basically viewed as belonging to a wider network of holy sites that gave shape to a sort of intermediary, liminal space, separating Christian Europe from Christ's native land. It can be assumed that in some cases new forms of worship made their appearance to answer the pilgrims' wishes to anticipate the site-bound holiness they expected to experience at their final destination.

Relics and memorial sites associated with Eastern Mediterranean saints

One of the most common strategies used to transform anonymous places into holy sites was the promotion of cultic phenomena centered upon an image to be a lifetime portrait of the Virgin Mary and Christ: the Venetian sea routes to the Holy Land were dotted with shrines housing images attributed to Saint Luke.[33] Another was the public exhibition of bone fragments pertaining to saints whom the pilgrims expected to worship at the end of the voyage. In the cultic landscape of Palestine the shrines associated with saints were rare and could not compete with the sacred attractions represented by biblical sites – a notable exception being the stone on which Saint George was beheaded, preserved in his church at Lydda.[34] Yet a number of prominent cult places associated with universally venerated martyrs were worshipped in the neighboring countries and were part of the pilgrimage experience because of their location on the way to Jerusalem. Besides Lydda, relics and mementos of Saint George could be seen in the neighborhoods of Beirut (where the episode of the saint fighting the dragon was localized),[35] in the Monastery of Stavrovouni on Cyprus,[36] in the sacred collection of the Hospitallers in Rhodes,[37] and, obviously, also in Venice itself.[38] Pilgrims wishing to visit Saint Barbara's tomb in Old Cairo (Figure 8.5)[39] and perhaps to bring a painted ex-voto there such as the Catalan work now preserved in the Coptic Museum (Figure 8.6)[40] could prepare for this intense experience by praying before her arm in Venice,[41] before her head in the Latin cathedral in Candia,[42] or in front a column located in Beirut, whose purple veins were said to have been produced by the blood pouring from her neck at the moment of her beheading.[43]

There was a special attraction in the relics of Saint Catherine, whose bodily remains preserved in the Sinai Monastery represented a main stop on the trip from Egypt to Jerusalem. Those pilgrims who managed to reach the summit of Mount Horeb to see the site of God's revelation to Moses climbed up to the highest peak of Djebel Katrin, where they could kiss and sprinkle with tears the holy rock where the angels laid the blessed corpse of the martyr and guarded it for five centuries. In order to see the latter and take advantage of its extraordinary virtues they had to move into the interior of the Katholikon, next to the site of the Burning Bush, and approach the sarcophagus that contained her remains and the miraculous liquid, the so-called *Myron*, that poured forth from her body. After going through the ritual devotions, they were allowed to pin a pilgrimage token shaped like the cogged rod to their dress, which was the main iconographical attribute of the saint in the late Middle Ages.[44]

If one was not able to embark on the very dangerous pilgrimage from Cairo to Sinai, there were other possibilities for manifesting one's devotion to the great martyr and miracle worker. The foreigners who boarded the Venetian ships could embrace the Sinaitic tradition by worshipping an ampulla filled with the holy liquid that poured from her body, preserved with great honor in the Church of Saint Daniel.[45] Once they

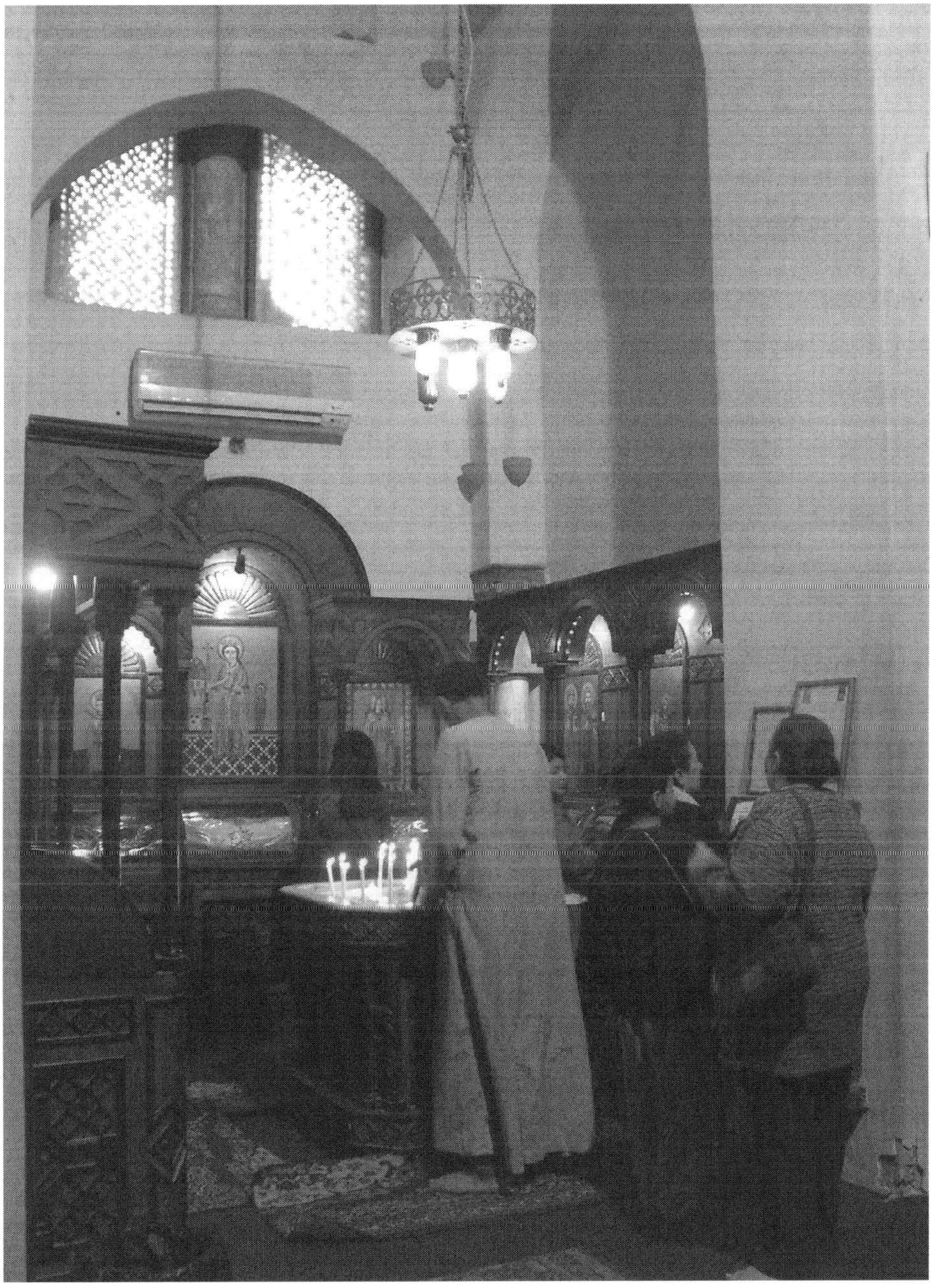

Figure 8.5 Pilgrims to the Tomb of Saint Barbara, Old Cairo, Church of Sitt Barbara. Photo credit: Michele Bacci.

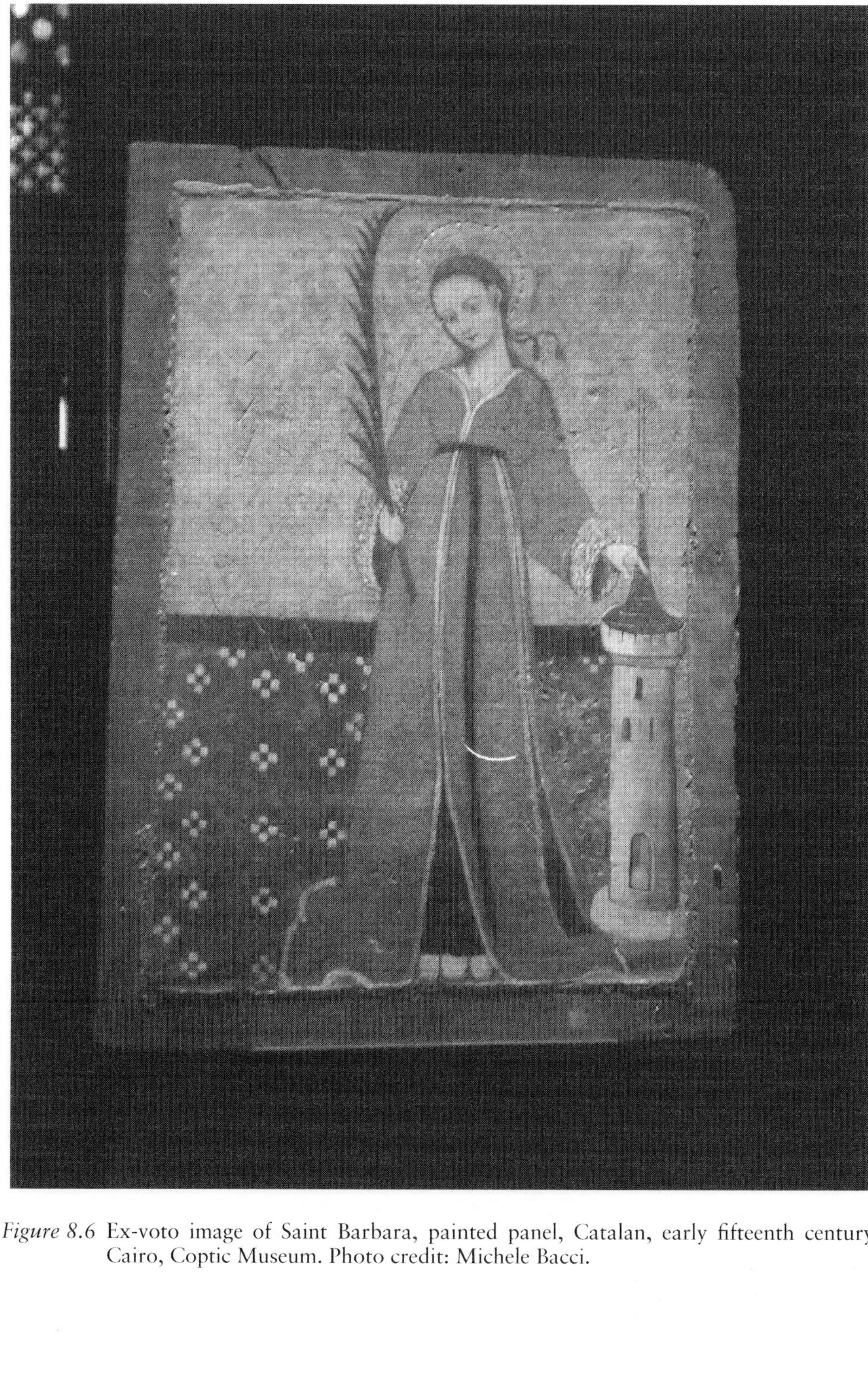

Figure 8.6 Ex-voto image of Saint Barbara, painted panel, Catalan, early fifteenth century. Cairo, Coptic Museum. Photo credit: Michele Bacci.

had crossed the Adriatic, the Ionian, and the Aegean seas, they could finally meet with the Alexandrian saint during their stay in Rhodes within the palace chapel of the Great Master, just before starting the voyage to the Levantine Sea. This building was dedicated to Saint Catherine herself and a number of luxurious tapestries illustrated the major episodes of her life. Her left arm was exposed for public viewing within a precious reliquary.[46]

By such means the Hospitallers could intercept and attract the extraordinary devotion that the pilgrims nourished for this holy figure. Analogous concerns were shared in the two successive landing places. After crossing the huge Gulf of Antalya, the ships finally berthed at Famagusta, on Cyprus, and the pious devotees, who had to spend some days in the most remote stronghold of Christianity in the Levant, strolled up to the ruins of the old Roman town of Konstantia. From the fourteenth century on it was described by their local guides as Saint Catherine's true native town, contending that the saint was associated with Alexandria only because she had moved there when she was thirteen years old and because it was there that she had been martyred. That her father was named Costus in some late versions of her life was used as an indication of her association with Konstantia and, according to fanciful etymology, the town's name, "Famagusta," was sometimes interpreted as "Fama Costi," that is, "Costus' renown." The visitors were shown a ruined building not far from the early Christian basilica of Epiphanius in Salamis, which was described either as the saint's house or her school. Near this spot, within a large necropolis, a tomb dating to the seventh century BCE was identified with the site where the mystical marriage of Saint Catherine had taken place (Figure 8.7).

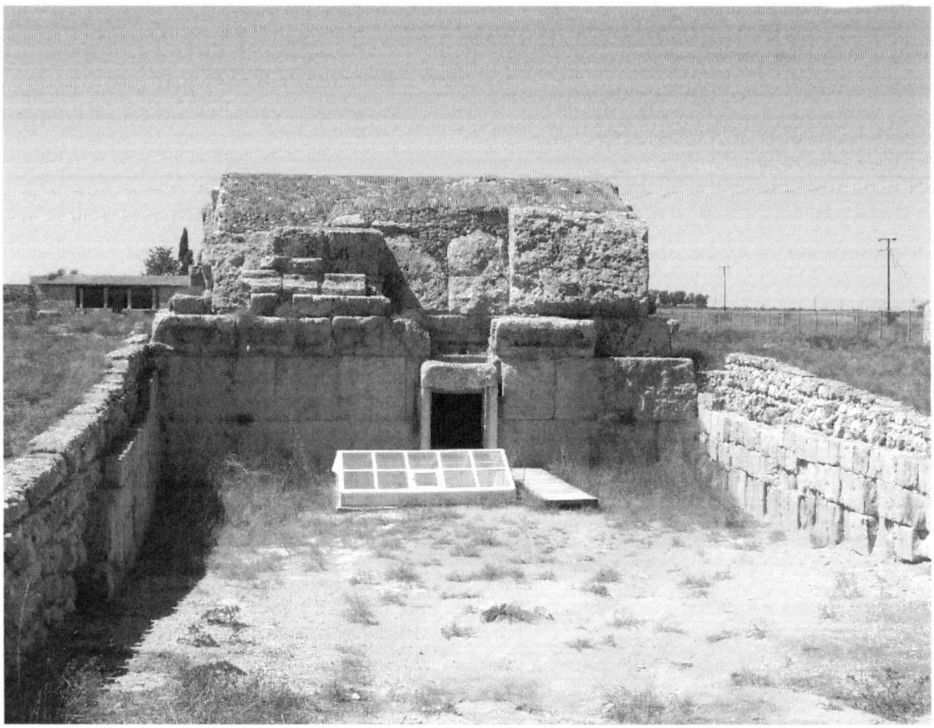

Figure 8.7 Cypro-archaic tomb, identified in the late Middle Ages as either the site of Saint Catherine's mystical marriage or Saint Catherine's Prison. Photo credit: Michele Bacci.

In order to enhance this connection, a painting displaying this theme was hung in the interior. From the late fifteenth century on, the place was already worshipped as the saint's prison, regardless of the fact that, according to all existing hagiographical sources, all the events of the martyrdom occurred in Alexandria.[47] The tomb-chapel was reportedly filled with murals scattered chaotically on its walls that featured the many coats-of-arms left there by Western pilgrims. Today only a small cross sculpted over the door to the inner grave remains as a testimony to the site's centuries-old cultic use.

Clearly, the Famagustans wanted to promote their town as an important stop for Holy Land pilgrims by creating an alternative holy topography. This was partly achieved, when the late fifteenth century the visitors to Saint Catherine's prison were allowed to pin half a rod to their garments – evidence that half of the indulgences connected to the pilgrimage to Saint Catherine had been already earned.[48] This did not prevent pilgrims from understanding that the most heroic and celebrated actions of the martyr took place in Alexandria. There it was possible to see the true prison, a very small underground room with a small window, which, according to legend, had been opened by the angel who daily brought food to the saint. There, two columns were said to have served as supports to the cogged rod. Another fragment of column marked the place where the saint was beheaded, as evidenced by its white color with red veins, which it was said had been made by the blood and milk that had spurted out of the saint's neck. In the same place the Greco-Melkite Church of Saint Sabas, where the icon painted by Saint Luke was preserved, marked the site of the house where she had lived.[49] Notwithstanding the overlaps and misunderstandings underlying all these traditions about the topography of Saint Catherine, the pilgrims felt extremely rewarded if they managed to visit all the sacred memorials scattered among Famagusta, Alexandria, and Sinai. Toward the end of the fourteenth century the Italian notary Nicola de' Martoni was especially pleased by the grace that enabled him to see with his own eyes *omnia facta beate Catherine et gesta per eam*,[50] all the sites connected with Saint Catherine and the mementos of all her actions.

It may be tempting to speak of the holy sites of Saints George, Barbara, and Catherine as shaping a specific sacred topography of the outskirts of the Holy Land. The shrines located on the Lebanese coasts, in Egypt, or even on Cyprus stood out for their distinctive nature as compared to those in more distant places. Whereas uncorrupted bodies, burials, and body parts prevailed in the cultic landscape of the ports of call along the Venetian sea routes to the Eastern Mediterranean, as the pilgrims approached their final goal these were substituted by sections of landscape and man-made structures associated with events described in the hagiographic narratives, which functioned in much the same way as most of the biblical *loca sancta*. Notable exceptions were the tombs of Saint Barbara in Cairo and of Saint Catherine on Mount Sinai, but even these experiences could be combined with visits to sites implying meditational practices about the deeds and vicissitudes of the saints: on Mount Sinai, the ascension to the top of Djebel Katrin was deemed to be as spiritually rewarding as the visit to the Mount of Revelation, Horeb, and in pilgrims' accounts was given much the same emphasis as Catherine's tomb. In a way, the locative experience of the holy prevailing in Palestine was extended to the neighboring areas by establishing an international network of memorial sites that evoked the hagiographies of three key figures of late medieval sainthood.

Extending the Holy Land network vs. imitating the architectural frame: Christological holy sites on Cyprus and Rhodes

The pilgrims' experiences of the places associated with a saint's birth, martyrdom, and burial corresponded to a meditation exercise, not unlike the one that, in the Palestinian holy sites, was expressed by means of a mental and emotionally charged evocation of the main Gospel events and especially of the passion. Compared with Saint George's or Saint Catherine's stories, which could be much more easily manipulated or even distorted, the memorial sites that evoked particular moments of Christ's passage on Earth could hardly be associated with sites outside the Holy Land without contradicting Scripture. Therefore, in order to evoke the memory of the Son of God's deeds it was essential to work out completely different strategies of cultic promotion. An original expedient was found by the canons regular of the Latin cathedral of Saint Sophia in Nicosia, who promoted the public worship for an ancient jasper sarcophagus that was said to have been made out of the very stone of the aniconic idol of Venus worshipped in the neighborhoods of Paphos by the ancient Cypriots. According to this odd legend, the Cypriots had heard that the Jews were going to cause Jesus' death on the cross; upon Jesus' refusal to follow them to Cyprus, they carved the idol in order to have it ready to be used at his burial site, but this pious intention was rendered vain by the resurrection. Nonetheless, it would have been appreciated as a "missed sepulchre" and, given that its creators' intention had been genuine, it proved to be venerable like an empty sepulchre in Jerusalem.[51]

This "missed burial" of Christ can be read as an absolutely original attempt at involving a non-Palestinian site in the topographic network of the Holy Land. It was probably too ambitious to prove successful, and was quickly sent into oblivion. Other regions made efforts to intercept Holy Land pilgrims by promoting forms of locative worship associated with indirect, that is, analogical or mimetic, evocations of Jerusalem. There are some indications that the Hospitallers tried to invest their town, Rhodes, with a Jerusalem symbolism by building a copy of the Holy Sepulchre and by attributing the topographic function associated with the "stations" of Christ's Way to Calvary in Jerusalem with their principal churches. A replica of the aedicule is mentioned by an early sixteenth-century pilgrim as being located in the cemetery of the Hospice of Saint Catherine, the structure where Jerusalem pilgrims were provided with hospitality: its funerary associations indicate that it served as a sort of surrogate Jerusalem, enabling those unlucky travelers who died there without reaching their final goal to be symbolically buried close to the Son of God's life-giving tomb.[52]

The location of replicas of the aedicule in cemeteries is paralleled by contemporary practice in Western Europe, where it often also served as the final station of a *Kalvarienberg* or *chemin de croix*, that is, a sequence of pillars or other structures commemorating the most important stops along Jesus' path to Mount Golgotha, which were used as visual supports for the performance of meditational exercises on Christ's passion, as in the notable case of Adam Kraft's Way of the Cross in Nuremberg.[53] The existence of analogous structures and their possible role as models for Western European complexes of the Way of the Cross is indirectly evidenced by a 1516 document that notes that the *chemin de croix* built at the beginning of the sixteenth century in Fribourg on the initiative of the Hospitaller knight Pierre d'Englisberg was modeled on a Rhodian archetype and that it replicated the authentic distances separating the holy places in Jerusalem. Indeed, the Way of the Cross in Fribourg consisted of seven pillars, located along the road leading from the cemetery of the Hospitaller Church of Saint-Jean to the Marian Church of Bourguillon through the hill of Montorge. It was deemed to be so authoritative as to be replicated in turn in yet another place, Romans-sur-Isère by Lyons.[54]

Evocations by synecdoche in Rhodes and Crete

The Rhodian complex was probably meant to meet the devotional needs of both Western pilgrims and locals. In any case, the mimetic replication of the architectural shape of this most worshipped site, combined with the topomimetic, analogical evocation of Jerusalem's sacred landscape, remained unparalleled along the sea routes of the Eastern Mediterranean. More often, visitors were reminded of the Holy City, the ultimate goal of their trip, by means of *pars pro toto* strategies with installations of mementos of Jerusalem. A case in point is the Franciscan Church of Saint Francis in Candia, which in the fifteenth century was enriched with a fragment of the flagellation column and a stone of the Golden Gate through which Christ had entered Jerusalem. In this way, the friars transplanted some of the holy stones that shaped the Holy City's material body onto Cretan soil.[55]

The Hospitallers, in their efforts to rival the renowned treasures of Constantinople, Rome, and Paris, presented their capital as a new Jerusalem by amassing a rich collection of passion relics, including reliquaries of the Holy Cross and venerable cross-shaped items made out of a material sanctified by contact with Christ's body. The Church of Saint John of Kollakion, in the general quarters of the Hospitallers on Rhodes, had an exceptional collection, including several fragments of the True Cross, a cup that was said to have been owned by Saint Martha and probably used by Jesus himself, a precious cross, which had allegedly been made out of the basin used by the Savior to wash his disciples' feet at the Last Supper, and one of Judas' thirty coins. Medieval pilgrims described it as a very old coin, plausibly an ancient Rhodian one, and it was used in the same way as the so-called *Agnus Dei*, small tokens that were distributed by the pope in Saint Peter's in Rome on Easter, a practice designed to rival Roman customs. The form of the coin was imprinted on small wax roundels, which were given to the devotees and later used as efficacious protection against sea storms and fires. The treasure also included a thorn from Christ's crown, but that was not considered to be as holy as that preserved in the Great Master's private chapel. The latter was no doubt more precious, being one of those that had wounded and perforated Christ's flesh. This special association afforded it the power to work the annual miracle by which it flourished for some hours on Holy Fridays and produced amazingly nice white flowers.[56]

In 1484, the Hospitaller knights received a very precious diplomatic gift from the emissaries of the Ottoman sultan Bayazet II – the right arm of Saint John the Baptist once worshipped in the Peribleptos Church in Constantinople. This was celebrated not only as a telling identity symbol, reinforcing the order's special devotion to the precursor, but also as a valuable addition to their collection of Christological relics, given that, as was often stressed, it consisted of the very finger that had pointed to Christ in order to demonstrate his role as the Lamb of God.[57] More or less in the same way, other relics that were eyewitnesses to the sacred events could also be described as contact relics and Gospel mementos. There was fierce competition between Venice and Zara about the ownership of the body of the wise Simeon, who had held the infant Jesus in his own arms at the moment of his presentation in the Temple. Two sepulchres, both considered as preserving his relics, faced each other across the Adriatic

Sea,[58] whereas a bit more to the south, inhabitants of the port of Ragusa/Dubrovnik, deprived of worshipping a third body, were eager to possess the cloth, presumably embroidered by the Virgin Mary, upon which the newborn Jesus had lain at the very moment of his presentation in the Temple. An arm of the same saint was allegedly preserved in the Church of Saint Francis in Candia.[59]

Overlapping topographies: Christ and Saint Francis on Crete

The Minor friars in the principal town of Crete made efforts to foster public worship in their church. As I have noted, different types of holy objects were offered to believers, including relics, miraculous icons, and stones from Jerusalem. Such objects were combined with mementoes of saints associated specifically with Western Europe and the history of their religious order, and especially of their founder, Saint Francis of Assisi. In Italy, the worship of the Poverello as the *alter Christus*, the new Christ, was expressed not so much by the spread of images of Saint Francis, but rather by the establishment of a specific "sacred topography," marked by the most important "mnemotopoi" of the saint's spiritual itinerary in Assisi, la Verna, Greccio, and elsewhere.[60] According to a widespread view, Assisi stood for Francis's holy sepulchre, whereas la Verna, the site of his stigmatization was compared to the Mount of Calvary, and the many fissures in the mountain were seen as a hint of this parallelism, as they reproduced the fissure produced by Christ's blood on the rock of Golgotha. The key role played by locative forms of religious experience became so prominent in Franciscan tradition that the cult of Saint Francis is the only one to have been promoted by establishing a topomimetic replica of his holy places, the Sacro Monte of Orta, Piedmont.[61]

In this respect, the memorial sites connected with Francis's voyage to Egypt and his visit to the Mamluke sultan during the Fifth Crusade could be profitably integrated into the wider network of Mediterranean holy sites and could work as an "extension" of Francis's own topography. This possibility was especially exploited in Candia, probably after the Observant branch of the order had taken possession of the convent around the mid-fifteenth century. Pilgrims to Candia were invited to visit the friars' garden and to look with reverence at a well that was said to have been built by Francis, the Poor of Assisi, with his own hands.[62] Worship was thus directed toward a material object, sanctified by its prestigious provenance and, at the same time, imbued with memorial meaning, given that it bore witness to the saint's physical presence on Crete during the Fifth Crusade. This allowed the Minor friars to present their convent as the only place where the holy topography of the *alter Christus* intersected with the network of venerable places that dotted the Venetian sea route to the Holy Land.

Site-bound holiness in a fluid setting

To summarize, it is worth stressing some of the outstanding characteristics of the new holy sites visited by late medieval pilgrims during their voyage to the Holy Land. As anthropologists have pointed out, one specificity of pilgrimage sites is their "liminality," that is, their perception as thresholds between the earthly and the divine dimensions.[63]

Pilgrims perceived their sea voyage as an introduction to the experience of the Jerusalem loca sancta for which they were longing. In the liminal context of coastal navigation, they tended to project their desire to see the sites of Christ's life onto the places they happened to visit along the route. Reminders of Jerusalem were to be found in various forms in each of the ports where they stopped. Christological relics, stones, and mementos of different kinds were housed in churches belonging to the various Christian denominations (Latin, Greek, and even Coptic and Maronite, as in Alexandria and Beirut), but this was not regarded as problematic. Their involvement in the Jerusalem pilgrimage, which started in the fourteenth century when Venice had a sort of monopoly in the organization of voyages to Palestine, contributed to invest these places with new meanings, to transform them into internationally recognized holy sites, and to present them not only as simple stops, but also as spiritually rewarding stages in the pilgrims' gradual approach to the Holy Land.

Different types of cultic strategies emphasized such a symbolic relationship with Jerusalem. The pilgrims' ultimate goal was evoked by fostering worship for holy figures mentioned in the Scriptures, such as John the Baptist and Simeon, as well as for saints whose cults were rooted either in Palestine or in the neighboring countries of the Eastern Mediterranean. Bodily remains and images were confined to the lands between Venice and Rhodes, whereas memorial sites – of Saint Barbara, Saint Catherine, Saint George, the apostles, and of Christ himself, as in the case of his "missed" sepulchre in Nicosia – became prominent as the ship got closer to the Holy Land. Evocations by synecdoche, in the form of both portions of Jerusalem's hallowed ground and collections of Christological relics, were also very widespread. On the other hand, analogical, topomimetic, and mimetic replicas of the loca sancta were rare and basically associated with issues of commemorative and individual piety, as in the case of the Rhodian copy of the Holy Sepulchre in the cemetery of the Hospice of Saint Catherine.

Late medieval pilgrims were responsible for integrating the holy sites located on the Dalmatian coasts, the Greek islands, Cyprus, and Egypt into a wider sacred topography of the Venetian sea routes to the Holy Land, which was essentially shaped by their own, individual experiences. It is difficult to judge, however, to what extent the association of each site with Jerusalem was the outcome of the pilgrims' desire to recognize material traces and anticipation of the experience of the loca sancta they were longing for or an intentional strategy on the part of local institutions to make their sites more attractive in the eyes of foreign visitors. Probably the truth lies somewhere in between, but there is no doubt that the extraordinary status of the holy sites was basically expressed by fostering worship for material objects and portions of ground deemed to imbibe, to some extent, the sanctity that imbued the land of Palestine. Indeed, no kind of mimetic or analogical evocation could rival cult phenomena that, even in a minimal sense, shared in – or aimed at sharing in – the same hallowed substance of the stones of Jerusalem.

Notes

* A preliminary and abridged version of this chapter was published in Italian as "La moltiplicazione dei luoghi sacri lungo le vie d'acqua per Gerusalemme nel tardo Medioevo," in *Peregrino, ruta y meta en las peregrinationes maiores: VIII Congreso internacional de estudios jacobeos (Santiago de Compostela, 13–15 Octubre 2010)*, eds. P. Caucci von Saucken and R. Vázquez (Santiago de Compostela: Xunta de Galicia, 2012), 179–194. The present chapter relies on data collected partly in the frame of the research project *Von Venedig zum Heiligen Land: Ausstattung und Wahrnehmung heiliger Orte an den mediterranen Küsten (1300–1550)*, financed by the Swiss National Fund, 2013–2016.

1 Richard Krautheimer, "Introduction to an 'Iconography of Mediaeval Architecture'," *Journal of the Warburg and Courtauld Institutes* 5 (1942): 1–33. Cf. the remarks of Paul Crossley, "Medieval Architecture and Meaning: The Limits of Iconography," *Burlington Magazine* 130 (1988): 116–121.

2 Among the most important contributions in the extensive literature are: Gustaf Dalman, *Das Grab Christi in Deutschland* (Leipzig: Dieterich, 1922); Jean Hubert, "Le Saint-Sépulchre de Neuvy et les pèlerinages en Terre-Sainte au XIe siècle," *Bulletin monumental* 87 (1927): 91–100; W. Eugene Kleinbauer, "Charlemagne's Palace Chapel at Aachen and Its Copies," *Gesta* 4 (1965): 2–11; Wolfgang Götz, *Zentralbau und Zentralbautendenz in der gotischen Architektur* (Berlin: Mann, 1968); Damiano Neri, *Il S. Sepolcro riprodotto in Occidente* (Jerusalem: Franciscan, 1971); Geneviève Bresc-Bautier, "Les imitations du Saint Sépulcre de Jérusalem (IXe – XVe siècle): Archéologie d'une dévotion," *Revue d'histoire de la spiritualité* 50 (1974): 319–342; Robert Ousterhout, "The Church of Santo Stefano: A 'Jerusalem' in Bologna," *Gesta* 20 (1981): 311–322; Otfried Ellger, *Die Michaelskirche zu Fulda als Zeugnis der Totensorge: Zur Konzeption einer Friedhofs- und Grabkirche im karolingischen Kloster Fulda* (Fulda: Parzeller, 1989); Matthias Untermann, *Der Zentralbau im Mittelalter: Form – Funktion – Bedeutung* (Darmstadt: Wissenschaftliche Buchgesellschaft, 1989); Sergio Gensini, ed., *La 'Gerusalemme' di San Vivaldo e i sacri monti in Europa* (Pisa: Pacini, 1989); Robert Ousterhout, ed., *The Blessings of Pilgrimage* (Urbana: University of Illinois Press, 1990); Stefanie Dathe, "Die Kirche la Vera Cruz in Segovia: Untersuchungen zur Bedeutung des romanischen Zentralbaus," *Mitteilungen der Carl Justi Vereinigung* 5 (1993): 92–121; Lieselotte Kötzsche, "Das Heilige Grab in Jerusalem und seine Nachfolger," in *Akten des XII: internationalen Kongresses für christliche Archäologie* (Münster: Aschendorff, 1995), 272–290; Nikolas Jaspert, "Vergegenwärtigungen Jerusalems in Architektur und Reliquienkult," in *Jerusalem im Hoch- und Spätmittelalter*, eds. Dieter Bauer, Klaus Herbers, and Nikolas Jaspert (Frankfurt am Main: Campus Verlag, 2001), 219–270; Jan Pieper, Anke Naukojat, and Ank Kappler, eds., *Jerusalemkirchen: Mittelalterliche Kleinarchitekturen nach dem Modell des Heiligen Grabes* (Aachen: Geymüller, 2003); Michael Rüdiger, *Nachbauten des Heiligen Grabes in Jerusalem in der Zeit von Gegenreformation und Barock: ein Beitrag zur Kultgeschichte architektonischer Devotionalkopien* (Regensburg: Schnell & Steiner, 2003); Piero Pierotti, Carlo Tosco, and Caterina Zannella, eds., *Le rotonde del Santo Sepolcro: Un itinerario europeo* (Bari: Edipuglia, 2005); Colin Morris, *The Sepulchre of Christ and the Medieval West* (Oxford: Oxford University Press, 2005); Annabel Jane Wharton, *Selling Jerusalem: Relics, Replicas, Theme Parks* (Chicago: University of Chicago Press, 2006); A. Barbero and G. Roma, eds., *Di ritorno dal pellegrinaggio a Gerusalemme: Riproposizione degli avvenimenti e dei luoghi di Terra Santa nell'immaginario religioso fra XV e XVI secolo* (Ponzano Monferrato: Atlas, 2008); Renata Salvarani, *La fortuna del Santo Sepolcro nel Medioevo: Spazio, liturgia, architettura* (Milan: Jaca Book, 2008); Ursula Röper and Martin Treml, eds., *Heiliges Grab: Heilige Gräber: Aktualität und Nachleben von Pilgerorten* (Berlin: Lukas Verlag, 2014).

3 On the *loca sancta* as memorial sites cf. Maurice Halbwachs, *La topographie légendaire des évangiles en Terre Sainte*, ed. Marie Jaisson (Paris: Presses universitaires de France, 2008). On the different forms of conceptualization and evocation of Jerusalem see especially Bianca Kühnel, ed., *The Real and Ideal Jerusalem in Jewish, Christian, and Islamic Art*, Special double issue of the annual Jewish Art 23, no. 24 (Jerusalem: Centre for Jewish Art, 1997/1998); Robert Ousterhout, "Flexible Geography and Transportable Topography," in *The Real and Ideal Jerusalem*, 393–404; idem, "'Sweetly Refreshed in Imagination': Remembering Jerusalem in Words and Images," *Gesta* 48 (2009): 153–168; Annette Hoffmann and Gerhard Wolf, eds., *Jerusalem as Narrative Space/Erzählraum Jerusalem* (Leiden: Brill, 2012); Bianca Kühnel, Galit Noga-Banai, and Hanna Vorholt, eds., *Visual Constructs of Jerusalem* (Turnhout: Brepols, 2014).

4 Bianca Kühnel, "Jerusalem between Narrative and Iconic," in *Jerusalem as Narrative Space*, eds. G. *Wolf* and A. *Hoffmann*, 105–123.

5 'Imad al-Din, as quoted in Carole Hillenbrand, *The Crusades: Islamic Perspectives* (New York: Routledge, 2000), 318.

6 Michele Bacci, "Il Golgotha come simulacro," in *Synergies in Visual Culture/Bildkulturen im Dialog: Festschrift für Gerhard Wolf*, eds. Manuela De Giorgi, Annette Hoffmann, and Nicola Suthor (Munich: Fink, 2013), 111–122.

7 Friar Antonio de Reboldi of Cremona, "Itinerarium ad Sepulchrum Domini," *Zeitschrift des deutschen Palästina-Vereins* 13 (1890): 153–174. "In Bethlehem est ecclesia in loco, ubi Christus fuit natus, quae dicitur Sancta Maria, tam pulcra, quod nunquam vidi tam pulcram, tam curiosam, tam sculptuosam in columpnis et picturis tam magnam, sicut est ista Bethleemitica toto orbe terrarum veneranda. Narrare siquidem seriose et sigilatim ipsius per totum mondum venerandae ecclesiae magnitudinem, latitudinem, longitudinem et diversorum lapidum marmoreorum ornatum, ordinem mirabilium et multiplicum colompnarum marmorearum, picturarum varietatem, ordinem et curiositatem et pavimentum miro lapide tabulatum, tectum metallo plombeo copertum, nimis esset longum enarrare. Set temporalia transeamus et solum, quae sunt in ipsa sacratissima ecclesia spiritualia, dicamus."

8 Jacopo da Verona, *Liber peregrinationis*, in *Liber Peregrinationis di Jacopo da Verona*, ed. Ugo Monneret de Villard (Rome: Tipografia dello Stato, 1950), 35: "Sciendum est itaque, quod de descriptione Jherusalem, quantum ad edificia antiqua et muros et municiones et portas, longum esset describere et nullius profectus. Cum civitas sit mutata et pro maiori parte dirupta, ego circuivi eam pluribus vicibus, et vix videntur in aliquibus locis porte et edificia antiqua, et ideo solum specificabo loca notabilia et venerabilia et memorie commendanda tam de veteri Testamento quam de novo."

9 Cf., among others, Herbert Thurston, *The Stations of the Cross: An Account of Their History and Devotional Purpose* (London: Burns & Oates, 1906, reprint 1914 and 2010); Karl Alois Kneller, *Geschichte der Kreuzwegandacht von den Anfängen bis zur völligen Ausbildung* (Freiburg im Breisgau: Herder, 1908); Paul Wilhelm von Keppler, *Die XIV Stationen des heiligen Kreuzwegs: Eine geschichtliche und kunstgeschichtliche Studie* (Freiburg im Breisgau: Herder, 1904); Amedee Teetaert de Zedelghem, *Saggio storico sulla devozione alla Via Crucis* [1949] (Crea: Centro di Documentazione dei Sacri Monti, Calvari e Complessi Devozionali Europei/ Ponzano: Regione Piemonte, 2004); Sylvia Schein, "La Custodia Terrae Sanctae franciscaine et les Juifs de Jérusalem à la fin du Moyen Âge," *Revue des études juives* 141 (1982): 370–371; Albert Storme, *The Way of the Cross: A Historical Sketch* (Jerusalem: Franciscan, 1984); Sandro Sticca, "The Via Crucis: Its Historical, Spiritual, and Devotional Context," *Mediaevalia* 15 (1989): 93–126; Nicky Zwijnenburg-Tönnies, "Die Kreuzwegandacht und die deutschen Pilgertexte des Mittelalters," in *Fünf Palästina-Berichte aus dem 15: Jahrhundert*, eds. R. Herz, D. Huschenbett, and F. Sczesny (Wiesbaden: Reichert, 1998), 225–260; Mitzi Kirkland-Ives, "Alternate Routes: Variation in Early Modern Stational Devotions," *Viator* 40 (2009): 249–270.

10 Leonhard Lemmens, *Die Franziskaner im Hl. Lande*, vol. 1: *Die Franziskaner auf dem Zion (1336–1551)* (Münster: Aschendorff, 1919), 164–187 and *passim*; Béatrice Dansette, "Les pèlerinages occidentaux en Terre Sainte: une pratique de la "Dévotion moderne" à la fin du Moyen Age?," *Archivum Franciscanum Historicum* 72 (1979): 106–133, 330–428; Nine Miedema, "Following in the Footsteps of Christ: Pilgrimage and Passion Devotion," in *The Broken Body: Passion Devotion in Late-Medieval Culture*, eds. A.A. MacDonald, H.N.B. Ridderbos, and R.M. Schlusemann (Groningen: Forsten, 1998), 73–92.

11 Francesco Suriano, "Trattato di Terra Santa e dell'Oriente," in *Il Trattato di Terra Santa e dell'Oriente di frate Francesco Suriano, missionario e viaggiatore del secolo XV*, ed. G. Golubovich (Milan: Artigianelli, 1900), 10–12. "[. . .] quelle cosse che tochorono uno pocho Christo recevetero molto la sua virtù et gratia. Adunque chi più se li acostò, più gratia ricevete. E chi maximamente se li acostò, recevete molto più virtù. Ma questa benedecta terra maximamente se li acostò per tacto sopra tute le altre cosse del mondo, adunque tuta è piena de divine virtude, et è facta abitatione sanctissima. [. . .] Unde sono sancti li fructi, sancti li arbori, sancte le legne, sancti li logumi, sancte le erbe, sancto lo pane, sancte le aque, sancte le pietre, sancta omni altra cossa, et piene de virtude."

12 Jeaninne Guérin Dalle Mese, *Égypte, la mémoire et le rêve: Itinéraires d'un voyage, 1320–1601* (Florence: L.S. Olschki, 1991), 396–400.

13 On such bad usages see the disapproval expressed by Felix Fabri, in *Evagatorium in Terrae Sanctae, Arabiae et Egypti peregrinationem*, vol. 2, ed. K.D. Hassler (Stuttgart: Societas Litteraria Stuttgardiensis, 1843–1849), 294.

14 See Jaroslav Folda, *The Art of the Crusaders in the Holy Land, 1098–1187* (Cambridge: Cambridge University Press, 1995), 136–137, 522–523; Denys Pringle, *The Churches of the*

Crusader Kingdom of Jerusalem: A Corpus, vol. 3 (Cambridge: Cambridge University Press, 1993–2009), 411.

15 On Wey's pilgrimage and chapel see, Pnina Arad, "Pilgrimage, Cartography and Devotion: William Wey's Map of the Holy Land," *Viator* 43, no. 1 (2012): 301–322.

16 William Wey, *Itinerarium* [1458], in *The Itineraries of William Wey, Fellow of Eton College, to Jerusalem, 1458*, ed. G. Williams (London: J. B. Nichols and Sons, 1857), 52–55.

17 On the holy topography of late medieval sea routes see Michele Bacci and Martin Rohde, eds., *The Holy Portolano: The Sacred Geography of Navigation in the Middle Ages* (Berlin: De Gruyter, 2014).

18 On the sanctity of Echmiadzin and its perception as 'alternative Jerusalem' see Armen Kazaryan, " 'Novyy Ierusalym' v prostranstvennykh kompleptsyakh i arkhitekturnykh formakh srednevekovoy Armenii," in *Novye Ierusalymi. Ierotopiya i ikonografiya sakralnykh prostranstv*, ed. Alexey Lidov (Moscow: Indrik, 2009), 520–543; idem, *Kafedralnyi sobor Surb Echmiadzin i vostochnokhristianskoe zodchestvo IV – VII vekov* (Moscow: Russian Academy of Architecture and Building Sciences, 2007).

19 For such traditions see Otto Meinardus, *Christian Egypt Ancient and Modern* (Cairo: American University in Cairo Press, 1965); idem, "The Itinerary of the Holy Family," *Studia Orientalia Christiana Collectanea* 7 (1962): 2–48; idem, *Coptic Saints and Pilgrimages* (Cairo: The American University in Cairo Press, 2002), 83–86. Cf. also Nabil Selim Atalla, *The Escape to Egypt According to Coptic Tradition* (Cairo: Lehnert & Landrock, 1993).

20 The sites of interest are Loutró by Agia Roumeli and Kaloi Liménes, as well as the island of Gávdos, which was eventually identified with the site of Saint Paul's shipwreck: cf. Cristoforo Buondelmonti, "Descriptio insulae Cretae," in *Creta sacra*, ed. Flaminio Corner (Venice: Pasqual, 1755), 1–18, 77–109, esp. 5 and 85. Paul's association with Crete was referred to by Jacopo da Verona, *Liber peregrinationis*, 16. Some Western pilgrims described a hermitage near Zeus' cave on Mount Ida, where the apostle was supposed to have written some of his letters: cf. Gaudenz, Herl of Kirchberg, *Jerusalemfahrt* [1470], ed. Reinhold Röhricht, "Jerusalemfahrt des Grafen Gaudenz von Kirchberg, Vogtes von Matsch (1470), nach der Beschreibung seines Dieners Friedrich Steigerwalder," *Forschungen und Mitteilungen zur Geschichte Tirols und Vorarlbergs* 1 (1904), 97–152, esp. 123; Peter Fassbender, "Pilgerfahrt [1492]," in *Deutsche Pilgerreisen nach dem heiligen Lande*, eds. Reinhold Röhricht and Heinrich Meisner (Berlin: Weidmann, 1880), 246–277, esp. 253. The small monastery that Jacques Le Saige visited in 1518 probably corresponds to Kaloi Liménes: M. Duthillœul, *Voyage de Jacques Le Saige de Douai à Rome, Nostre Dame de Lorette, Venise, Jérusalem et autres saints lieux* (Douai: D'Aubers, 1851–1852), 83–86. On the shrine at Agia Roumeli cf. Lucia Nixon, *Making a Landscape Sacred: Outlying Churches and Icon Stands in Sphakia, Southwestern Crete* (Oxford: Oxbow Books, 2006), 62, 64–65.

21 The earliest references date from the late fifteenth and early sixteenth centuries, cf. Johannes Graf zu Solms, "Beschreibung der Reyse und Wallfahrt des Johannes Graff zu Solms [1483]," in *Reissbuch des heyligen Lands*, ed. Sigmund Feyerabend (Frankfurt: I. Feyerabendt, 1584), ff. 50r–98r, esp. 56v; Ambrosius Zeebout, "Tvoyage van Mher van Joos van Ghistele [1485]," in *Ambrosius Zeebout: Tvoyage van Mher Joos van Ghistele*, ed. R. J. G. A. A. Gaspar (Hilversum: Verloren, 1998), 273; Spanish anonymous, *Viaje de Terra Santa* [c. 1520], in *Viajeros españoles a Tierra Santa (siglos XVI y XVII)*, ed. Joseph R. Jones (Madrid: Miraguano, 1998), 143.

22 The cultic history of the site has been only partially and insufficiently investigated: cf. Francesco Polese, *S. Piero a Grado e la sua leggenda* (Livorno: Giusti, 1905); *Nel segno di Pietro: la basilica di San Piero a Grado da luogo della prima evangelizzazione a meta di pellegrinaggio medievale*, ed. Maria Luisa Ceccarelli Lemut (Ospedaletto: Felici, 2003); Emilio Tolaini, *Lo sposalizio del mare e altri saggi su San Piero a Grado* (Pisa: ETS, 2004). The date of the church is still a matter of debate; based on archeological investigations carried out in the 1980s, it is now assumed that the present building was erected in the tenth century on the site of an older church and restored again in the twelfth and fourteenth centuries; see Fabio Redi, "Basilica di San Piero a Grado: Le strutture, gli scavi e la cronologia," in *Terre e paduli: reperti, documenti, immagini per la storia di Coltano*, ed. Renzo Mazzanti (Pontedera: Bandecchi & Vivaldi, 1986), 216–228.

23 Kateřina Kubínová, *Imitatio Romae: Karel IV. a Řím* (Prague: Artefactum, 2006), 280–282; Helena Soukupová, "Pisanský oltář apoštola Petra v Chrámu Sv: Petra a Pavla na Vyšehradé", in *Královský Vyšehrad IV: Sborník příspesků ze semináře 940-let královské kolegiátní kapítuly Sv: Petra a Pavla na Vyšehradě* (Prague: Královská kapitula Sv: Petra a Pavla na Vyšehradě, 2012), 284–308.

24 On such issues see Neta Bodner, "Earth from Jerusalem in the Pisan Camposanto," in *Between Jerusalem and Europe: Essays in Honour of Bianca Kühnel*, eds. Renana Bartal and Hanna Vorholt (Leiden: Brill, 2015), 74–93; and Lucy Donkin's chapter in this volume.

25 Alfons Maria Schneider, "Römische und byzantinische Bauten auf dem Garizim," *Zeitschrift der deutschen Palästina-Vereins* 68 (1946–1951): 211–234, esp. 230–231.

26 On such themes, see Yamit Rachman-Schrire, "Evagatorium in Terrae Sanctae [. . .]: Stones Telling the Story of Jerusalem," in *Jerusalem as Narrative Space*, eds. Gerhard *Wolf* and Annette *Hoffmann*, 353–366.

27 The expression *appendicia exteriora* was used in the twelfth century: Thiophridus of Epternach, *Flores epytaphii sanctorum*, 3, 1, in *Thiofridi abbatis Epternacensis Flores epytaphii sanctorum*, ed. Michele Camillo Ferrari (Turnhout: Brepols, 1996), 57. Notwithstanding the theologians' generally negative attitude, some bodily relics, including Christ's blood, teeth, hair, and foreskin, were offered to public worship in Western Christianity from the eleventh–twelfth centuries onward, and were occasionally the object of criticism: see Caroline Walker Bynum, *The Resurrection of the Body in Western Christianity, 200–1336* (New York: Columbia University Press, 1995), esp. 140. This book provides the most valuable information on the theological debates about notions of the resurrected body in late antique and medieval traditions. Cf. also idem, *Christian Materiality: An Essay on Religion in Late Medieval Europe* (New York: Zone Books, 2011).

28 For a survey of Christological collections, see Holger A. Klein, "Sacred Things and Holy Bodies: Collecting Relics from Late Antiquity to the Early Renaissance," in *Treasures of Heaven: Saints, Relics, and Devotion in Medieval Europe*, eds. Martina Bagnoli, Holger A. Klein, C. Griffith Mann, and James Robinson (New Haven and London: Yale University Press, 2011), 55–67, with bibliography.

29 As first witnessed by the twelfth-century "Descriptio Lateranensis ecclesiae," in *Codice topografico della città di Roma*, vol. 3, eds. Roberto Valentini and Giuseppe Zucchetti (Roma: Istituto storico italiano per il Medio Evo, 1940–1953), 336–339.

30 Michele Bacci, "Relics of the Pharos Chapel: A View from the Latin West," in *Vostochnochristianskie relikvii/ Eastern Christian Relics*, ed. Alexey Lidov (Moscow: Progress-Traditsiya, 2003), 234–248, esp. 241 and 244.

31 Michele Bacci, "Patterns of Church Decoration in Famagusta," in *Famagusta: Art and Architecture*, ed. Annemarie Weyl Carr (Turnhout: Brepols, 2014), 203–276, esp. 239–240.

32 See especially Franco Cardini, "La devozione a Gerusalemme in Occidente e il "caso" sanvivaldino," in *La 'Gerusalemme' di San Vivaldo*, ed. Sergio Gensini, 55–102; Luciano Vaccaro and Francesca Ricardi, eds., *Sacri Monti. Devozione, arte e cultura della Controriforma* (Milan: Jaca Book, 1992); Amilcare Barbero, ed., *Atlante dei Sacri Monti, Calvari e complessi devozionali europei* (Novara: Istituto Geografico De Agostini, 2001); Barbero and Roma, *Di ritorno dal pellegrinaggio a Gerusalemme*.

33 Michele Bacci, "La moltiplicazione dei luoghi sacri lungo le vie d'acqua per Gerusalemme nel tardo Medioevo," in *Peregrino, ruta y meta en las peregrinationes maiores: VIII Congreso internacional de estudios jacobeos (Santiago de Compostela, 13–15 Octubre 2010)*, eds. P. Caucci von Saucken and R. Vázquez (Santiago de Compostela: Xunta de Galicia, 2012), 179–194, esp. 185–187.

34 Pringle, *The Churches*, vol. 2, 9–27.

35 By the present-day cave of Saidet el-Bazaz (Our Lady of the Seven Mammas): cf. Nina Jidejan, *Beirut through the Ages* (Beirut: Dar el-Machreq, 1996), 88–89; Victor Somma, "Des saints heroïques vénérés au Liban," *Proche Orient chrétien* 41 (1991): 258–288, esp. 272–275; Nada Hélou, "Les lieux sacrés de Beyrouth au Moyen Âge: Les deux églises de Saint-Georges," in *The Holy Portolano*, eds. Bacci and Rohde, 73–93.

36 As witnessed by Nicola de' Martoni, *Liber peregrinationis ad loca sancta* [1394–1395], ed. Louis Le Grand, *Relation du pèlerinage à Jérusalem de Nicolas de Martoni, notaire italien (1394–1395)*, *Revue de l'Orient latin* 3 (1895): 566–669, esp. 636.

37 For the earliest sources cf. ibid., 641; Fabri, *Evagatorium in Terrae Sanctae*, vol. 3, 287–288.
38 A fragment of the saint's skull was worshipped in the Church of San Giorgio Maggiore from 1462 on: cf. Roberto D'Antiga, *Guida alla Venezia bizantina: Santi, reliquie e icone* (Padua: Casadei libri, 2005), 39–40; Agathangelos Bishop of Fanar, Chryssa Maltezou, and Enrico Morini, *Ἱερά λείψανα Ἁγίων τῆς καθ'ἡμᾶς Ἀνατολῆς στή Βενετία* (Athens: Apostolikē Diakonia, 2005), 127–140.
39 Charalambia Coquin, *Les édifices chrétiens du Vieux-Caire* (Cairo: Institut français d'archéologie orientale du Caire, 1974), 129–130.
40 Paul Van Moorsel, Mat Immerzeel, and Linda Lange, *Catalogue général du Musée Copte: The Icons* (Cairo: Institut français d'archéologie orientale, 1994), 141, n. 158.
41 D'Antiga, *Guida alla Venezia bizantina*, 43–44; Agathángelos, Maltézou, Morini, *Ἱερά λείψανα*, 103–118.
42 Maria Georgopoulou, *Venice's Mediterranean Colonies: Architecture and Urbanism* (Cambridge: Cambridge University Press, 2001), 112–113; idem, "The Holy Sites of Candia," 146. The identification with the Egyptian martyr Saint Barbara was questioned by some pilgrims, such as Felix Fabri, *Evagatorium in Terrae Sanctae*, vol. 3, 281–282, and Friederick II of Liegnitz and Brieg, in *Deutsche Pilgerreisen nach dem heiligen Lande*, eds. Röhricht and Meisner, 120.
43 Ogier VIII, Lord of Anglure, "Ogier VIII, Lord of Anglure," in *Le saint voyage de Jherusalem* [1395], eds. François Bonnardot and Auguste Longnon (Paris: F. Didot, 1878), 10–11: "Item, l'eglise de Sainte Barbe; et au dehors en une estroicte rue a ung petit pillier de marbre de plusieurs couleurs, sur lequel saincte Barbe ot coppée la teste; et est le dit pillier jusques au jourd'uy couloures de son sang."
44 See especially Mahfouz Labib, *Pèlerins et voyageurs au Mont Sinaï* (Cairo: Institut français d'archéologie orientale, 1961); Anastasia Drandaki, "Through Pilgrims' Eyes: Mt. Sinai in Pilgrim Narratives of the Thirteenth and Fourteenth Centuries," *Δελτίον της χριστιανικής αρχαιολογικής αρχαιολογικής εταιρείας*, ser. IV, 27 (2006): 491–503; David Jacoby, "Christian Pilgrimage to Sinai until the Late Fifteenth Century," in *Holy Image, Hallowed Ground: Icons from Sinai*, eds. Robert S. Nelson and K.M. Collins (Los Angeles: J. Paul Getty Museum, 2006), 78–93; Nancy P. Ševčenko, "St. Catherine of Alexandria and Mount Sinai," in *Ritual and Art: Byzantine Essays for Christopher Walter*, ed. P. Armstrong (London: Pindar, 2006), 129–143; idem, "The Monastery of Mount Sinai and the Cult of Saint Catherine," in *Byzantium: Faith and Power (1261–1557): Perspectives on Late Byzantine Art and Culture*, ed. S.T. Brooks (New York: Metropolitan Museum of Art, 2006), 118–137; Michele Bacci, "Llocs de la memòria de Caterina d'Alexandria," in *La princesa sàvia: Les pintures de santa Caterina de la Seu d'Urgell*, eds. Manuel Castiñeiras and J. Verdaguer (Barcelona: Museu nacional d'art de Catalunya, 2010), 49–56; Jaš Elsner and Gerhard Wolf, "The Transfigured Mountain: Icons and Transformations of Pilgrimage at the Monastery of St. Catherine at Mount Sinai," in *Approaching the Holy Mountain: Art and Liturgy at St Catherine's Monastery in the Sinai*, ed. Sharon E.J. Gerstel (Turnhout: Brepols, 2010), 37–71.
45 Cf. Antonio da Crema, *Itinerario al Santo Sepolcro 1486*, ed. G. Nori (Pisa: Pacini, 1996), 138: "In Sancto Daniel: licore di sancta Katerina"; Santo Brasca, in *Viaggio in Terrasanta* [1480], ed. A.L. Momigliano Lepschy (Milan: Longanesi, 1966), 51.
46 Giacomo Bosio, *Dell'istoria della sacra religione et illustrissima militia di S. Giovanni gerosolimitano*, vol. II (Venice: Facciotti, 1695), 111, 363, 385, 485; G. Sommi Picenardi, *Itinéraire d'un chevalier de Saint-Jean de Jérusalem dans l'île de Rhodes* (Lille: Desclée, de Brouwer & Cie, 1900), 79–80; I. Kóllias, *Η μεσαιωνική πόλη της Ρόδου και το Παλάτι του Μεγάλου Μαγίστρου* (Athens: Tameio archaiologikōn porōn kai apallotriōseōn/ T.A.P.A., 1994), 157–160.
47 On this theme see especially Lorenzo Calvelli, *Cipro e la memoria dell'antico fra Medioevo e Rinascimento: La percezione del passato romano dell'isola nel mondo occidentale* (Venice: Istituto Veneto di Scienze, Lettere ed Arti, 2009), 157–245.
48 This usage was referenced in 1485–1486 by Georges Lengherand, *Voyage de Georges Lengherand, Mayeur de Mons en Haynaut, à Venise, Rome, Jérusalem, Mont Sinaï & le Kayre – 1485–1486*, ed. Marquis de Godefroy Ménilglaise (Mons: Masquillier & Dequesne, 1871), 110, and Konrad Grünenberg, as excerpted in Gilles Grivaud, *Excerpta Cypria nova*

(Nicosia: Centre de Recherches Scientifiques, 1990), 126. Cf. also Philipp von Hagen, *Vier rheinische Palaestina-Pilgerschriften*, ed. Conrady, 278.

49 Detailed descriptions of Catherine's sites in Alexandria were provided by Ludolph von Südheim, "De itinere Terrae Sanctae [1335–1341]," in *Ludolphi, rectoris ecclesiae parochialis in Suchem, de itinere Terrae Sanctae liber: Nach alten Handschriften berichtigt herausgegeben*, ed. F. Deycks (Stuttgart: Litterar. Verein, 1851), 36; Niccolò da Poggibonsi, "*Libro d'oltramare* [1346]," in *Pellegrini scrittori: Viaggiatori toscani del Trecento in Terrasanta*, eds. A. Lanza and M. Troncarelli (Florence: Ponte alle Grazie, 1990), 112; Lionardo Frescobaldi, "Viaggio in Terrasanta [1384]," in ibid., 177; Giorgio Gucci, "Viaggio ai luoghi santi," in ibid., 261; Martoni, *Liber peregrinationis ad loca sancta*, 588; Sebaldt Rieter [1462], in *Das Reisebuch der Familie Rieter*, eds. R. Röhricht and H. Meisner (Tübingen: Litterar. Verein in Stuttgart, 1884), 128; Anselmo Adorno, "Itinerarium Terrae Sanctae [1470–1471]," in *Itinéraire d'Anselme Adorno en Terre Sainte (1470–1471)*, eds. J. Heers and G. de Groer (Paris: Éditions du Centre National de la Recherche Scientifique, 1978), 164; Rhenish Pilgrim [c. 1473], *Pilgerschrift*, ed. Conrady, *Vier rheinische Palaestina-Pilgerschriften*, 168; Fabri, *Evagatorium in Terrae Sanctae*, vol. 3, 158–159; Arnold van Harff [1496–1499], in *The Pilgrimage of Arnold von Harff*, trans. and ed. M. Letts (London: Hakluyt Society, 1946), 95; Jean Thenaud [1511], in *Le voyage d'outre-mer (Égypte, Mont Sinay, Palestine) de Jean Thenaud*, ed. Charles Schefer (Paris: E. Leroux, 1884), 27; Domenico Trevisan, *Relazione dell'ambasciata* [1511], ibid., 174; Faostino da Toscolano, *Itinerario di Terra Santa* [1654], ed. W. Bianchini (Spoleto: Centro Italiano di Studi sull'Alto Medioevo, 1992), 92–93.

50 Martoni, *Liber peregrinationis ad loca sancta*, 632.

51 On this object cf. Fabri, *Evagatorium in Terrae Sanctae*, vol. 3, 230–234, and Johannes Graf zu Solms, *Beschreibung der Reyse und Wallfahrt des Johannes Graff zu Solms*, f. 57r. Cf. also Calvelli, *Cipro e la memoria dell'antico fra Medioevo e Rinascimento*, 43–44.

52 Barthélémy de Salignac, *Itinerarium Hierosolymitanum* [1522], vol. 3, 5, ed. Rainerius Reineccius Steinhemius, *Descriptio Terrae Sanctae et regionum finitimarum, auctore Borchardo, monacho Germano, familiae Dominicanae, quem vixisse accepimus circa annum Iesu Christi MCCXXCIII, item Itinerarium Hierosolymitanum Bartholomaei de Saligniaco, equitis et ic. Galli, idem argumentum pertractans* (1522), pars II, Magdeburg 1587, without a page number: "Et est alia Latina ecclesia in honorem Divae virginis et martyris Catharinae erecta, in cuius coemeterio sepulchrum extat instar sepulchri dominici."

53 See especially Susanne Wegmann, "Der Kreuzweg des Adam Kraft im Spiegel spätmittelterlicher Frömmigkeit," in *Adam Kraft*, ed. F.M. Kammel (Nürnberg: Germanisches Nationalmuseum, 2002), 295–306.

54 The connection of the *chemin de croix* in Fribourg with an archetype in Rhodes is referred to in a document of the town municipality dated November 10, 1516: cf. Kneller, *Geschichte der Kreuzwegandacht*, 202–204; Ludovic Viallet, *Bourgeois, prêtres et cordeliers à Romans (v. 1280 – v. 1530)* (Saint-Étienne: Presses de l'Université de Saint-Etienne, 2001), 462; Jyri Hasecker, *Die Johanniter und die Wallfahrt nach Jerusalem (1480–1522)* (Göttingen: V&R Unipress, 2008), 96; Verena Villiger, "Monter à Bourguillon: La mise-en-scène du sacré aux portes de Fribourg," *Annales fribourgeoises* 72 (2010): 31–42; Stephan Gasser, Katharina Simon-Muscheid, and Alain Fretz, *Die Freiburger Skulptur des 16: Jahrhunderts: Herstellung, Funktion und Auftraggeberschaft*, vol. 1 (Petersberg: Michael Imhof Verlag, 2011), 337–339.

55 See *Friedrich II of Liegnitz and Brieg*, eds. Röhricht and H. Meisner, 120, and Spanish anonymous, *Viaje de Terra Santa*, 130.

56 Picenardi, *Itinéraire d'un chevalier de Saint-Jean*, 89–93; Charles Oman, "The Treasure of the Conventual Church of St. John at Malta," *The Connoisseur* 173 (1970): 101–107, 177–185, 244–249; Mario Buhagiar, "The Treasure of Relics and Reliquaries of the Knight Hospitallers in Malta," in idem, *Essays on the Knights and Art and Architecture in Malta 1500–1798* (Santa Venera: Midsea Books, 2009), 29–54; idem, "The Treasure of the Knight Hospitallers in 1530: Reflections and Art Historical Considerations," ibid., 55–72.

57 This relic was sent as a diplomatic gift to the Great Master Pierre d'Aubusson: cf. Sebastiano Paoli, *Codice diplomatico dell'Ordine gerosolimitano* (Lucca: Marescandoli, 1733–1737), 554; the solemn ceremonies on the occasion of its arrival in Rhodes are described in detail

by Guillaume Caoursin in his *De translatione Sacrae Dextrae sancti Johannis Baptistae Christi praecursoris ex Constantinopoli ad Rhodios commentarium*, published as an appendix to his *Obsidionis Rhodiae urbis descriptio* (Ulm: Johann Reger, 1496). See Ida Sinkević, "Afterlife of the Rhodes Hand of St John the Baptist," in *Byzantine Images and Their Afterlives: Essays in Honor of Annemarie Weyl Carr*, ed. Lynn Jones (Farnham: Ashgate, 2014), 125–141.

58 Daniele Forlati, *Illyricum sacrum*, vol. 5 (Venetiis: Apud Sebastianum Coleti, 1775), 81–84, 608–610; Charles Seymour, "The Tomb of Saint Simeon the Prophet in San Simeon Grande in Venice," *Gesta* 15 (1976): 193–200; Agathángelos, Maltézou, Morini, *Ἱερά λείψανα*, 329–337.

59 The Dubrovnik relic is frequently mentioned in fifteenth century pilgrims' travelogs: cf. Lochner, *Beschreibung der Pilgerfahrt*, 210; Gabriele Capodilista, *Itinerario* (1458), in *Viaggio in Terrasanta di Santo Brasca*, ed. A. L. Momigliano Lepschy, 169; Peter Rot,"Hans und Peter Rot's Pilgerreisen 1440 und 1453," ed. A. Bernoulli, *Beiträge zur vaterländischen Geschichte*, n. ser., 1 (1882): 331–408, esp. 401; Roberto da Sanseverino, *Viaggio alla Terra Santa*, ed. G. Maruffi (Bologna: Commissione per i testi di lingua, 1858), 33–34; Wilhelm Tzewers, *Itinerarius Terrae Sancte*, ed. G. Hartmann (Wiesbaden: Harrassowitz 2004), 96; Santo Brasca, ed., *Momigliano Lepschy, Viaggio*, 57; *Friedrich II of Liegnitz and Brieg*, eds. Röhricht and Meisner, "Die Pilgerfahrt des Herzogs Friedrich II", 117. The arm relic in Candia is mentioned in Johann Meisenheimer's description of the Holy Land pilgrimage of Alexander of Palatinate and Georg-Ludwig of Nassau-Saarbrücken, ed. J. Karbach, "Die Reise Herzog Alexanders von Pfalz-Zweibrücken und Graf Johann Ludwigs von Nassau-Saarbrücken ins Heilige Land, 1495–1496, nach dem Bericht des Johann Meisenheimer," *Zeitschrift für die Geschichte der Saargegend* 45 (1997): 11 118, esp. 59.

60 Michele Bacci, "Immagini sacre e pietà 'topografica' presso i Minori," in *Le immagini del Francescanesimo: Atti del XXXVI Convegno internazionale, Assisi, 9–11 ottobre 2008* (Spoleto: Centro italiano di studi sull'alto medioevo, 2009), 31–57.

61 Anna Maria Morello, ed., *Il Sacro Monte d'Orta e san Francesco nella storia e nell'arte della Controriforma* (Orta San Giulio: Pubblicazione a cura della Regione Piemonte, 1983); A. Marzi, ed., *Orta San Giulio. La Fabbrica del Sacro Monte: Conoscenza Progetto Restauro* (Turin: Istituto Bancario San Paolo, 1991).

62 Jacques Le Saige [1518], in *Voyage de Jacques Le Saige de Douai à Rome, Nostre Dame de Lorette, Venise, Jérusalem et autres saints lieux*, ed. M. Duthillœul (Douai: Adam d'Aubers, 1851), 81. Cf. also Georgopoulou, "The Holy Sites of Candia," 147–150.

63 First introduced by Victor and Edith Turner, *Image and Pilgrimage in Christian Culture: Anthropological Perspectives* (Oxford: Blackwell, 1978)

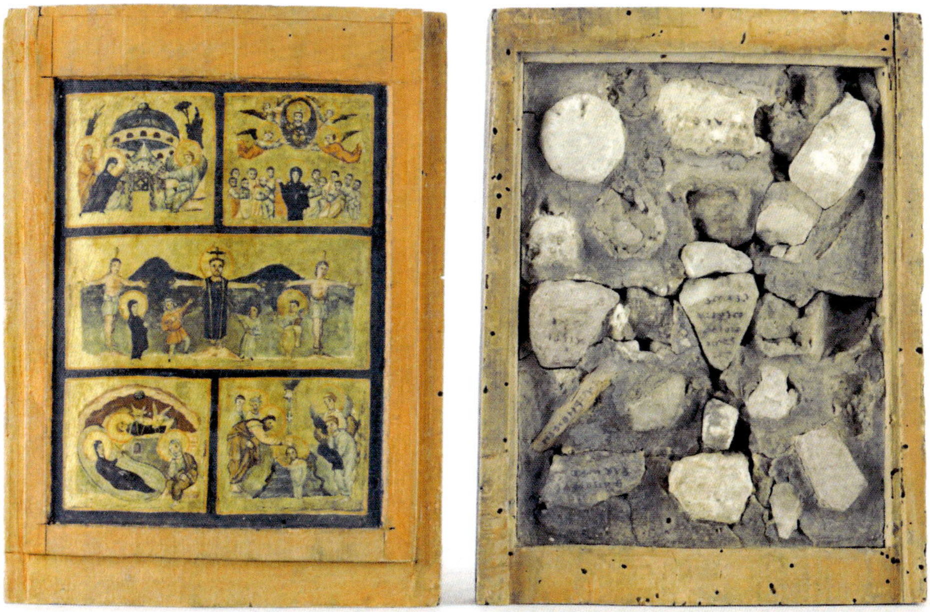

Plate 1 Reliquary box with stones and wooden fragments from the Holy Land, 24 × 18.4 × 3 cm, lid 1 cm thick, sixth/seventh century, from the Sancta Sanctorum treasure, Rome, Vatican, Musei Vaticani, Museo Sacro.

Image source: Martina Bagnoli et al., eds, *Treasures of Heaven: Saints, Relics and Devotion in Medieval Europe* (London: The British Museum Press, 2010), Cat. No. 13.

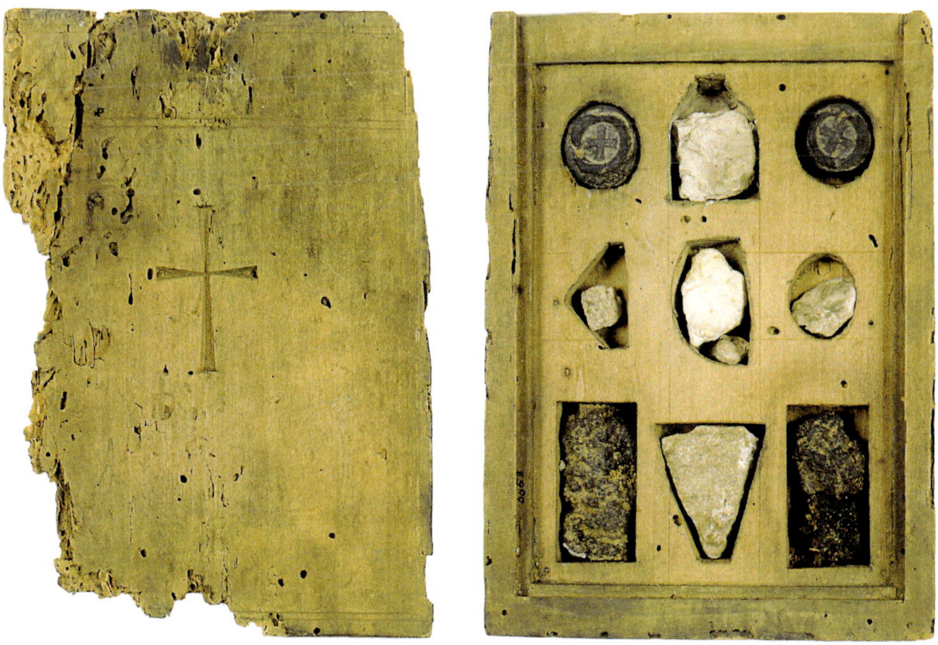

Plate 2 Wooden reliquary box with stones, fragments of carbon and pilgrim's tokens, 19 × 14.5 × 2.5 cm, Palestine, early medieval, from the Sancta Sanctorum treasure, Rome, Vatican, Musei Vaticani, Museo Sacro.

Photo credit: © Vatican Museum.

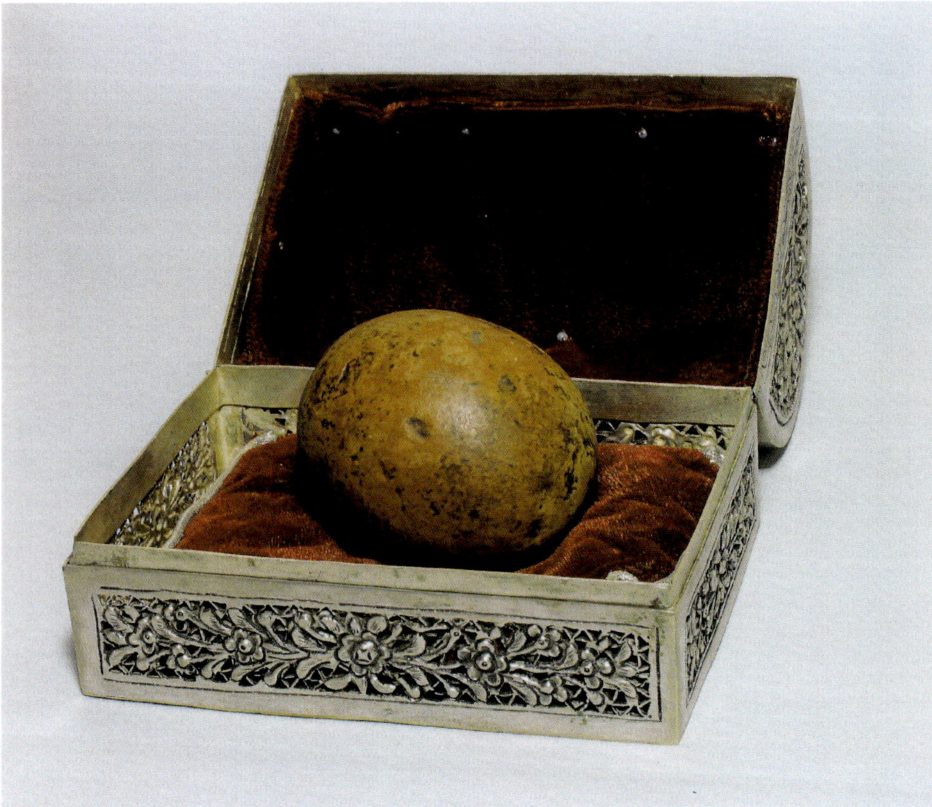

Plate 3 The Stone of Grace, Trinity Cathedral, Tbilisi, Georgia.
Photo credit: Zaza Skhirtladze.

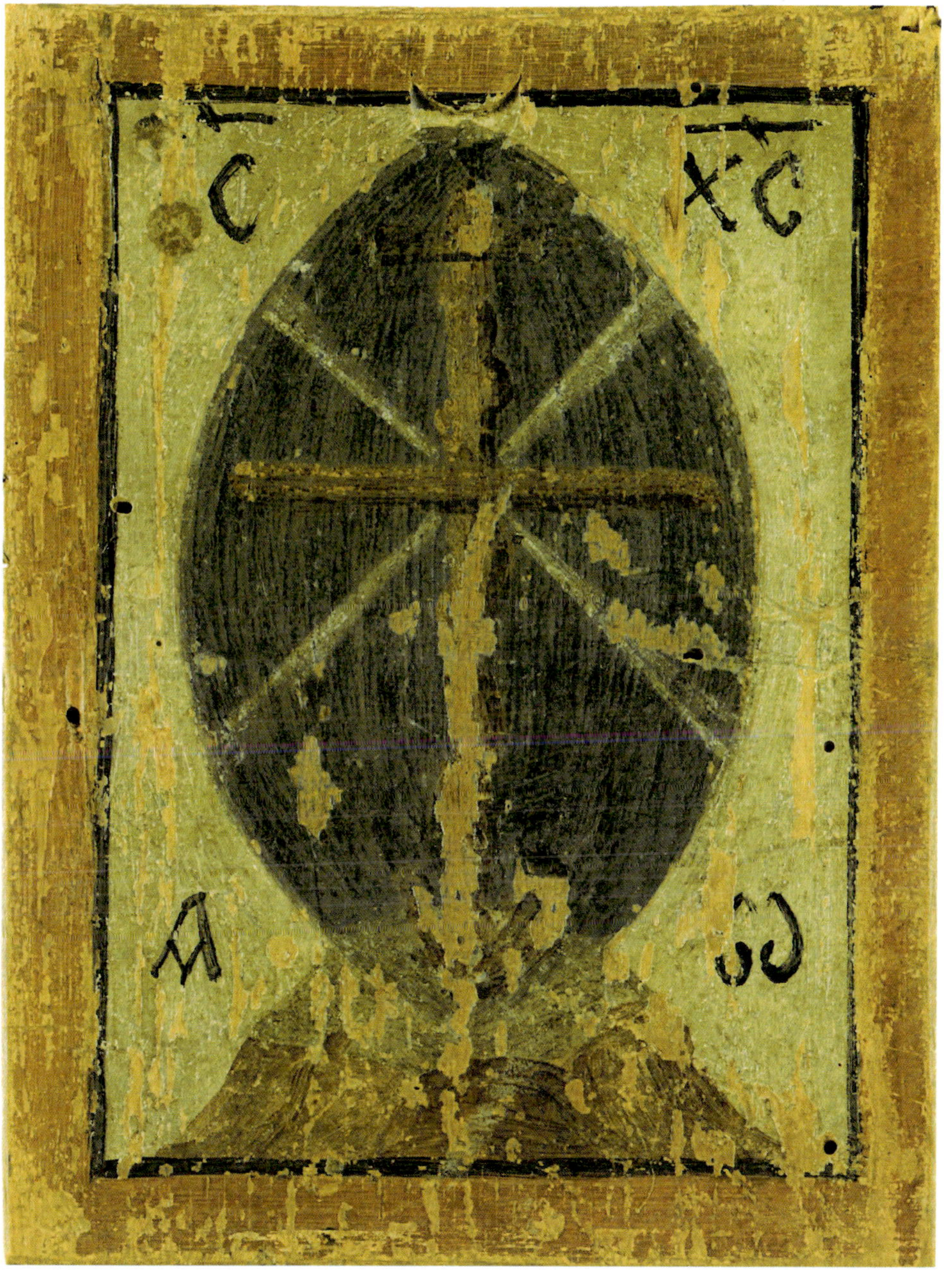

Plate 4 Obverse of lid, the Sancta Sanctorum reliquary box with stones, wood and cloth, sixth century, Syria or Palestine, painted wood, stones, wood fragments, and plaster, 24 × 18.4 × 3 cm, lid 1 cm thick.

Photo credit: Public domain.

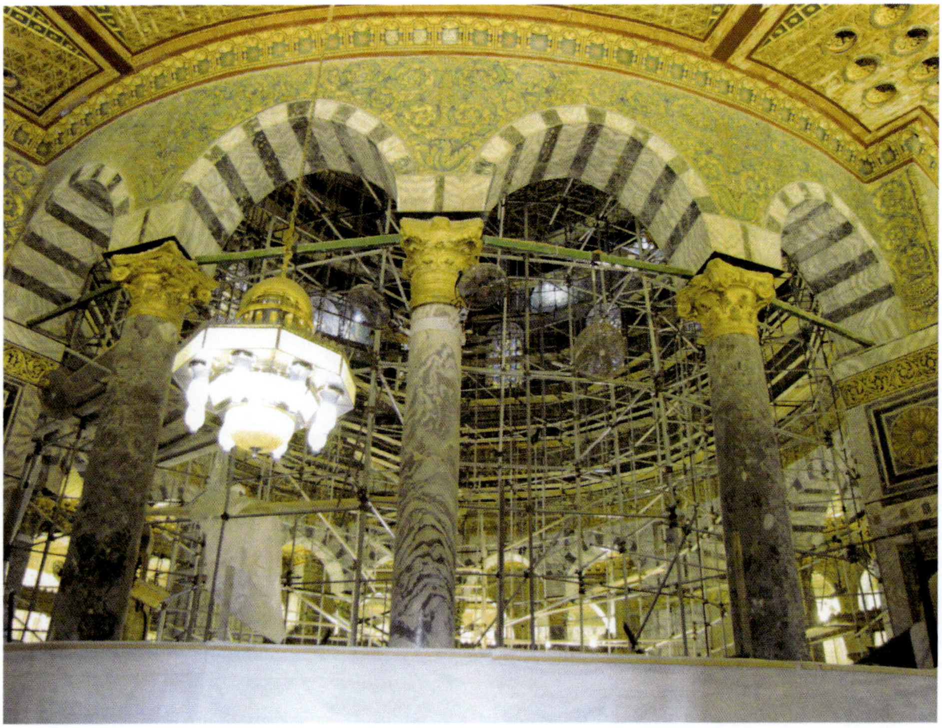

Plate 5 Dome of the Rock, interior, view of inner colonnade on south side.
Photo credit: Lawrence Nees.

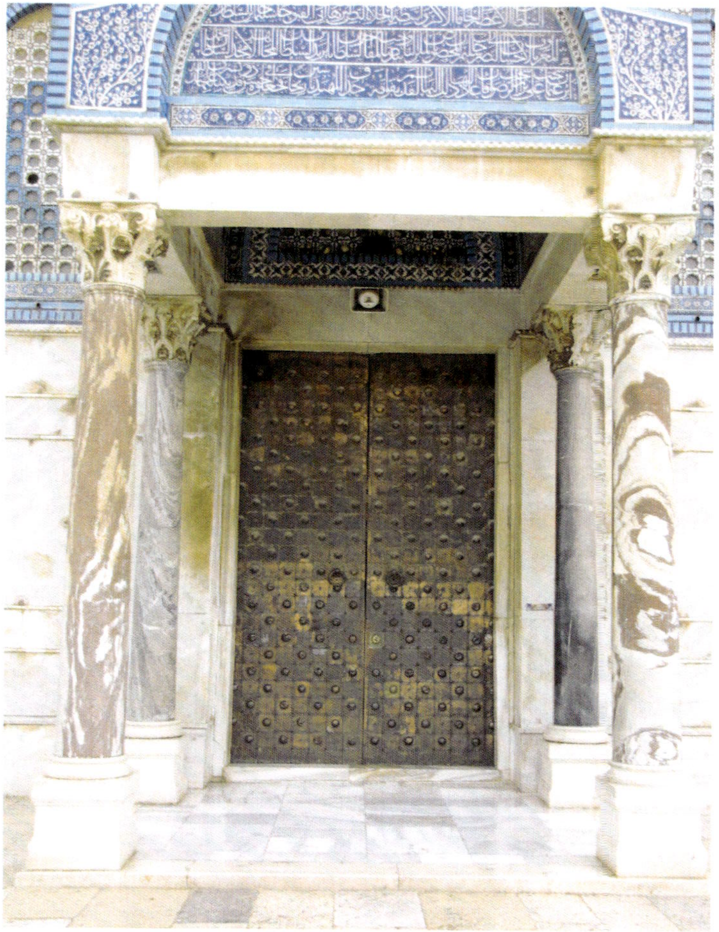

Plate 6 Dome of the Rock, exterior, north porch, 2012.

Photo credit: Lawrence Nees.

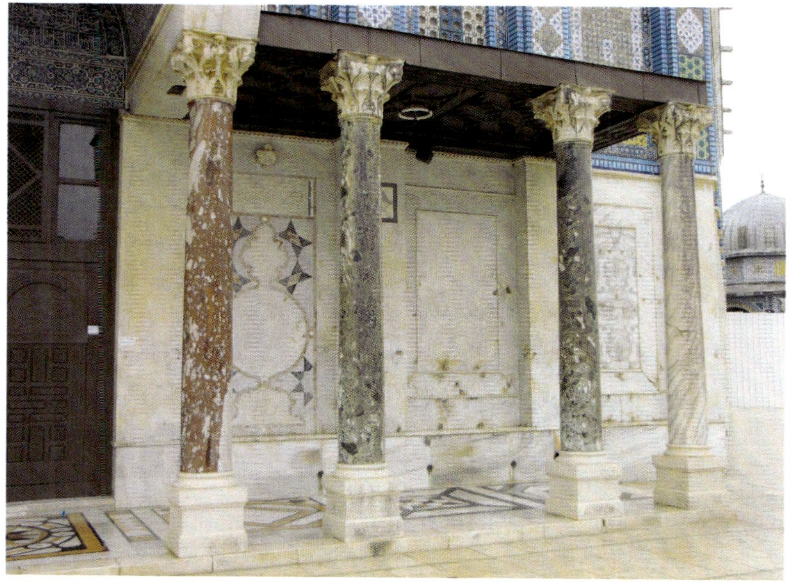

Plate 7 Dome of the Rock, exterior, south porch, columns to the right of the door.

Photo cred t: Lawrence Nees.

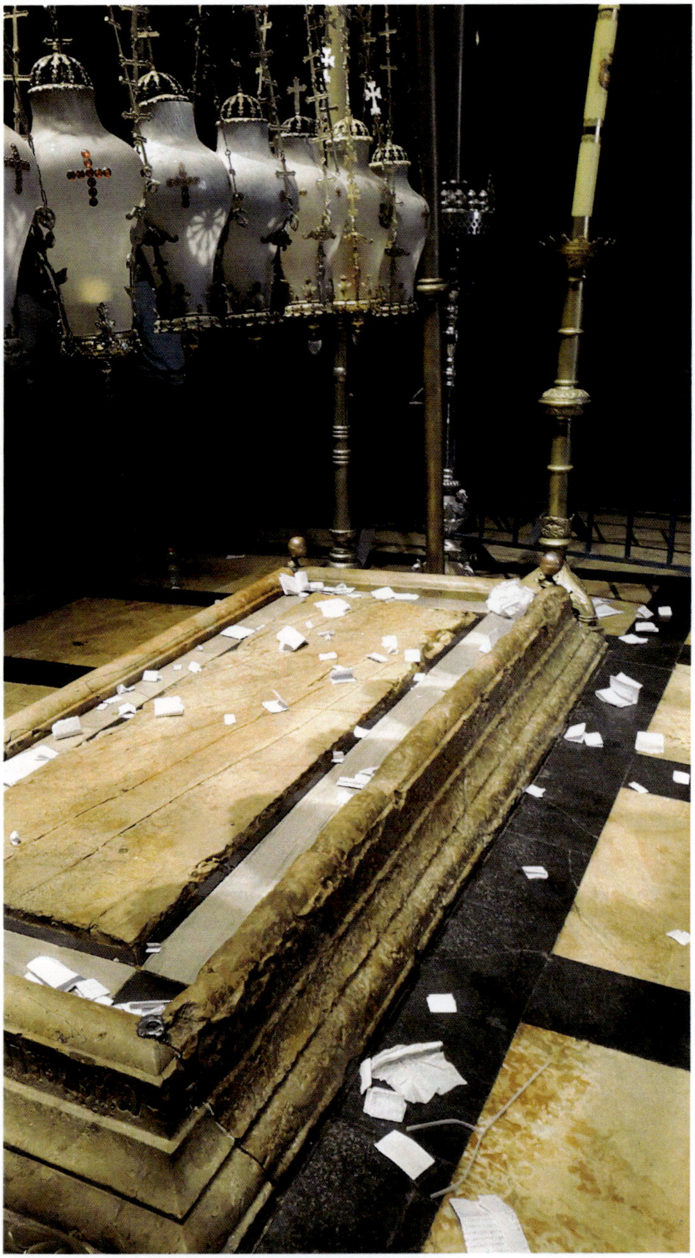

Plate 8 The Stone of the Unction, Church of the Holy Sepulchre, Jerusalem.

Photo credit: Netta Amir.

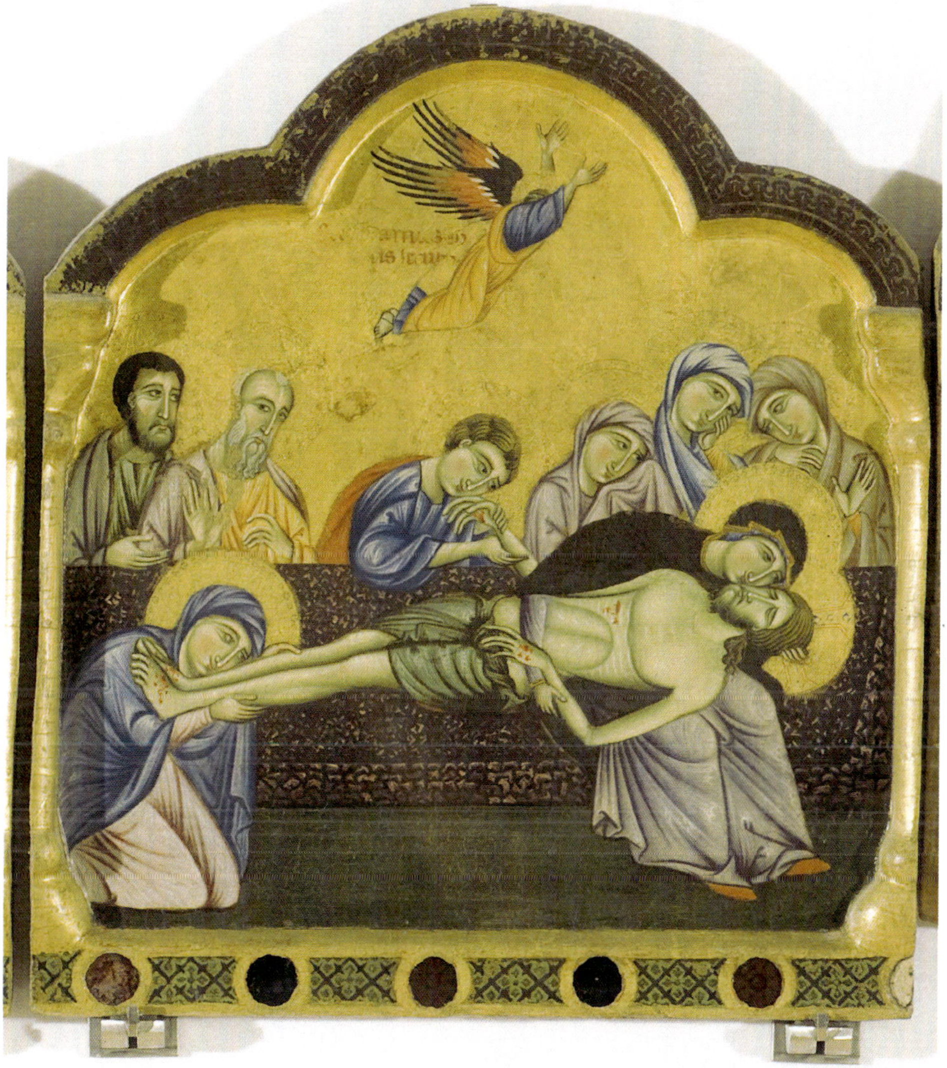

Plate 9 Maestro di San Francesco, Lamentation, between 1262 and 1272, tempera on wood, a panel from a double-sided polyptych, chiesa di San Francesco al Prato, Perugia, Italy.

Photo credit: Galleria Nazionale dell'Umbria.

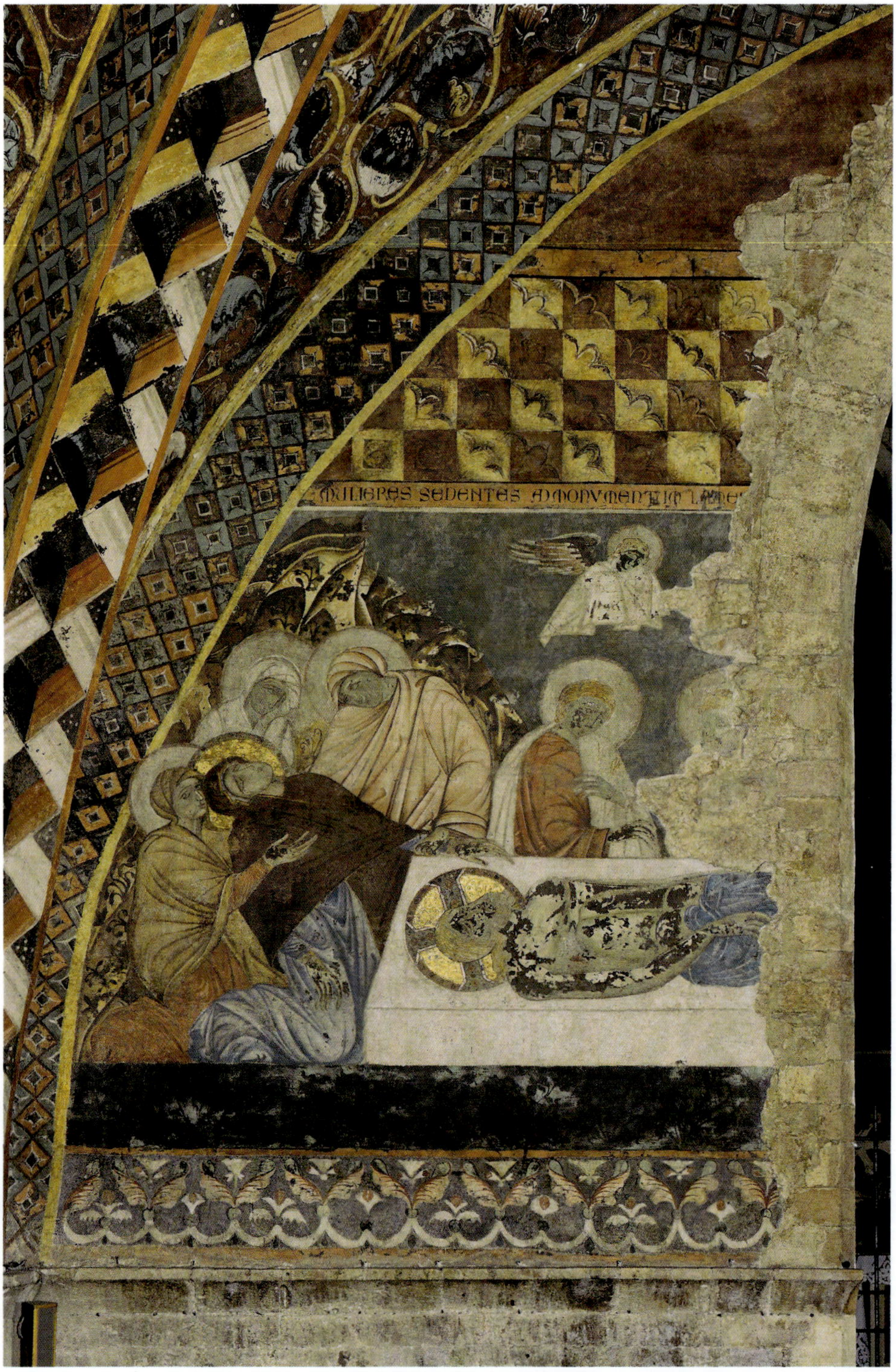

Plate 10 Maestro di San Francesco, Lamentation, before 1265, fresco, nave of the lower church of Saint Francis, Assisi.

Photo credit: Foto di Marcello Fedeli, Spoleto – 2014; ©Archivio fotografico del Sacro Convento di S. Francesco in Assisi, Italia.

III

Instillation and enactment

9 Rocks of Jerusalem

Bringing the Holy Land home

Elina Gertsman and Asa Simon Mittman

Introduction: "Feelings for sacred *things*"

In his seminal work, *The World of Late Antiquity*, published in 1971, Peter Brown characterizes the difference between late Roman polytheism and the earliest Christianity as a difference between *things* and *people*. He writes:

> Traditional paganism had expressed itself through forms as impersonal as the universe itself: it mobilized feelings for sacred *things* – for ancient rites, for statues, for oracles, for vast beloved temples. The "new mood," by contrast, threw up men – raw individuals who believed that they were the agents of vast forces.[1]

Toward the end of his study, Brown makes a finer distinction. The devotion to holy *men* only lingered as long as those men lived, as long as the initial age of martyrs and other early Christian heroes endured. Eventually, as was inevitable, these people were outlasted by things they left behind, by pieces of their bodies, objects they touched, and places they lived:

> They had lasted while mere men came and passed away. The new devotion was an upsurge of loyalty to holy *things*, while the enthusiasm of previous centuries had concentrated on holy *men*. In Rome and in Gaul, the relic and the martyr's grave totally ousted the living holy man in the popular imagination.[2]

Brown generalizes, of course, but his observations on the role of objects in changing patterns of devotion encapsulate the main themes we explore here. Our focus is a remarkable object – or, rather, a collection of objects, in turn housed within another object, which bears on it representations of yet other things: a reliquary box, once held in the treasury of the Sancta Sanctorum in the Lateran Palace, containing bits of stone, wood, and cloth, labeled with locations from the "Holy Land" (Plate 1).[3] The box, now in the Museo Sacro in the Vatican, has been credited to sixth-century Syria or Palestine.[4] Its sliding lid (obverse) bears a painting of the cross intersected by what appear to be a lance and a reed with a sponge, forming a schematic Christogram, inscribed within a mandorla and placed on the Golgotha hill; Christ's initials are inscribed in the upper corners, and *alpha* and *omega* are painted on either side of the hillock (Plate 4). On the reverse side, which faces the relics enshrined in the box, is a series of images narrating scenes from the life of Christ: the Nativity, baptism, crucifixion, Holy Women at the Tomb, and ascension. These five scenes portray an encapsulation of Christ's earthly experience with great economy. Altogether, the contents of the box, the monogram, and the paintings function as a threefold conjuring of Christ: in image, name, and material remains made sacred by contact with Christ.[5]

Building on Brown's observations, we consider this box and its contents through the lenses of thing theory and theories of memory. We pay particular attention to the tensions between the individual objects and the ways their arrangement and proximity create a collective; their invocation of distant locales; and their agentic potential. The objects housed in this box look to most modern viewers like no more than what they literally are – rocks, splinters of wood, scraps of textile. Yet for their collector, and for subsequent medieval viewers – including the popes who had exclusive access once the box was in the Sancta Sanctorum – these *things* would have served as a complex mnemonic map, punctuated with a series of visual, material, and cognitive triggers that potentially encouraged an ersatz, performative pilgrimage.[6] We inquire into the possible patterns and processes of an imaginative, transcendent journey, made possible by these base fragments.

Pilgrimages far and near

Christian pilgrimage, which originated in late antiquity, was popular until 638, when the Holy Land was incorporated into the vast Umayyad Empire.[7] The sacred sites in and around Jerusalem were of great importance to early pilgrims, who collected various natural and man-made objects considered to have therapeutic and transcendent properties, which allowed the travelers to retain a material connection to these locales once they returned home. Indeed, Jaś Elsner argues that "[t]hese places, themselves [are] a collection to be experienced by pilgrims."[8] However, the Holy Land's centrality to pilgrimage dwindled even before the fall of Jerusalem. As Elsner and Joan-Pau Rubiés note:

> By the end of the fourth century, the entire landscape of the Roman empire would be dotted with martyria, shrines and churches marking the specific spots where important saints had lived, the places where their bodies finally came to rest, and the location of significant artefacts associated with them, such as the True Cross.[9]

From this point on, travel to the Holy Land was decreasing in irreplaceability as the distributed network of sacred sites expanded throughout the Mediterranean and European world.[10] As Brown writes, Gregory of Tours found that "relics were everywhere, scattered throughout the entire Christian world. In every region, there were specks of dust unlike all other specks of dust, fragments of bone unlike other fragments, tombs unlike other tombs."[11] Our present subject was, then, not only a node within this network, nor only a model of the sacred geography from which it came, but also a microcosm of both the Holy Land it explicitly references and the Western European world, whose many relics – which reconfigured the landscape into the body of Christ – were being indexed by the assemblage of fragments, pebbles, and splinters arranged within the box.

Spiritual pilgrimages challenged the necessity of physical journeys to the Holy Land through the proliferation of relics.[12] To discourage pilgrimage, in the fifth century

Abbot Shenoute of Atripe related his vision of Christ, who inspired Shenoute to "glorify Jerusalem in your monastery, which you have dedicated to my name together with those who will hear and obey you, as equals of the angels. . . . You must know that my Cross is everywhere for whoever desires to repent."[13] That is, although there was certainly value in the sites of the Nativity and the Passion, the importance of visiting them was on the decline by the time the Sancta Sanctorum reliquary was assembled. This was not because they were unimportant. Quite the contrary: it was because they were seen as transportable and because "my Cross" – embodied in each of its splinters – was already everywhere. Indeed, even for those visiting Jerusalem in the flesh, the spiritual pilgrimage might eventually displace the physical journey. For example, Peter the Iberian, a highly influential fifth-century Palestinian bishop, was "an enthusiastic pilgrim," who on one visit to Jerusalem fell down in full *proskynesis*, crawled the final stretch of the route, and then "repeatedly touched the holy ground with [his] lips and eyes."[14] Nonetheless, on a subsequent visit, Peter "refrained from visiting the holy places in Jerusalem" even though he was just outside the city.[15] As his *Vita* reads:

> [T]he blessed one returned to the brethren in the plain. When he went, some were indignant in their souls and said, "How, when he abode all these days beside Jerusalem, did the blessed one not desire greatly to enter the Holy City, even if by night, and venerate the worshipful places, and especially the holy Golgotha and the life-giving tomb?"[16]

A monk "who was very simple and innocent" then recounts a vision in which Peter guides him through all the major sites. Together, they travel to the Martyrium of Saint Stephen and the cave with his tomb. Peter – in the vision – then "ran down to the holy Golgotha and the holy Tomb," to the churches of Pilate and the Paralytic, Gethsemane, the holy Ascension, the house of Lazarus, Bethlehem, the tomb of Rachel, and "the rest of the temples and houses of prayer on the road," and to Siloam, and to Zion. In the vision, they "completed a holy course and had worshipped the Lord in every place."[17] All this travel, following the standard pilgrimage route, was accomplished without *actually going* to these places, either for Peter, who found himself just outside the city, or for the "simple" monk stationed in Beth Tafsha, a few miles away.[18] During the period in which the present reliquary box was made, the worshipper's physical presence in the original landscape was decreasing in importance, just as its sites were transported throughout the Christian world in visions and in relics, among which were the so-called *benedictiones* or eulogiae (blessed objects), stored within the Sancta Sanctorum box.[19] By the seventh century – after the Islamic conquest – the city of Rome, filled with the very objects that were seen as facilitating spiritual pilgrimages, had superseded Jerusalem as the primary pilgrimage destination for those farther west.[20] Elsner writes of a box of relics found in the crypt of the Basilica of Saint Columban, Bobbio, containing "ampullae . . .[,] earth, dust and various small objects or tokens in wood, metal and terracotta."[21] These objects "evoke the totality of the Holy Land,"[22] an effect amplified by the arrangement of relics of place within the Sancta Sanctorum box. Once collections of relics, such as the Sancta Sanctorum box and its contents, were transported to the West, they brought Jerusalem with them.

Remainders and reminders

The mnemonic function of the box – for the original collector and later viewers familiar with its landscape of origin and the stories set therein – plays out in multifold ways. Most essentially, the gathering of eulogiae within strives to grasp what Christ left behind after the resurrection. Fragments of cloth, wood, and stones are keyed to the locations where Christ experienced his terrestrial existence: the Mount of Olives, for instance, and Bethlehem; one is marked "from Zion"; another identifies its provenance "from the life-giving [site of] Resurrection," recalling the reference in the *Vita* of Peter to "the life-giving tomb," and also the longing for curative effects.[23] Each eulogia, therefore, was capable of conjuring not only the specific place but also the specific narrative attached thereto.[24] Further, each could function as an experiential mechanism in the larger network of significations attached to the very material that was made to evoke these narratives and link them together typologically.

The bit marked "Bethlehem," for instance, is simply a thin piece of wood, but it is rich in heuristic potential. It could have been a fragment *from* Bethlehem, a souvenir of the place where Christ was born, and the kind of material available to the pilgrim. Why not a stone, which was likely just as readily obtainable? Certainly, because it comes from the place associated with Christ's birth, it would reference Christ's crib, the manger originally made of wood, which was kept in the crypt of the Nativity Church in Bethlehem, and mentioned by Origen ca. 247:

> If anyone wishes to have further proof to convince him that Jesus was born in Bethlehem besides the prophecy of Micah and the story recorded in the Gospels by Jesus' disciples, he may observe that, in accordance with the story in the Gospel about his birth, the cave in Bethlehem is shown where he was born and the manger in the cave where he was wrapped in swaddling clothes.[25]

The wood from the ostensible manger – the five sycamore planks – was likely brought from Palestine to Rome in the mid-seventh century, and is preserved in Santa Maria Maggiore (Figure 9.1).[26] Thus, the wood splinter embedded in the reliquary box was ontologically associated with the wood of the manger, pointing to it without necessarily coming from it. These fragments – the manger as a piece of the assemblage of the barn and the splinter as a piece if not of the manger then of the place from whence it came – were related in their material, source location, and, at least in part, in their recollection of and resonance with nativity narratives.

In addition to the crib the Bethlehem eulogia recalls – we argue, purposefully – another relic associated with Christ: the cross. The cross's many splinters, multiplied by the relic-hungry pious, circulated widely throughout "Christendom" and, when stripped of the expensive reliquaries that enshrined them and hid their irregularities, look precisely like the humble shard inserted into paste in the Syrian box (Figure 9.2).[27] Moreover, as one of the main *brandea* of the passion, the cross not only stood as a reminder of the event but indexed the body of Christ itself. In general, as a propagative matter, the wood was semantically and conceptually likened to flesh: it is not an accident that the Greek *ὕλη* means both "wood" and "matter."[28] The analogy between the cross and the body stretched on it was articulated by numerous theologians throughout the Middle Ages, anxious to explain that the worship of the cross was, in essence, like the worship of Christ himself.[29] In other words, the spiritual history

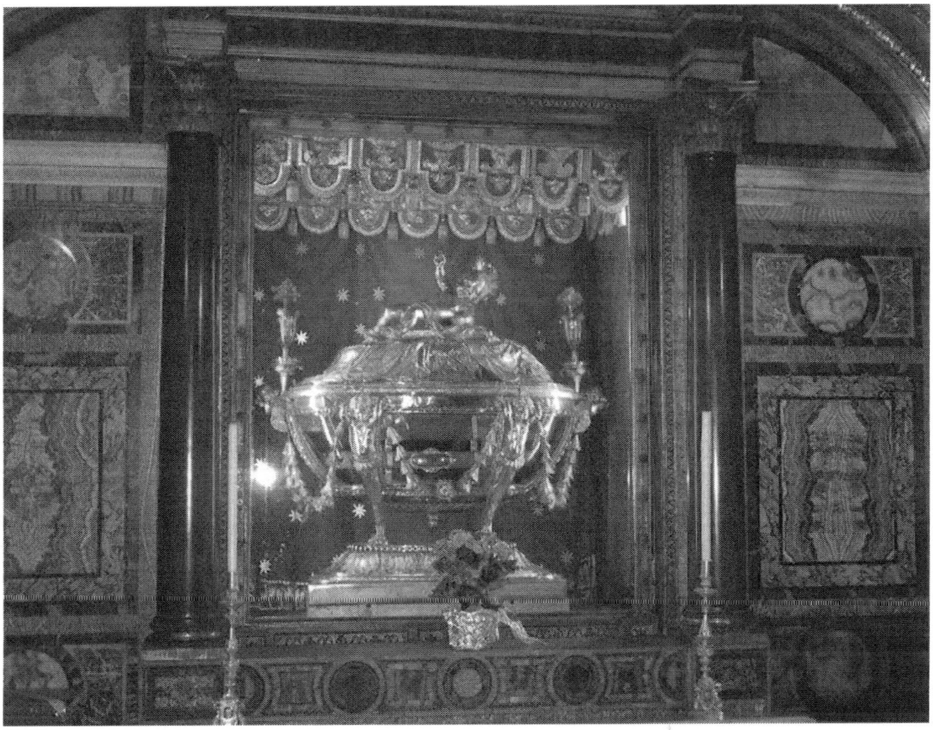

Figure 9.1 Giuseppe Valadier, Reliquary of the Holy Crib (sacra *culla*) (1802, Rome), Santa Maria Maggiore Basilica, Rome. Photo credit: Riccardov.

of the wood inflected the meaning of the splinter visually formulated as a devotional object, so the fragment marked with "Bethlehem" would also evoke Christ's infancy and his Passion.[30] It would also bear resonances of the earthly sites of these events, and the heavenly home of their protagonist. As suggested by the *Life of Desiderius of Cahors* – written shortly after his death in 655 – any relic acted as "a fragment of Paradise."[31] In this case, the wood fragment – as other objects in the box – would have a dual function: for a devout Christian in search of material sacredness (or sacred material), it would transport something of heaven, as other relics, but also something of the present Holy Land as well. The pilgrim gathering sacred splinters and stones and locking them away in a box would essentially enact Matthew 13:44, which compares the Kingdom of Heaven to a treasure, first hidden in the field, then unearthed, and then hidden again. This mnemonic evocation would have been echoed by the image of the Nativity painted on the lid, which figures the Christ Child not in the wooden manger but atop a stone altar and above a *fenestella* – a niche for the relic.[32] This Eucharistic image models the meditation on the Nativity that merges Christ's birth, sacrifice, and resurrection, offering an *espèce* of a visual guide for contemplation of the wood fragment.

There is no doubt that the associative net structured by the raw materials of the eulogiae would have been complicated by the relationships between and among the objects within the box.[33] Although we have no way of knowing what most of them

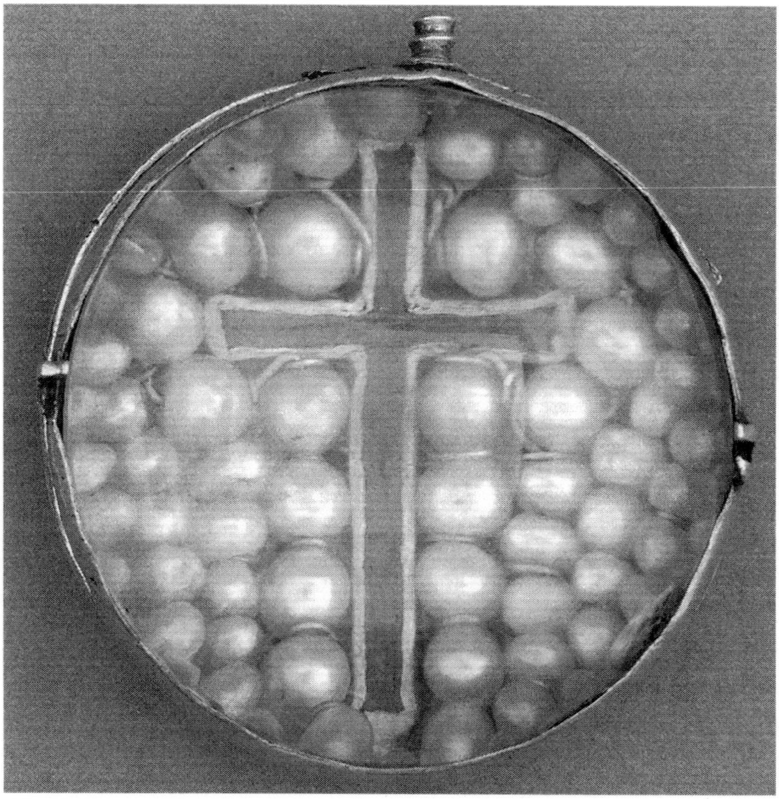

Figure 9.2 True Cross reliquary pendant (ca. 1200, Scotland), gold, wood, rock crystal, pearls, 5.5 × 5.2 × 2.8 cm, The British Museum, London. Photo credit: © Trustees of the British Museum.

index – there are very few surviving identifications – the arrangement of the objects into a loose Christogram suggests that they were meant to function not only as discrete mementos, but also as a coherent whole, mapping the topology of the Holy Land as Christ's very *nomen sacrum*, and so his body. Derek Krueger speculates that the arrangement of the stones might not be original to the box because some are now too high for the lid to slide into place over them.[34] He suggests that they might have originally been placed in a bag, and in turn placed in the box. However, as Krueger notes, the paintings on the inside of the lid did not suffer the abrasions that would likely have resulted from the contents shifting around, even in a bag. Similarly, the delicate bit of wood would probably have been destroyed by being jostled by the sharp-edged rocks. It seems more likely that the arrangement is original, but that some of the stones were reset over time, as the compound dried out and some of the contents loosened. At the fulcrum of the Christogram sits a triangular stone, marked as an object from the place of the resurrection. Its form is pregnant with associations: for Plato, the triangle stands at the heart of each body and each element; for Boethius, it is the "principle and element of all forms"; and it certainly evokes the Trinity.[35] The material is significant as well: it recalls a stone sepulchre in which Christ was buried and from which he arose, and it evokes the stone used to seal the tomb and then rolled away to announce the resurrection.[36] Placed at the center, the stone becomes a pivot on which the narratives unfolding through other eulogiae turn: its function as a visual and theological nexus, for example, activates the numerous typological parallels between the Nativity and the Resurrection and strengthens the association between Christ's life and death already inherent in the small wooden fragment marked with "Bethlehem."[37]

The centrality of the resurrection eulogia is echoed in the prominence of its inscription: it is the only one – at least the only extant one – that describes the locus sanctus from which it comes, offering a characterization of the place it comes from as "life-giving." Others serve rather as abbreviated, practical markers with which to anchor the objects and the memories. The stress on locality is significant inasmuch as it highlights the manifold metonymic function of the eulogiae: just as their arrangement into a Christogram symbolizes and therefore embodies Jesus as the Messiah, so too, their very nature as pieces of certain locations makes them, in fact, their locations.[38] The importance of these locations are also evident in the four narrative images on the lid, which moor the beholder's imagination to particular, lovingly rendered places: the Nativity cave, the River Jordan, Golgotha, and Christ's sepulchre as it looked before its transformation in the subsequent century.[39] For viewers who had been on pilgrimage, including the original collector, this specificity would facilitate a vivid recall of the visited sites. On the other hand, the images' consistent and unifying gold backgrounds translate these worldly and perhaps personally observed locations into a heavenly realm. Even for those who had not physically traveled to the sites, they would still serve as an aid for imaginary re-enactments of the episodes from Christ's life, a devotional exercise that would aid prayer and meditation. Indeed, the inside of the lid functions as a narrative icon, particularly effective because of its very layout, familiar most immediately from ancient and medieval treatises on memory.[40]

Perhaps the most germane source on medieval mnemonic theory is the so-called *Rhetorica ad Herennium*, a treatise formerly attributed to Cicero, extant in more than 100 exemplars, cited by Rufinus and Grillius in the fifth century and by Priscian in the sixth.[41] Book III of the treatise outlines strategies for successful construction of artificial or trained memory, which would facilitate subsequent recall of whatever information was required. The treatise suggests the construction of so-called *loci*, or backgrounds, into which *imagines*, or images, are to be inserted. Images are distinctive figures; *loci* are spaces; together, they constitute a mnemonic vehicle for heuristic recall of various events. Scholars have long recognized the value of the treatise for unpacking a wide variety of medieval images, especially those suggestive of a diagrammatic structure especially resonant with what *Ad Herennium* proposes.[42] It suggests that *loci* be arranged in distinctive series; not be crowded; be distinct from one another; and be properly, but not too vibrantly, illuminated. The treatise acknowledges that the variety of backgrounds presupposes "a relatively large experience," although anyone can do it through the use of one's imagination: "For the imagination can embrace any region whatsoever and in it at will fashion and construct the setting of some background."[43] In turn, the images ought to be of the kind that "can adhere longest in the memory": that is, colorful and brilliant, at times violent, and striking in their palette and their configuration. They are magnets for the process of recollection, visually interesting, filled with memorable details. Once the loci are populated, they have to be "again and again run over rapidly in the mind . . . in order to refresh the images."

The inner lid of the reliquary, in other words, would make an ideal schema for such a *loci*-and-*imagines* construction. Set within a symmetrical grid, uniformly gilded, and enclosed in distinctive frames, Christological narratives inhabit five separate compartments, each a locus for *memoria rerum* that is configured as a summary image of a complex event. Moreover, each compartment is clearly outlined, with two grouped atop and two below the central locus – a suggestion we find in later medieval treatises on memory, such as Thomas Bradwardine's *De Memoria Artificiali* – all interrelated through the presence of Christ and thus arranged in a well-defined series.[44] Images are uniformly eye-catching, visceral: one revels in the vivid rendition of the Sepulchre, for example, whereas another foregrounds the violence inflicted upon Christ's body. This is not to say that the painters and users of the box referred to the treatise as a source, but, rather, that they utilized commonly understood mnemotechnic strategies, also evident in contemporary codices, such as the sixth-century Italian copy of the Gospels of Saint Augustine, which Mary Carruthers has described as "a complete set . . . of '*imagines rerum*' for the events of the Passion" and posited as a near embodiment of "medieval mnemotechnic pedagogy."[45] The eulogiae container thus becomes a quintessence of the *thesaurus*, the strongbox – a common medieval metaphor for trained memory – that contains, stores, and organizes visual and material prompts for remembering.[46]

The organizing principle of this prompt is the cross. Painted on the exterior of the lid, it takes center stage on its interior with the crucifixion episode, painted twice the size of the other scenes. As the cross nearly disappears behind Christ's body, the man on the cross becomes, in essence, the cross itself. Christ's figure, moreover, suggests a metonymic correspondence between the cross and the format of panels on the lid: his, and the thieves', outstretched arms emphasize the horizontal stretch of the image that functions visually as a *patibulum*, while his blue *colobium* is continued in the vertical lines that divide the top and bottom scenes, implying the *stipes*. Within, the arrangement of the objects echoes the outer image of the lid: the "X" form simultaneously presents the *chi* and evokes the crossed lance and reed, the vertical line appears both as the stem of the *rho* and the *stipes*, while a corresponding horizontal run of stones figures the *patibulum*. The lid's interior thus serves a dual purpose. On the one hand, it functions as an intermediary between the symbolic Christogram painted on the outside and another Christogram inlaid with stones within: the crucified Christ appears at the exact center of the panel, and of the box as a whole, and, when the lid is closed, his painted body is nestled between the center of the cross on the outside and the stone that marks the resurrection on the inside. Christ's death on the cross, then, at once alludes to the dogma of the Resurrection and to a perpetual Passion. On the other hand, because the design on the outside of the lid is repeated with objects within, the lid becomes a transparency, revealing rather than concealing the contents beneath, as they, in turn, reveal the locales and the events they index through mnemonic chains of association, and through their own vibrant presence.

Imaginary worlds, material remains

This vibrancy asserts itself to this very day. The collection of eulogiae is quietly arresting. Perhaps much of its force comes from the juxtaposition of the beautiful paintings, which have frequently and justly been compared with the renowned illuminations of the Rabbula Gospels,[47] and the apparent banality of the objects they cover, celebrate, and ennoble. The fragments recall Jane Bennett's observations about the surprisingly attractive contents of a sewer drain ("one large men's black plastic work glove, one dense mat of oak pollen, one unblemished dead rat, one white plastic bottle cap, one smooth stick of wood"). Therein, she found:

> [A]n energetic vitality inside each of these things, things that I generally conceived as inert. In this assemblage, objects appeared as things, that is, as vivid entities not entirely reducible to the contexts in which (human) subjects set them, never entirely exhausted by their semiotics.[48]

Bennett continues, a bit further on: "the items on the ground that day were vibratory – at one moment disclosing themselves as dead stuff and at the next as live presence: junk, then claimant; inert matter, then live wire."[49] The contents of the reliquary are quite similar in their presence – indeed, much of what we have argued here is predicated on *medieval* ideas about the power and presence of these objects. At once vital and sacred, this gathering of meaningless detritus vibrates with yet greater presence when pulled together. In a sense, this is the operative principle of all visible relics, which look like what they are – bits of bone, locks of hair, scraps of fabric, splinters of wood – and simultaneously like things of great importance and potential.

What is created within (and through) this assemblage? In his *Building Imaginary Worlds*, Mark Wolf presents many strategies by which authors, artists, filmmakers, and others construct fictional worlds of various types. What occurs with all of these overlaps and imbrications is akin to his notion of diegetic braiding, "[t]he condition that occurs when multiple stories or narrative threads set in the same world share the same locations, objects, characters, and other details."[50] Each object in the reliquary is a piece of the Holy Land, and therefore is part of a series of interlocking stories.

The members and images of this complex object (or complex of objects) construct an imaginary world by conjuring, wherever it is, the earthly and heavenly Jerusalem. In so doing, it/they generate(s) an interaction with the viewers. As Wolf writes, "imaginary worlds invite audience participation in the form of speculation and fantasies."[51] In the appendix to his book, Wolf provides a list of imaginary worlds that contains many entries familiar to medievalists – the Land of the Arimaspi (from Herodotus), Thule (from Pytheas's *On the Ocean*), "Blemmyae Land" (Pliny),[52] and so on, but he also lists Eternal Jerusalem, citing Augustine's *City of God*.[53]

The world encased in the reliquary box is reminiscent of Wolf's diegetic braiding in a second sense: for Augustine, following Psalm 87, "Jerusalem was the 'City of our God' of which 'glorious things are spoken,'" and so, as Brown articulates, "Jerusalem stood for heaven, the distant home of the saved."[54] The landscape of the present, earthly Jerusalem (and its immediate environs) was something of a stage set for the performance of visionary experiences of the City of God to come. This might also be seen as an example of an overlaid world, "[a] fictional diegesis in which an existing, Primary World location is used" – that is, the present Jerusalem – "with fictional characters and objects appearing [in] it, but without enough invention to isolate it from the Primary World into its own separate secondary world."[55] Just as many relics were seen at once as the fragmentary remains of dead people and as active, living presences of saints in heaven, so the landscape of Jerusalem was dually resonant. The box, in turn, contains fragments of the landscape that, once removed from their points of origin and reconfigured (obscurely, loosely) into the *nomen sacrum*, the implements of the Passion, and the body of Christ, bring the earthly Jerusalem to distant Rome. As

Elsner writes, "[t]he power of relics – and in particular of *collections* of relics – lies in this special combination of tangible presence and particularity with highly generalized and suggestive evocation of a greater and more sacred past."[56] The objects betray a longing for Jerusalem at the same time as they render its physical presence unnecessary, even obsolete. The pilgrimage and the return journey thus establish an ersatz path for the pilgrim: the objects carried from the Holy Land and eventually deposited in Rome allowed the viewer to be transported, in a sense, to the earthly Jerusalem – obviating the need for actual pilgrimage – and, from there, onward to heaven.

The objects in the box thereby not only serve to recall a past pilgrimage but also to create a present one, granting the box a role beyond its original user.[57] The vitality of the assemblage is rooted in the perceived power of the eulogiae to generate – to quote McKenzie Wark, writing about a different context – a "virtual geography, the experience of which doubles, troubles, and generally permeates [the] experience of the space" actually inhabited by the viewer.[58] Several scholars have written about "spiritual" or "imagined pilgrimage," a process facilitated by medieval *mappaemundi*, relics, and other objects. Daniel Connolly posits Cassiodorus – who lived for a time at Constantinople – as a possible point of origin for this notion, and one that is roughly contemporaneous with the Sancta Sanctorum reliquary.[59] This sixth-century historian and founder of the Monastery of Vivarium, with its highly influential scriptorium, sought to encourage his readers to use geographical texts as a way to know the world. He noted:

> [I]f a noble concern for knowledge has set you on fire, you have the work of Ptolemy, who has described all places so clearly that you judge him to have been practically a resident in all regions, and as a result you, who are located in one spot, as is seemly for monks, traverse in your minds that which the travel of others has assembled with very great labor.[60]

Cassiodorus' text echoes the dynamic we see at play in the reliquary box, likewise "assembled with very great labor." Like the geographies of Ptolemy, the assemblage of relics and the paintings with which they are accompanied were the starting point for mental and spiritual travels. These could be rooted in memories of the actual pilgrimage during which these objects were collected or in memories of other pilgrimages taken by other travelers. So too, though, these memories could be rooted in – and borrowed from – the rich complex of biblical and exegetical texts describing the events of the Nativity, Life, and Passion of Christ, as well as other works of art, literature, and liturgy. After all, travelers to the Holy Land did not merely see what was before them, but saw – in visions, in their minds' eyes – the long-past events that granted these otherwise ordinary places their extraordinary significance.

* * *

Encased in and contained by their box, the small *benedictiones* – much of the significance of which is rooted in their materiality, in their wood, their stone – bear witness to what Brown characterizes as "an upsurge of loyalty to holy *things*."[61] This upsurge proved to be tenacious: to this day, one may purchase a box filled with just such holy things, neatly arranged and carefully labeled, brought all the way from Jerusalem to be worshipped at home (Figure 9.3).

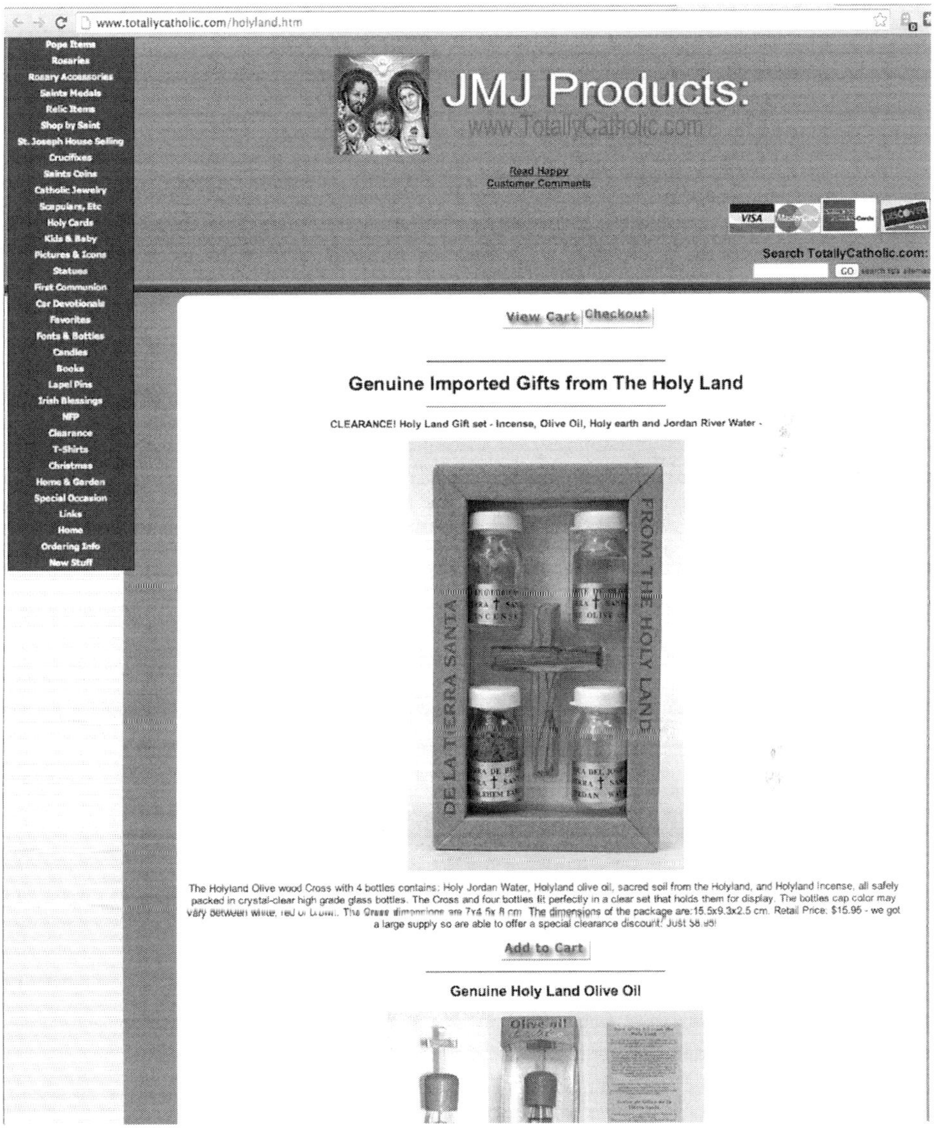

Figure 9.3 JMJ Products, "Genuine Imported Gifts from The Holy Land," *Totally Catholic* <www.totallycatholic.com/holyland.html> (accessed June 2014).

Or just, perhaps, to be put on display: the transparency of the lid in this contemporary box of relics suggests, more than anything else, constant scopic consumption. Not so in the Sancta Sanctorum container, whose engagement with sacred topography is predicated on enclosure. For here, the *benedictiones* were granted the primary view of the sites from whence they came, as depicted on the *inside* of the lid, which slides into place over them, concealing both the objects and the images from human spectatorship. That is, for the majority of their post-removal existence, these active objects, vibrant with potential and redolent of their sacred sites of origin, have been granted a view of the lavish gold-and-polychrome images that recall their autochthonous homes. When the box is closed, we are given a view of the hybrid image uniting Christogram and crucifixion: a pair that implies the way God came to Earth, and also the route by which to follow him to the City of God. When it is open, we are greeted with a palimpsestic pair of assemblages – painted and gathered – each of which dually presents and signifies the landscape of the earthly Jerusalem of early Christian pilgrimage and the longed-for landscape of the heavenly Jerusalem at the heart of its viewers' aspirations.

Notes

1 Peter Brown, *The World of Late Antiquity: AD 150–750* (London: Thames and Hudson, 1971, repr. 1974), 52. Emphasis is original.

2 Brown, *Antiquity*, 182. For more on Christian pilgrimage to holy men, see Jaś Elsner and Ian Rutherford, "Introduction," in *Pilgrimage in Graeco-Roman and Early Christian Antiquity: Seeing the Gods*, eds. J. Elsner and I. Rutherford (Oxford: Oxford University Press, 2006), 26, and Georgia Frank, *The Memory of Eyes: Pilgrims to Living Saints in Christian Late Antiquity* (Berkeley: University of California Press, 2000). For more on the shift toward pilgrimage to holy things, see Elsner and Rutherford, *Pilgrimage in Graeco-Roman and Early Christian Antiquity*, 29, and Hippolyte Delehaye, *Les origines du culte des martyrs* (Brussels: Société des Bollandistes, 1912), 60–119; Delehaye, *Sanctus: Essai sur le culte des saints dans l'antiquité* (Brussels: Société des Bollandistes, 1927), 196–207; and Delehaye, "Loca Sanctorum," *Analecta Bollandiana* 48 (1930), 5–64.

3 For more on groups of relics as collections, see Jaś Elsner "Replicating Palestine and Reversing the Reformation: Pilgrimage and Collecting at Bobbio, Monza and Wahingham," *Journal of the History of Collections* 9, no. 1 (1997): 117–130.

4 For the most recent catalog entry, see Martina Bagnoli, Holger A. Klein, C. Griffith Mann, and James Robinson, eds., *Treasures of Heaven: Saints, Relics, and Devotion in Medieval Europe* (Baltimore: The Walters Art Museum, 2010), #13, 36, entry by Cristina Pantalella. See also Anton Legner, *Ornamenta Ecclesiae: Kunst und Künstler der Romanik: Katalog zur Ausstellung des Schnütgen-Museums in der Josef-Haubrich-Kunsthalle*, vol. 3 (Köln: Stadt Köln, 1985), #H8, 80, entry by Anton Legner.

5 *Treasures of Heaven*, 36.

6 Derek Krueger, "The Religion of Relics in Late Antiquity and Byzantium," in *Treasures of Heaven*, eds. Bagnoli et al., #11, characterizes this as "a virtual pilgrimage to the Holy Land. The subject of early Christian pilgrimage is vast; we cite here three studies, all published by Oxford University Press, that offer fundamental overviews of this practice: Edward D. Hunt, *Holy Land Pilgrimage in the Later Roman Empire, AD 312–460* (1982); Peter W. L. Walker, *Holy City, Holy Places? Christian Attitudes to Jerusalem and the Holy Land in the Fourth Century* (1990); and Joan E. Taylor, *Christians and the Holy Places: The Myth of Jewish-Christian Origins* (1993).

7 Krueger, "The Religion of Relics," 11. Pilgrimage increased again after the Crusader conquests and endured through the Renaissance, despite Islamic rule.

8 Elsner, "Replicating Palestine," 117.

9 Jaś Elsner and Joan-Pau Rubiés, "Introduction," in *Voyages and Visions: Towards a Cultural History of Travel* (London: Reaktion Books, 1999), 17–18.

10 Brouria Bitton-Ashkelony, *Encountering the Sacred: The Debate on Christian Pilgrimage in Late Antiquity*, The Transformation of the Classical Heritage 38 (Berkeley: University of California Press, 2005), 200. For a list of the great variety of pilgrimage types in the period, see Elsner and Rutherford, 28–30. Although pilgrimage to the Holy Land did continue, it was no longer the only type available.

11 Peter Brown, *The Rise of Western Christendom: Triumph and Diversity, A.D. 200–1000*, rev. ed. (Chichester: John Wiley and Sons, 2013), 164.

12 Bitton-Ashkelony, *Encountering the Sacred*, 200.

13 Cited in Bitton-Ashkelony, *Encountering the Sacred*, 202–203. For source text, see E.C. Amélineau, ed. and trans., *Vita Sinuthii: Monuments pour servir a` l'histoire de l'Egypte chrétienne*, vol. 1 (Paris: Ernest Leroux, 1888), 333ff. Shenoute was abbot of the "White Monastery," on the outskirts of Atripe (modern-day Sohag), north of Thebes, and "the most important Coptic writer of Late Antiquity": Pablo Alvarez, *Rediscovering Shenoute of Atripe [ca. 348–465]*, a Digital Project, University of Michigan Library [01/30/2013]. Digital facsimile: http://www.lib.umich.edu/special-collections-library/rediscovering-shenoute-atripe-ca-348–466-digital-project (accessed November 2013). According to Heike Behlmer, "Visitors to Shenoute's Monastery," in *Pilgrimage and Holy Space in Late Antique Egypt*, ed. David Frankfurter (Leiden: Brill, 1998), 341, "[T]he monastery grew [under Shenoute] to be the most important religious center of the area, numbering as many as several thousand inhabitants, male and female," with a "large monastic library."

14 Cornelia B. Horn, *Asceticism and Christological Controversy in Fifth-Century Palestine: The Career of Peter the Iberian* (Oxford: Oxford University Press, 2006), 247, and John Rufus, *The Lives of Peter the Iberian, Theodosius of Jerusalem, and the Monk Romanus*, eds. and trans. Cornelia B. Horn and Robert R. Phenix, Jr. (Leiden: Brill, 2008), 52–53.

15 Bitton-Ashkelony, *Encountering the Sacred*, 185–186, quotation from John Rufus (?), *Vita Petri Iberi*, ed. and trans. R. Raabe (Leipzig: J.C. Hinrichs'sche Buchhandlung 1895), 98–100.

16 *Peter the Iberian*, 195.

17 *Peter the Iberian*, 196–197.

18 See Horn, *Asceticism*, 245–249.

19 See Gary Vikan, "Icons and Piety in Early Byzantium," in *Sacred Images and Sacred Power in Byzantium* (Aldershot: Ashgate, 2003), 572–574. For a brief discussion of topography vis-à-vis the box and its semantic relationship to the work of Robert Smithson, see Alexander Nagel, *Medieval Modern: Art out of Time* (New York: Thames and Hudson, 2012), 116–121.

20 Diana Webb, *Pilgrims and Pilgrimage in the Medieval West* (London: I.B. Tauris, 2001), 11ff.

21 Elsner "Replicating Palestine," 119.

22 Elsner "Replicating Palestine," 122.

23 "απ ζωοποιον 'Αναστάσεως." ["From the Life-Giving (place of) Resurrection."] Transcription and translation from C.R. Morey, "The Painted Panel from the Sancta Sanctorum," in *Festschrift zum Sechzigsten Geburtstag von Paul Clemen*, ed. Richard Klapheck (Düsselfdorf: L. Schwann, 1926), 151, who thanks Pio Franchi de' Cavalieri for assistance therewith.

24 Jas Elsner, "Replicating Palestine," 126, writes, "the collections are of 'relics' whose link with the origin provides a direct, metonymic evocation of the whole from which the fragments have come."

25 Origen, *Contra Celsum*, i.51, trans. Henry Chadwick (Cambridge: Cambridge University Press, 1953), 47–48. On the nativity crib, see Kurt Weitzmann, "Loca Sancta and the Representational Arts of Palestine," *Dumbarton Oaks Papers* 28 (1974): 31–55, 36; note to *Presepe Domini, exornatum insuper argento atque auro*, in T. Tobler, *Itinera et Descriptiones Terrae Sanctae*, 1 (Geneva: J.G. Fick, 1877), 53, par. IX. See also H. Vincent and F.-M. Abel, *Bethléem: Le Sanctuaire de la Nativité* (Paris: J. Gabalda, 1914), 136.

26 Sible Lambertus de Blaauw, *Cultus et decor: Liturgie en architectuur in laatantiek en middeleeuws Rome: Basilica Salvatoris, Santae Mariae, Sancti Petri* (Delft: Eburon, 1987), 197–198.

27 On the veneration of the cross and cross reliquaries, see, entries 27, 32, 34, 35, 36, 37, 45, 46, 47, 48,, 49, 62, 77, 85, 87, 89, and 90, and Guido Cornini, "'Non Est Toto Sanctior Orbe Locus': Collecting Relics in Early Medieval Rome," in *Treasures of Heaven*, eds. Bagnoli et al., esp. 71–72.

28 See Isidore of Seville, *Isidori Hispalensis episcopi Etymologiarum sive originum libri XX*, ed. W.M. Lindsay, 2 vols (Oxford: E Typographeo Clarendoniano, 1911), vol. 2, 321–322, book 19, sections 19.3–19.5 for the suggestion that "all wood is called matter" and the association between the words "matter" (*materia*) and "mother" (*mater*).

29 As early as the second century, Justin Martyr formulated the cross as the defining aspect of Jesus' identity as Christ, as the son of God (see *Dialogue with Trypho*, trans. Thomas

Falls [Washington: Catholic University of America Press, 2003], esp. 89–96); this trope, which was discussed in its many variants throughout the Middle Ages, received a definitive formulation by Thomas Aquinas, who argued that because the cross touched Christ's body and was drenched in his blood, and because it "represents to us the figure of Christ extended thereon," the cross must be "worshipped with the same adoration as Christ, viz. the adoration of '*latria*.' And for this reason also we speak to the cross and pray to it, as to the Crucified himself." Thomas Aquinas, *Summa theologica*, pt. 3, question 25, article 4, "Of the Adoration of Christ."

30 For the importance of materials in medieval art, see Herbert Kessler, *Seeing Medieval Art* (Toronto: University of Toronto Press, 2004), ch. 20, and Thomas Raff, *Die Sprache der Materialen: Anleitung zu einer Ikonologie der Werkstoffe* (Berlin: Waxmann Verlag, 1994).

31 "acsi partem paradysi": *Vita Desiderii ep. Cadurcae urbis*, ed. B. Krusch, eMGH SS rer. Merov. 4, 1902, 563–602. For a brief discussion, see Brown, *Christendom*, 162.

32 Weitzmann, "Loca Sancta," 36, writes: "The structure on the lid . . . is a massive cube with a niche in the center. This is the early form of the *altare fixum*, a so-called block altar, with a fenestella, a niche meant to enclose a relic, as seen in a sixth-century altar in the church of Bagnocavallo near Ravenna."

33 See Bianca Kühnel, "The Holy Land as a Factor in Christian Art," in *Christians and Christianity in the Holy Land, from the Origins to the Latin Kingdoms*, eds. Ora Limor and Guy G. Stroumsa (Turnhout: Brepols, 2006), 463–504.

34 Derek Krueger, "Liturgical Time and Holy Land Reliquaries in Early Byzantium," in *Saints and Sacred Matter: The Cult of Relics in Byzantium and Beyond*, eds. Holger Klein and Cynthia Hahn (Washington: Dumbarton Oaks, 2015), 111-113.

35 Plato, *Timaeus*, in *Collected Dialogues of Plato, Including the Letters*, ed. Edith Hamilton (New York: Pantheon, 1961), 53C–55C; Boethius, *De institutione arithemtica libri duo*, ed. Gottfried Friedlein (Leipzig: Teubner, 1867), 104; for the discussion of these sources and the exploration of the meaning of the triangle, see Rebecca Zorach, *The Passionate Triangle* (Chicago: University of Chicago Press, 2011), esp. ch. 1.

36 Mark 15:46, Vulgate Bible, emphasis ours: "Ioseph autem mercatus sindonem, et deponens eum involvit sindone, et posuit eum in monumento, *quod erat excisum de petra*, et advolvit *lapidem* ad ostium monumenti"; Mark 16:3–4: "Et dicebant ad invicem: Quis revolvet nobis *lapidem* ab ostio monumenti? Et respicientes viderunt revolutum *lapidem*."

37 For a detailed exploration of these many prefigurations, see Lynne Hilton Wilson, "Jesus' Atonement Foretold through His Birth," in *To Save the Lost*, eds. Richard Neitzel Holzapfel, and Kent P. Jackson (Provo, UT: Religious Studies Center, Brigham Young University, 2009), 103–126.

38 Elsner, "Replicating Palestine," also takes up the metonymic function of relic collections.

39 The outside of the lid is also evocative of place; the abstracted symbols of the Passion are planted into the hillock of Golgotha.

40 On narrative icons, see Paroma Chatterjee, "Archive and Atelier: Sinai and the Case of the Narrative Icon," in *Approaching the Holy Mountain: Art and Liturgy at St. Catherine's Monastery in the Sinai*, eds. Sharon E.J. Gerstel and Robert S. Nelson (Turnhout: Brepols, 2011), 319–344, esp. 329–334.

41 Harry Caplan, *Introduction to Rhetorica ad herennium* (Cambridge: Harvard University Press, 1954), xxxiv–xxxv.

42 See *Ad C. Herennium: De Ratione Dicendi*, 209–225. On the importance of *Ad Herrenium* for later medieval audiences, see, in addition to the work of Mary Carruthers, e.g., *The Book of Memory: A Study of Memory in Medieval Culture*, 2nd ed. (Cambridge: Cambridge University Press, 2008), Paolo Rossi's *Logic and the Art of Memory: The Quest for a Universal Language*, trans. Stephen Clucas (London: Athlone, 2000), and Frances Yates's *The Art of Memory* (London: Routledge, 1999 [1966]); art historians from Lina Bolzoni to Georges Didi-Huberman have long recognized its importance for the study of visual imagery.

43 *Ad Herennium*, III, XIX.32.

44 Thomas Bradwardine, *De Memoria Artificiali (On Acquiring a Trained Memory)*, trans. Mary Carruthers, in Carruthers, *The Book of Memory*, Appendix C, 281–288.

45 Carruthers, *The Book of Memory*, 249.

46 Ibid., 33–45.
47 Florence, Biblioteca Medicea Laurenziana, cod. Plut. I, 56 (586 CE). This resemblance was noted as early as 1926: Morey, "Painted Panel," 163, who in turn cites his reliance on observations in Oskar Konstantin Wulff, *Altchristliche und byzantinische Kunst*, vol. 1 (Berlin: Akademische Verlagsgesellschaft Athenaion, 1914). See also Bagnoli et al., *Treasures of Heaven*, 36; and Kurt Weitzmann, "Loca Sancta," 41.
48 Jane Bennett, *Vibrant Matter: A Political Ecology of Things* (Durham, NC: Duke University Press, 2009), 5.
49 Ibid., 5.
50 Mark Wolf, *Building Imaginary Worlds: The Theory and History of Subcreation* (New York: Routledge, 2012), 367.
51 Ibid., 13.
52 Perhaps "Blemmyae Land" was not as far away from the world of late antique pilgrimage as we usually imagine. Shenoute helped shelter refugees from the historical Blemmye incursions, and then tried to convert the Blemmyes, as documented in several works by and about him: Behlmer, "Visitors to Shenoute's Monastery," 344–345).
53 Wolf, *Building Imaginary Worlds*, 290–291.
54 Peter Brown, *The Rise of Western Christendom: Triumph and Diversity, A.D. 200–1000*, 1st ed. (Oxford: B.H. Blackwell, 1996), 52. This line is not present in the revised second edition, otherwise cited throughout.
55 Wolf, *Building Imaginary Worlds*, 379.
56 Elsner, "Replicating Palestine," 126.
57 As Elsner, "Replicating Palestine," 118, notes, "fragments of material from distant holy *places* . . . [make] the past *present* (for the believer), and in doing so [sanctify] the present with the full holy effect of the past." Emphasis is original.
58 McKenzie Wark, *Virtual Geography: Living with Global Media Events* (Bloomington: Indiana University Press, 1994), vii. Wark is writing about the contemporary world and global media events, but the discussion is quite resonant with the present material. Recent studies in cognitive science suggest the creation of the so-called blended space, which brings together past and present sites and re-creates one through the use of the other: Gilles Fauconnier and Mark Turner, *The Way We Think: Conceptual Blending and the Mind's Hidden Complexities* (New York: Basic Books, 2002); for the application of the concept to medieval theater, see Jill Stevenson, *Performance, Cognitive Theory, and Devotional Culture: Sensual Piety in Late Medieval York*, Cognitive Studies in Literature and Performance Series (New York: Palgrave, 2010); for its potential use for art and performance historians, see Elina Gertsman, "Image and Performance: An Art Historian at the Crossroads," in *Research on Medieval and Renaissance Drama*, 51, special issue: "Opportunities in Medieval and Renaissance Drama," eds. Mario Longtin and Jill Stevenson (2013): 5–13.
59 Daniel K. Connolly, *The Maps of Matthew Paris: Medieval Journeys through Space, Time and Liturgy* (Woodbridge: Boydell and Brewer, 2009), 28, 32. On Cassiodorus's time in Constantinople, see Cassiodorus, *Variae*, trans. S.J.B. Barnish (Liverpool: Liverpool University Press, 1992), liii.
60 "tum si vos notitiae nobilis cura flammaverit, habetis Ptolomei codicem, qui sic omnia loca evidenter expressit, ut eum cunctarum regionum paene incolam fuisse iudicetis, eoque fiat ut uno loco positi, sicut monachos decet, animo percurratis quod quiquorum peregrination plurimo labore collegit.": R.A.B. Mynors, ed., *Cassiodori Senatoris Institutiones* (Oxford: Clarendon Press, 1937), 66, no. 25. Translation from Cassiodorus, *An Introduction to Divine and Human Readings*, trans. Leslie Wever Jones (New York: Octagon Books, 1966), 125 "Divine Letters," no. 25.
61 Brown, *Antiquity*, 182.

10 Image, epigram, and nature in Middle Byzantine personal devotion*

Brad Hostetler

In *Nectar and Illusion*, Henry Maguire examines Byzantium's ambiguous relationship with nature in both art and literature. He demonstrates that after Iconoclasm, visual representations of the terrestrial world displayed in public settings were in "a constant tension between acceptance and denial," but "tended to flourish most abundantly in relatively inconspicuous locations," such as on small private objects.[1] I build upon Maguire's work by examining the ways in which nature was invoked, represented, and utilized through epigrams, images, and materials in personal devotional contexts in the Middle Byzantine period.

I focus on two reliquaries: one now in the treasury of the Protaton at Mount Athos, which contains a relic of the True Cross and four stones from the Holy Land, and another reliquary of the True Cross, now lost, formerly in the Grandmont Abbey in the Limousin region.[2] These two reliquaries have a great deal in common. They were approximately the same size.[3] The Protaton is, and the Grandmont was, a silver-gilt rectangular box with a sliding lid. Each is, or was, inscribed with a 12-syllable dedicatory epigram.[4] Their iconography was also similar: a repoussé image of the crucifixion with a portrait of the patron at the base of the cross – Zosimas for the Protaton and Alexios for the Grandmont. To my knowledge, only these two Middle Byzantine reliquaries are known to have included a so-called donor portrait.[5] The fact that Zosimas and Alexios are part of a biblical narrative scene – a rare iconographic feature in any medium of this period – makes these depictions all the more unusual.[6]

The Protaton and Grandmont reliquaries share many formal and iconographic similarities, yet they convey distinct and nuanced messages about the patrons and their relics. The epigrams are essential for understanding these messages, but they have not yet been subject to close art historical analysis.[7] I argue that for the Protaton, the structure, alliteration, and placement of the epigram emphasize Zosimas's ownership of the four stones. For the Grandmont, the epigram's innovative terminology articulates Alexios's rightful possession of the True Cross and conveys the ways in which he hoped the relic would protect him. I demonstrate that, for the patrons, both reliquaries functioned as holy sites, where they came into contact with the sacred matter, witnessed the biblical event that sanctified these materials, and harnessed their salvific properties.[8]

The Protaton reliquary

The Protaton reliquary is a composite work, constructed in the eleventh–twelfth centuries and altered in the eighteenth.[9] The crucifixion panel on the lid is part of the original reliquary (Figure 10.1). With the exception of Zosimas's portrait, it conforms

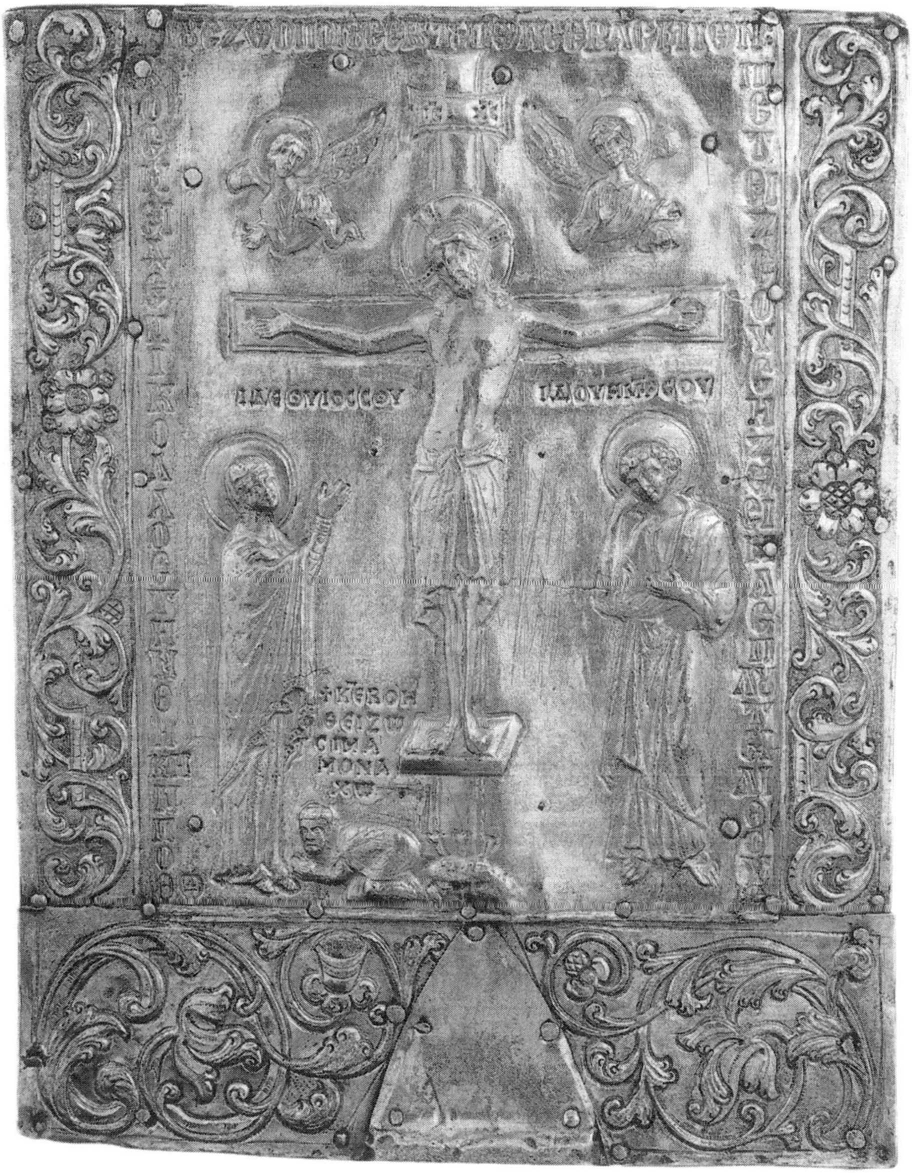

Figure 10.1 Protaton reliquary, lid, eleventh–twelfth centuries and 1758, silver and gilded silver on wood, 17.8 × 14.2 × 1.2 cm. Treasury of the Protaton, The Holy Community of Mount Athos. After Stelios Papadopoulos and Chrysoula Kapioldasi-Sotiropoulou, eds., *The Treasury of the Protaton*, vol. 1 (Mount Athos: The Holy Community of Mount Athos, 2001), Figure 10, page 58.

to traditional Middle Byzantine iconography. Christ is at the center, the Mother of God is on the left, John the Apostle is on the right, and two angels are positioned above them. The cross is mounted on Golgotha, represented as a small rock, and is supported by three stakes. Zosimas is shown in *proskynesis* between the Theotokos and Christ, and is identified by the inscription placed above him: "Lord help Zosimas the monk."[10] He has short curly hair, is bearded, and is wearing a tunic covered by a mantle. He is facing the Theotokos in left profile and is cupping his hands together in supplication. His feet are touching the rock of Golgotha.

An inscription frames the crucifixion panel on three sides. It identifies Zosimas and Nikolaos as the two patrons of the reliquary. This 12-syllable epigram is composed of three verses; each verse is placed along a different edge of the panel. It begins on the upper edge, continues down the right side, and concludes on the left.[11] The translation is best rendered by rearranging the order of the first two verses:

2 With welling faith Zosimas is rich with
1 the life-giving stones from venerable places;
and with desire Nikolaos adorns the reliquary.[12]

The "life-giving stones" mentioned in the epigram are inside the reliquary, placed in the four corners of the box (Figure 10.2).[13] Inscribed on the revetment next to each of these relics is an identifying inscription: "from the holy Tomb of Christ" (upper left), "from the holy Skull," that is, Golgotha (upper right), "from holy Bethlehem" (lower left), and "from holy Gethsemane" (lower right).[14] There was originally a gilded repoussé cross, fragments of which survive, in the center of the box.[15] This cross has an inscription, "*xylon*" (wood), which is divided in two parts, placed on either side of the original image.[16] The letter forms of the cross and relic inscriptions are the same as those of the dedicatory epigram on the lid, verifying that the lid and box are contemporary. Based on the presence of the word *xylon*, coupled with the repoussé image of the cross, Anatole Frolow suggests that the Middle Byzantine reliquary originally contained a relic of the True Cross.[17]

Given our limited knowledge concerning Zosimas and Nikolaos, it is not possible to identify them in contemporary texts or to locate Zosimas's monastery. Brigitte Pitarakis suggests that Zosimas may have acquired the stones while on pilgrimage to Jerusalem at some point after 1099, when the Crusaders captured the city and there was an increase in the number of Christian pilgrims to the Holy Land.[18] There is no evidence that either confirms or refutes this suggestion.

The alterations dating to 1758 are documented by the inscription placed on the frame of the box.[19] At this time the reliquary was enlarged to make room for an eighteenth-century enkolpion of the True Cross; it is not known whether or not this enkolpion contains the same wood fragment that was part of the original Middle Byzantine reliquary. It is housed within a cruciform cavity that was cut into the center of the box, destroying most of the Middle Byzantine repoussé cross. The lid was enlarged with silver strips decorated with rinceaux added to the left, right, and lower edges of the crucifixion panel.[20] My focus is on the original form of the Protaton reliquary. I offer a reconstruction of the Middle Byzantine appearance of the box based on the surviving components (Figure 10.3).[21]

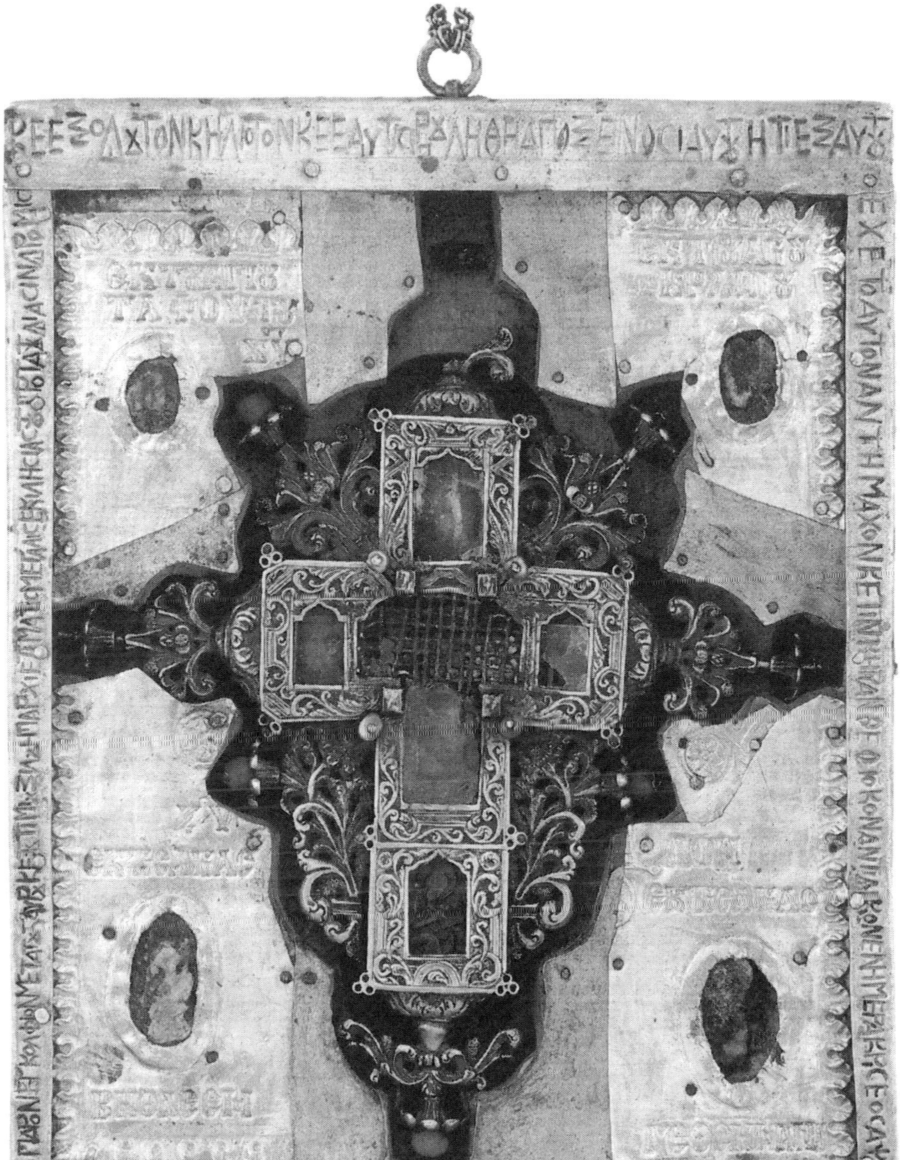

Figure 10.2 Protaton reliquary, box, eleventh–twelfth centuries and 1758, silver and gilded silver on wood with four stones, 19.3 × 16.1 × 3.0 cm. Treasury of the Protaton, The Holy Community of Mount Athos. After Stelios Papadopoulos and Chrysoula Kapioldasi-Sotiropoulou, eds., *The Treasury of the Protaton*, vol. 1 (Mount Athos: The Holy Community of Mount Athos, 2001), Figure 8, page 51.

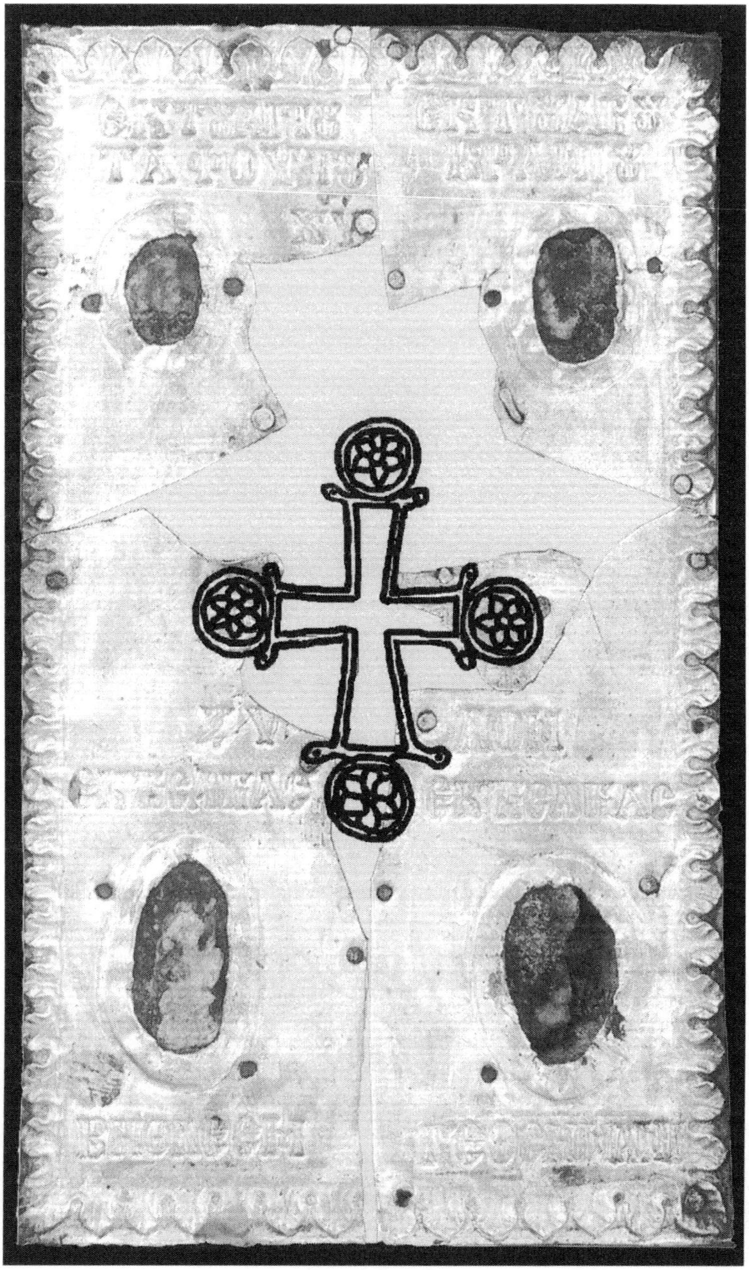

Figure 10.3 Protaton reliquary, reconstruction of the box, eleventh–twelfth centuries, silver and gilded silver on wood with four stones. Treasury of the Protaton, The Holy Community of Mount Athos. After Stelios Papadopoulos and Chrysoula Kapioldasi-Sotiropoulou, eds., *The Treasury of the Protaton*, vol. 1 (Mount Athos: The Holy Community of Mount Athos, 2001), Figure 8, page 51, with alterations by Brad Hostetler.

I argue that the epigram's compositional structure, the alliteration in the first two verses, and the placement of specific words were designed to privilege Zosimas over Nikolaos and to convey his ownership of the relics and his access to the holy sites. Each of the three verses focuses on a different aspect of the reliquary. The first references the stones, their life-giving power and their association with the "venerable places" of the Holy Land. The second verse identifies Zosimas as the owner of the relics. The third names Nikolaos, who adorned the reliquary. Grammatically, there are two subjects in the epigram: Zosimas and Nikolaos. Their names are in the nominative, and the words *lithous* (stones) and *thēkēn* (reliquary) are in the accusative. The first two verses form a unified thought that centers on Zosimas (the subject), who is rich with (the verb) the life-giving stones (the object). Through this structure, Zosimas is presented as the more important patron; he is named first and is the subject of two-thirds of the epigram.[22]

The use of alliteration draws attention to the first two verses.[23] When the epigram is recited aloud, the words *zōopoious* (life-giving), *zeousē* (welling), and *Zosimas* create an aural rhythm that emphasizes verses 1 and 2.[24] The choice of words to alliterate is also significant. *Zōopoious* and *Zosimas* are formed from the same root, *zōē*, meaning "life," and, in the Christian context, "eternal life."[25] Wordplay thus establishes the relationship between patron and relic. The salvation desired by *Zosimas* is accessed through the *life-giving* stones, which he possesses with *welling* faith. The relics are material proof that his hope for eternal life will be fulfilled.

The placement of specific words was also determined by Zosimas's desire to convey his ownership of the relics and thus of the holy sites. The phrase *ek topōn* (from places) is centered over the image of the cross at the upper edge of the panel (Figure 10.4, marked a). This inscribed phrase visually links with the image of the rock of Golgotha at the base through the vertical axis of the cross. The connection between word and image positions Calvary as a visual response to the phrase *ek topōn*, identifying the venerable place of one of the life-giving stones within the reliquary.

The placement of the word *lithous* is also significant (Figure 10.4, marked b). This word is positioned at the lower right-hand corner of the panel, where it is aligned with its visual analogue – the rock of Golgotha.[26] My suggestion that this pairing is deliberate is supported by the composition of the epigram. The word *lithous* is the last word of verse 2, and its corresponding article, *tous* (the), is the first word of verse 1. Andreas Rhoby notes that it is unusual for *lithous* to be separated from its article by two verses.[27] This seemingly odd syntax was necessary, as it allowed for the placement of *lithous* at the lower right-hand corner of the panel. This pairing of the inscribed word with the image of Golgotha visually reinforces Zosimas's contact with the rock.

In addition to placing Zosimas at the holy site, the crucifixion iconography also envisions the salvation granted to him through his contact with the life-giving rock of Golgotha. This is expressed visually by his position on the ground line at Calvary in the presence of Christ, the Theotokos, and John. He offers his prayer of invocation to the Lord (*Kyrie*), but he faces the Mother of God. His petition is therefore one of intercession, directed to the Theotokos, who relays it to her son through gesture and pose. Her body is turned in profile, her head is angled upward, and she is raising her right hand to him, recalling Byzantine iconography in which she is imaged as an intercessor.[28] Christ is tilting his head down toward her as if he is responding. The upright beam of the cross then directs the viewer's eye from Christ down to Zosimas, crouching next to the rock of Golgotha.[29]

If Frolow is correct that the Protaton reliquary originally contained a relic of the True Cross then it is unusual that the epigram lacks any reference to it. Frolow documents

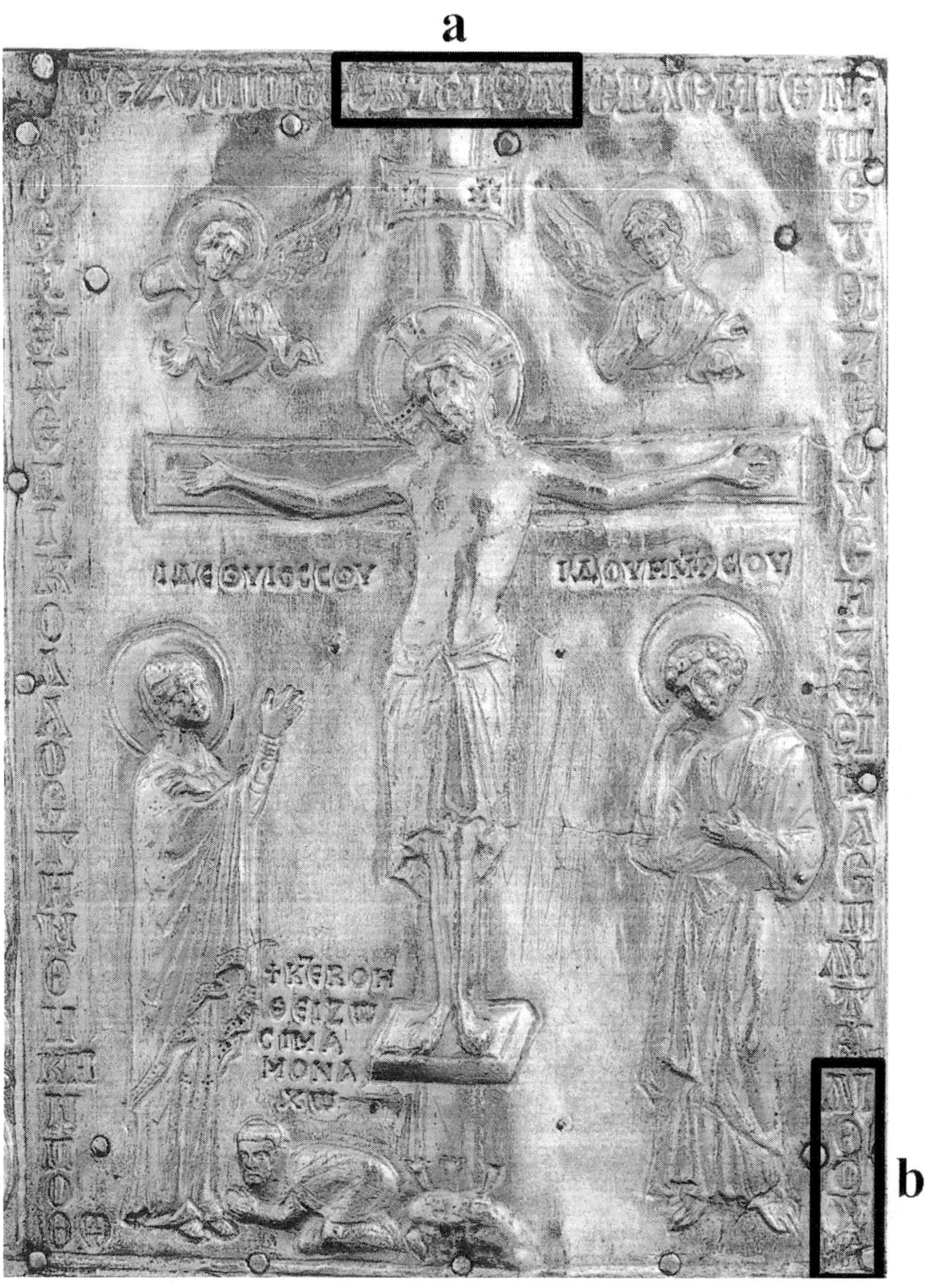

Figure 10.4 Protaton reliquary, detail of the lid marking the words (a) *ek topōn*, and (b) *lithous*, eleventh–twelfth centuries, silver and gilded silver on wood, 17.8 × 14.2 × 1.2 cm. Treasury of the Protaton, The Holy Community of Mount Athos. After Stelios Papadopoulos and Chrysoula Kapioldasi-Sotiropoulou, eds., *The Treasury of the Protaton*, vol. 1 (Mount Athos: The Holy Community of Mount Athos, 2001), Figure 10, page 58, with alterations by Brad Hostetler.

thirty Middle Byzantine True Cross reliquaries with epigrams; only three do not explicitly reference the wood.[30] I argue that by omitting any mention of the relic of the True Cross in the epigram and elevating the importance of the rocks in text and image, the Protaton reliquary encompasses the material whole of the Holy Land, both the earth and the wood. In this way, the reliquary elides geographical distance and topographical specificity, allowing the holy sites to be seen, touched, and experienced through their metonymic material references brought together in a personal devotional object.

The Grandmont reliquary

The twelfth-century Grandmont reliquary was, according to its dedicatory epigram, made for Alexios Doukas, grandson of Irene Doukaina and Emperor Alexios I Komnenos (r. 1081–1118). Unfortunately, it is not possible to know the precise identity of this Alexios because five of the emperor's grandsons had the same name.[31] The reliquary came into the possession of Amalric I, the Latin King of Jerusalem (r. 1163–1174), who donated it to the Abbey of Grandmont in 1174.[32] The reliquary was destroyed in the French Revolution, but its appearance is preserved in a description and two drawings published in 1658 by the French cleric François Ogier, supplemented by a description of the reliquary in the Grandmont Abbey inventory of 1666, published by the abbot Jacques Rémy Antoine Téxier.[33]

The Grandmont reliquary was an enkolpion in the form of a box with a sliding lid.[34] Like the Protaton, it featured a scene of the crucifixion with an image of the patron.[35] Christ was at the center, the Theotokos was on the left, John was on the right, and two angels were positioned above the cross. Alexios was at the foot of the cross, clasping his hands.[36] The interior and back of the reliquary differed in appearance from that of the Protaton. Ogier's engraving of the interior shows that the reliquary featured a double-arm cruciform receptacle for a relic of the True Cross, bordered by cabochons (Figure 10.5).[37]

Ogier also drew the reverse of the reliquary, which was inscribed with a 16-verse epigram written in majuscule letters with diacritics (Figure 10.6).[38] Each line of inscribed text on the reliquary was equal to one 12-syllable verse. A verse-by-verse translation reads:

> Having slept a short sleep on the three-part tree (*tridendria*)
> the all-king and divine-man Logos
> granted much grace to the tree (*dendrō*).
> For anyone inflamed with sickness is refreshed
> who fleeing for refuge to the branches of the three-part tree (*tridendrias*).
> But being ablaze right at the peak of midday
> I ran, I came, I sneaked to the branches.
> Receive and rightly shelter me with thy shade,
> O Tree (*dendron*), shading the whole earth,
> and instill the dew of Haërmon upon me,
> having descended from the beautiful-tree (*kallidendrias*) of the Doukai,
> whose root being the Empress Irene,
> the maternal grandmother, the glory of the kings,
> spouse of Alexios, ruler of the Ausones.
> Yea, yea, I beg (you) my sole protector
> your servant Alexios, from the Doukai family.[39]

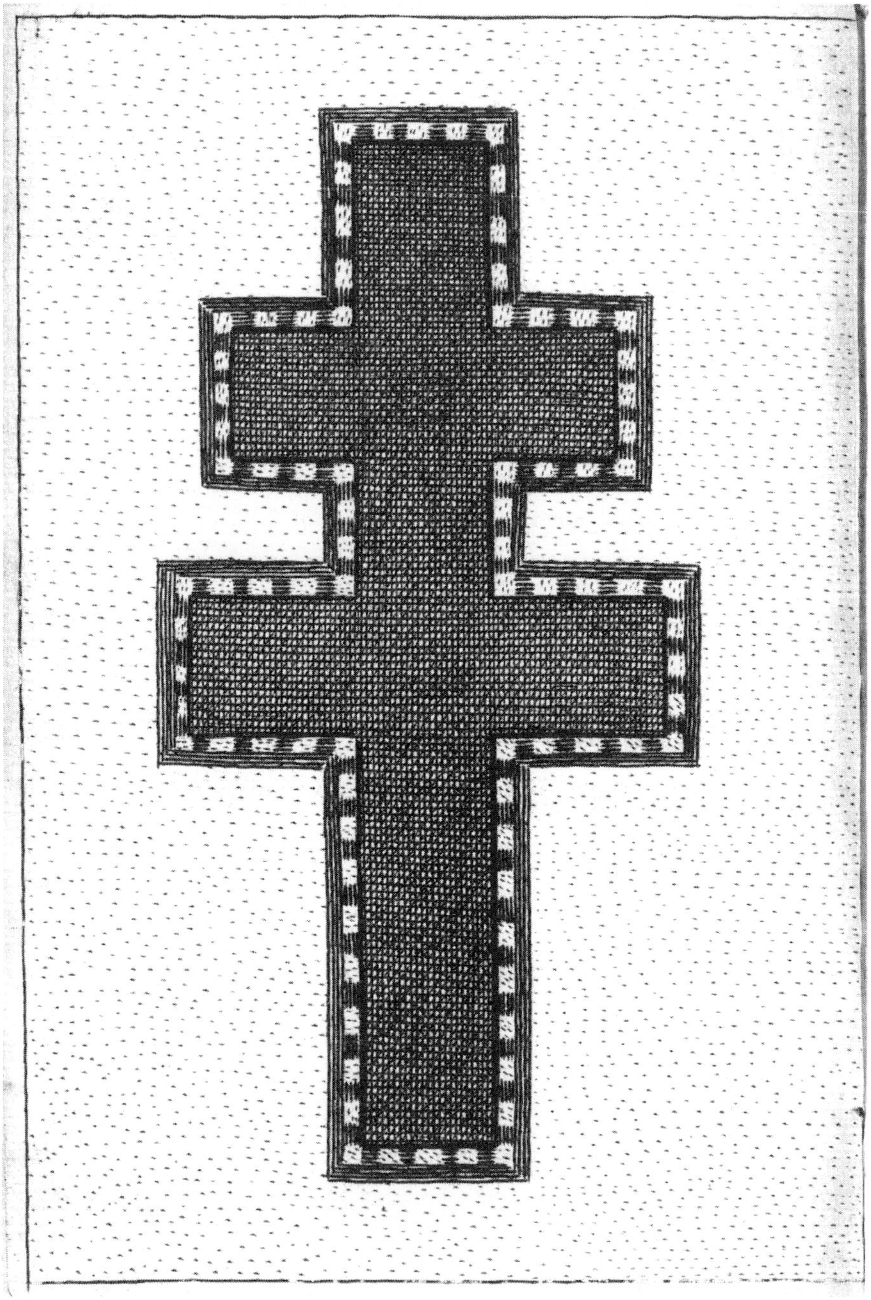

Figure 10.5 Grandmont reliquary, interior of box, twelfth century, drawing after François Ogier, *Inscription antique de la Vraye Croix de l'Abbaye de Grandmont, Avec un sermon de la Passion* (Paris: J. Hénault, 1658), plate I.

Source: BNF

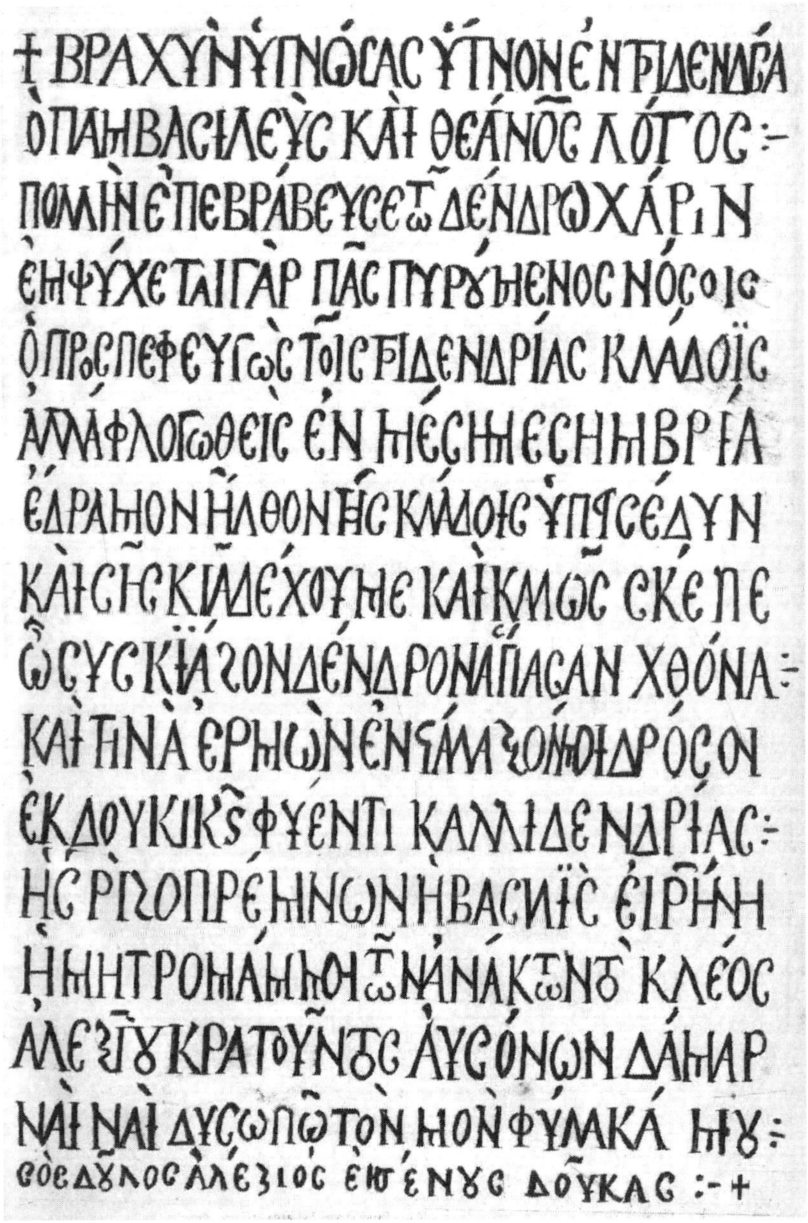

Figure 10.6 Grandmont reliquary, interior of box, twelfth century, drawing after François Ogier, *Inscription antique de la Vraye Croix de l'Abbaye de Grandmont, Avec un sermon de la Passion* (Paris: J. Hénault, 1658), plate II.

Source: BNF

The epigram is written in the first person, in the voice of Alexios. The first seven verses address the reader and describe the crucifixion and the restorative power of the True Cross. In the last nine verses, Alexios addresses the relic itself, urging it to protect him, and relaying his genealogy in order to justify his request. Frolow convincingly argues that, based on style and vocabulary, this epigram was composed by Nicholas Kallikles, the Komnenian court poet and physician. Kallikles also authored True Cross reliquary epigrams for Alexios's grandmother Irene and for her daughters Eudokia and Maria.[40]

It is noteworthy that the words *stauros* (cross) and *xylon* (wood) are not included in the epigram. Two-thirds of the epigrams on Middle Byzantine True Cross reliquaries that have been documented by Frolow use one or both of these terms.[41] Kallikles instead introduces various cognates of the word *dendron* (tree) in verses 1, 3, 5, 9, and 11.[42] This particular terminology is found in only one other True Cross reliquary epigram of this period, and it too was authored by Kallikles.[43]

The use of this new terminology is significant because it reflects the ways in which the True Cross was meant to benefit Alexios. The Grandmont epigram describes the cross as a three-part tree (*tridendria*), to which Alexios runs for shade.[44] The protection of its branches extinguishes the heat of the sun "at the peak of midday" – the time when Christ was crucified.[45] By using these words and phrases, the epigram portrays Alexios finding solace at the crucifixion, next to the cross – a sheltering tree. This description corresponds to the iconography in which Alexios was shown at Calvary, beneath the cross beam, as if shaded by the "branches" of the cross.[46]

The relic was also an instrument of healing. Alexios implores the tree to anoint and heal him because, according to verses 4 and 6, he is "inflamed with sickness."[47] The miracle-working nature of the True Cross is conveyed by verse 10, which references Psalm 132:3: "It is like the dew of Haërmon, which descends onto the mountains of Zion, because there the Lord commanded the blessing, life forevermore."[48] Early pilgrims to the Holy Land collected the dew from Mount Haërmon, which was renowned for its medicinal properties.[49] In Byzantium, the dew was also typologically associated with the *anargyroi* saints, Kosmas and Damianos, through their feast day on November 1.[50] According to the eleventh-century Synaxarion of the Euergetis Monastery in Constantinople, this was the only day of the year when Psalm 132:3 was read in the liturgy.[51]

The "*dendron* terminology" is also used to describe Alexios's genealogy. This functioned to convey his legitimate ownership of the relic of the True Cross. Verses 11 and 12 declare that Alexios is part of the beautiful tree (*kallidendrias*) of the Doukai family, and his grandmother, the Empress Irene, is the root (*rhizopremnon*) from which he sprang. The epigram also names three family members that make up his *kallidendria*, corresponding in number to the *tridendria* that form the True Cross.

Frolow notes that for genealogies, tree imagery was "official jargon" of Byzantine poets of the twelfth century, and that Kallikles frequently made use of it.[52] Although this metaphor may have been a topos, its application in the context of the Grandmont reliquary underscores his family's possession of, and Alexios's right to, the sacred wood. The True Cross was emblematic of Byzantine imperial power and identity.[53] Emperors invoked relics of the True Cross on ceremonial occasions and in battle, controlled their distribution to the Christian *oikoumene*, and housed them in richly decorated reliquaries.[54] Alexios's grandfather and namesake utilized relics of the True Cross to negotiate treaties with the Normans.[55] His grandmother Irene is known to have commissioned at least five reliquaries of the True Cross, and her daughters, Maria and Eudokia, each commissioned one.[56]

In this epigram, Alexios insistently proclaims his membership in this illustrious family. His connections thus granted him access to the precious wood and to the same court poet who authored True Cross reliquary epigrams for his family. By using "*dendron* terminology" to describe his relic and his lineage, Alexios reinforces the association between the True Cross and his patrimony, thereby establishing his continuity with the past and his entitled claim to the relic.

My argument that the *dendro*-centric language was deliberately crafted in order to convey these messages for Alexios is supported by the etymologies of the words *tridendria*, *kallidendria*, and *rhizopremnon*. Rhoby observes that the Grandmont epigram is the earliest known use of the word *tridendria*. He also notes that the words *rhizopremnon* and *kallidendria* are *hapax legomena*.[57] All three terms are compound words that were created to evoke tree imagery. The words *tridendria* and *kallidendria* combine the prefixes *tria* (three) and *kallos* (beautiful) with the word *dendron*.[58] *Rhizopremnon* is formed from the word *rhiza*, which can, in its literal sense, refer to roots of any type or, more broadly, to a race from which a family springs.[59] Combined with the term *premnon* (base of a tree), the word is given a more specific arboreal connotation.[60] This epigram, therefore, utilizes nontraditional vocabulary for the relic of the True Cross, combined with newly invented compound words. Word repetition and word innovation were used in order to articulate and personalize Alexios's devotion to the relic of the True Cross as his protector and healer.

The unusual iconography of the Grandmont can thus be understood in the context of the epigram's messages. The lid featured Alexios at the crucifixion, at the foot of the cross. Thus he was depicted beside his relic at the specific event when (at the peak of midday) and at the place where (beneath the "branches" of the cross), he ran to the tree for shade. The presence of the relic of the True Cross in his enkolpion made Alexios's hope for protection by the tree a material reality.[61]

* * *

Peter Brown demonstrates that for relics of the Early Latin West, the *praesentia* of the holy was made manifest through material remains. This allowed the faithful to "feed on the facts of distance and the joys of proximity."[62] Whereas pilgrimage was one way to overcome such distance, the Protaton and Grandmont reliquaries reveal how – in the Middle Byzantine period – the intimate link between patron and sacred matter could be imagined, verbalized, and experienced through a single object.[63] The Protaton reliquary, and its *litho*-centric messages, emphasizes Zosimas's ownership of the four life-giving stones and, in turn, his access to the holy sites. The Grandmont reliquary with its *dendro*-centric messages presented the wood of the True Cross as a tree that protected, sheltered, and healed Alexios. The patrons' choices of iconography, while unusual by Middle Byzantine standards, were visual expressions of what was conveyed by the epigrams. Their imagined presence at the crucifixion, in visual proximity to the earth and the wood, was realized through the inclusion of a piece of Golgotha layered beneath the portrait of Zosimas and a fragment of the cross beneath that of Alexios. The reliquaries thus functioned, for Zosimas and Alexios, as holy sites, where they accessed their relics, witnessed the biblical event that sanctified these materials, and harnessed their salvific properties.

Notes

* A version of this chapter was presented at the 48th International Congress on Medieval Studies in 2013 for the panel "Eco-Critical Approaches to Medieval Art: East and West." I thank the session chairs, Nancy Ševčenko and Anne Harris, as well as those in attendance, for their helpful comments. I also benefitted from the advice of Lynn Jones, Stratis Papaioannou, Ida Toth, and the anonymous reviewers. All translations are my own, unless otherwise noted.

1 Henry Maguire, *Nectar and Illusion: Nature in Byzantine Art and Literature* (Oxford: Oxford University Press, 2012), 105.

2 For the Protaton reliquary, see Gabriel Millet, Jeanne Fourier-Pargoire, and Louis David Petit, *Recueil des inscriptions chrétiennes de l'Athos*, vol. 1, Bibliothèque des écoles françaises d'Athènes et de Rome 91 (Paris: Albert Fontemoing, 1904), no. 26; Anatole Frolow, *La relique de la Vraie Croix: recherches sur le développement d'un culte*, Archives de l'Orient chrétien 7 (Paris: Institut Français d'Études Byzantines, 1961), no. 1120; Katia Loverdou-Tsigarida in *Treasures of Mount Athos*, ed. Athanasios Karakatsanis (Thessaloniki: Ministry of Culture, Museum of Byzantine Culture, 1997), no. 9.18; Brigitte Pitarakis and Yota Ikonomaki-Papadopoulos, "Decorative Arts," in *The Treasury of the Protaton*, vol. 1, eds., Stelios Papadopoulos, and Chrysoula Kapioldasi-Sotiropoulou (Mount Athos: The Holy Community of Mount Athos, 2001), 48–55; Brigitte Pitarakis, *Άγιον Όρος: Κειμήλια Πρωτάτου* (Thessaloniki: Hagioreitike Hestia, 2006), no. 67; Démétrios Liakos in *Le Mont Athos et l'Empire byzantin: trésors de la Sainte Montagne* (Paris: Paris-Musées, 2009), no. 124; Andreas Rhoby, *Byzantinische Epigramme in inschriftlicher Überlieferung* (hereafter, *BEIÜ*), vol. 2: *Byzantinische Epigramme auf Ikonen und Objekten der Kleinkunst* (Vienna: Verlag der Österreichischen Akademie der Wissenschaften, 2010), Me34; Brad Hostetler, "The Iconography of Text: The Placement of an Inscription on a Middle Byzantine Reliquary," *Eastern Christian Art* 8 (2011): 49–55; and idem, "The Function of Text: Byzantine Reliquaries with Epigrams, 843–1204 (PhD dissertation, Florida State University, 2016), 145–159, 172–173. For the Grandmont reliquary, see François Ogier, *Inscription antique de la Vraye Croix de l'Abbaye de Grandmont: Avec un sermon de la Passion* (Paris: J. Henault, 1658); Charles Dufresne Du Cange, "Explication des inscriptions de la vraye croix, qui est en l'abbaye de Grandmont, et de celle qui est au monastere du mont Saint-Quentin en Picardie," Dissertation XXVI in *Histoire de Saint Louis par Jean de Joinville* (Paris, 1668), reprinted in *Glossarium mediae et infimae latinitatis*, ed. Léopold Favre, vol. 10 (Niort: L. Favre, 1887), 90–93; Jacques-Paul Migne, *Dictionnaire d'épigraphie chrétienne*, vol. 1, Nouvelle encyclopédie théologique 30 (Paris: J.-P. Migne, 1852), cols. 672–675; Jacques Rémy Antoine Téxier, *Dictionnaire d'orfévrerie, de gravure et de ciselure chrétiennes* (Paris: J.-P. Migne, 1857), cols. 833–836; Ernst Curtius, and Adolph Kirchhoff, eds., *Corpus inscriptionum graecarum*, vol. 4 (Berlin: Vendit G. Reimeri Libraria, 1877), no. 8735; Anatole Frolow, "Deux inscriptions sur des reliquaires byzantins," *Revue Archéologique* 18 (1941): 233–239; idem, *La relique*, nos. 319, 365; Marie-Madeleine Gauthier, *Highways of the Faith: Relics and Reliquaries from Jerusalem to Compostela*, trans J. A. Underwood (Secaucus, NJ: Wellfleet Press, 1986), no. 28; Holger Klein, *Byzanz, der Westen und das 'wahre' Kreuz: die Geschichte einer Reliquie und ihrer künstlerischen Fassung in Byzanz und im Abendland* (Wiesbaden: Reichert, 2004), 219–224; *BEIÜ* 2:Me15; and Hostetler, "The Function of Text," 167–168.

3 The Protaton is a Middle Byzantine reliquary altered in the eighteenth century. Its current dimensions are 19.3 × 16.1 × 3.0 cm (overall), and 17.8 × 14.2 × 1.2 cm (lid). The original dimensions, based on my measurements of a scaled photograph of the crucifixion panel, were approximately 14.5 × 10.5 cm. The dimensions of the Grandmont reliquary were 15 × 10 cm; see note 34.

4 For an overview of Byzantine epigrams, see *BEIÜ* 1: 37–73.

5 Scholars have traditionally used the term *donor portrait* to refer to any contemporary figure depicted on a work of art. For problems associated with the scholarly use of this term, see Linda Safran, "Deconstructing 'Donors' in Medieval Southern Italy," in *Female Founders in Byzantium and beyond*, Wiener Jahrbuch für Kunstgeschichte 40/41, eds. Lioba Theis et al. (Vienna: Böhlau Verlag, 2011–2012), 135–151. I have chosen to use the word *patron* to refer to Zosimas and Alexios because it emphasizes their roles in commissioning the reliquaries. I assume, therefore, that as patrons, Zosimas and Alexios had to make choices about the reliquaries' form, decoration, and dedicatory epigrams that reflect their individual needs and desires. The Saint Demetrios ciborium reliquary, now in the Moscow Kremlin (MZ 1148), features portraits of the Emperor Constantine X Doukas (r. 1059–1067) and his wife, Eudokia Makrembolitissa. These are not donor images, but rather commemorative imperial portraits placed there by the patron named in the dedicatory epigram: the *mystographos* John Autoreianos; *BEIÜ* 2: Me99.

6 Brigitte Pitarakis notes that it is unusual for Zosimas to be shown at the crucifixion, but does not discuss its significance; "Decorative Arts," 50. Nancy P. Ševčenko, "The Representation of Donors and Holy Figures on Four Byzantine Icons," *Δελτίον τῆς Χριστιανικῆς ἀρχαιολογικῆς ἑταιρείας* 17 (1993–1994): 162–164, identifies three other examples of Middle Byzantine "donors" shown with biblical narrative scenes. To this list I add the unnamed figure shown with a scene of the Annunciation in the Cappadocian rock-cut church of Yusuf Koç Kilisesi, which dates to the first half of the eleventh century, as well as the two men named John Entalmatikos and Genethlios represented with the *Benediction of the Apostles* in the mid-eleventh-century rock-cut Church of Karanlık Kilise, also in Cappadocia; Lyn Rodley, *Cave Monasteries of Byzantine Cappadocia* (Cambridge: Cambridge University Press, 1985), 53–54, 156, figs. 10a, 29a. See also the processional cross at the Cluny Museum in Paris which features a monk next to an image of the Virgin being fed by an angel; in John A. Cotsonis, *Byzantine Figural Processional Crosses*, Dumbarton Oaks Byzantine Collection Publications 10 (Washington, DC: Dumbarton Oaks Research Library and Collection, 1994), Figure 14b. The Khakouli Triptych features an enamel image of a Byzantine empress in what appears to be a scene of the Annunciation; Ioli Kalavrezou, "Female Popular Beliefs and Maria of Alania," *Journal of Turkish Studies* 36 (2011): 85–101, at Figure 4. For Late Byzantine examples, see Fani Gargova, "The Meteora Icon of the Incredulity of Thomas Reconsidered," in *Female Founders in Byzantium and beyond*, Wiener Jahrbuch für Kunstgeschichte 40/41, eds. Lioba Theis et al. (Vienna: Böhlau Verlag, 2011–2012), 369–381.

7 The most important study of these epigrams is *BEIÜ* 2: Me15 and Me34. For the Grandmont, see also Frolow, "Deux inscriptions," 233–239.

8 The seminal work on the topic of portable *loca sancta* is Robert G. Ousterhout, ed., *The Blessings of Pilgrimage* (Urbana: University of Illinois Press, 1990). See also the unpublished dissertation by Laura Veneskey, "Alternative Topographies: 'Loca Sancta' Surrogates and Site Circulation in Late Antiquity and Byzantium" (PhD dissertation, Northwestern University, 2012).

9 The date of the Middle Byzantine components is based on style, iconography, and epigraphy; Frolow, *La relique*, 653; *Treasures of Mount Athos*, 341; Pitarakis, "Decorative Arts," 49; *Ἅγιον Ὄρος*, 119; and *Le Mont Athos*, no. 124.

10 Κ(ύρι)ε, βοήθει Ζωσιμᾷ μοναχῷ. *BEIÜ*, 2:202.

11 The inscription is complete; however, the upper left corner of the panel is missing. That portion of the panel likely featured a cross that marked the beginning of the inscription; see Hostetler, "Iconography of Text," 51.

12 Τοὺς ζωοποιοὺς ἐκ τόπων σεβασμίων
πίστει ζεούσῃ Ζωσιμᾶς πλουτεῖ λίθους·
κοσμεῖ δὲ Νικόλαος τὴν θήκην πόθῳ.
Ed. *BEIÜ* 2: 202

This translation revises my previous rendering in Hostetler, "Iconography of Text," 53.

13 On the topic of stone relics in the eleventh and twelfth centuries, see Sandrine Lerou, "L'usage des reliques du Christ par les empereurs aux XIe et XIIe siècles: le Saint Bois et les

Saintes Pierres," in *Byzance et les reliques du Christ*, eds. Jannic Durand and Bernard Flusin (Paris: Association des amis du Centre d'histoire et civilisation de Byzance, 2004), 159–182. Other examples of Middle Byzantine reliquaries with stones from the Holy Land are to be found in Frolow, *La relique*, nos. 332, 405, 473. See also Hostetler, "The Function of Text," 41–43.

14 ἐκ τοῦ ἁγίου Τάφου τοῦ Χ(ριστο)ῦ; ἐκ τοῦ ἁγίου Κρανίου; ἐκ τῆς ἁγίας Βηθλεέμ; ἐκ τῆς ἁγίας Γεθσημανί.

15 The finial rosettes on the left, right, and lower ends of the cross are extant. For a similar image of a cross with rosettes, see the twelfth-century gold and enamel enkolpion at the Natsionalen Arkheologicheski Muzei in Sofia (487) in *The Glory of Byzantium: Art and Culture of the Middle Byzantine Era, A.D. 843–1261*, eds. Helen C. Evans and William D. Wixom (New York: The Metropolitan Museum of Art and New Haven: Yale University Press, 1997), no. 574. On the Protaton reliquary, the right rosette was cut from the revetment, rotated clockwise 135 degrees, and reattached in approximately the same location.

16 [τίμιον] ξύλον. Frolow, La relique, 652.

17 Frolow, *La relique*, 652.

18 Pitarakis, "Decorative Arts," 52; and *Ἅγιον Ὄρος*, 120.

19 ΤΟ ΠΑΡΟΝ ΕΓΚΟΛΦΙΟΝ ΜΕΤΑ ΤΟΥ ΣΤΑΥΡΟΥ ΚΕ ΤΟΥ ΤΙΜΙΟΥ ΞΙΛΟΥ ΗΠΑΡΧΙ ΕΤΙΜΑ [ΚΤΗΜΑ] ΤΙΣ ΜΕΓΑΛΙΣ ΕΚΙΛΗΣΙΑΣ ΤΟΥ ΠΡΩΤΑΤΟΥ ΔΙΑ ΣΙΝΔΡΟΜΙΣ | ΚΕ ΕΞΟΔΟΥ ΤΟΝ ΚΗΛΙΟΤΟΝ ΚΕ ΕΑν ΤΙΣ ΒΟΥΛΗΘΗ ΑΠΟΞΕΝΟΣΙ ΑΥΤΟΥ Η ΤΙ ΕΞ ΑΥΤΟΥ | ΕΧΕΤΟ ΑΥΤΟΝ ΑΝΤΗΜΑΧΟΝ ΚΕ ΤΙΝ ΚΗΡΙΑΝ ΘΕΟΤΟΚΟΝ ΑΝΤΙΔΙΚΟΝ ΕΝ ΗΜΕΡΑ ΚΡΙΣΕΟΣ ΑΨΝ8. (This *enkolphion* (sic) with the cross and the True Cross is the possession of the great church of the Protaton, through the subvention and at the expense of the kelli-dwellers, and if anyone alienates this or anything from it, may he have the Lady Theotokos as his adversary in the Day of Judgment. 1758.) Transcription and translation by Yota Ikonomaki-Papadopoulos in "Decorative Arts," 49. See also Millet, *Recueil des inscriptions*, no. 26.

20 On these alterations, see Ikonomaki-Papadopoulos in "Decorative Arts," 52–55.

21 Pitarakis also published a reconstruction drawing in "Decorative Arts," 281, ill. 1, but it does not include a reconstruction of the repoussé cross.

22 Zosimas is further distinguished from Nikolaos through the use of the word *de* in verse 3, which creates a conceptual break between the second and third verses. It also serves as a conjunction, offsetting verse 3 from the rest of the epigram in order to present a new idea about a different individual; Herbert Weir Smyth, *Greek Grammar*, rev. Gordon M. Messing (Cambridge: Harvard University Press, 1956), no. 2834.

23 The alliteration was first noted by *BEIÜ* 2: 203.

24 On the practice of reading epigrams aloud, see Amy Papalexandrou, "Text in Context: Eloquent Monuments and the Byzantine Beholder," *Word & Image* 17 (2001): 259–283.

25 *A Patristic Greek Lexicon*, ed. G.W.H. Lampe (Oxford: Clarendon Press, 1961), s.v. "ζωή."

26 Hostetler, "Iconography of Text," 54.

27 *BEIÜ* 2: 203.

28 Pitarakis notes that this pose recalls the iconography of the Theotokos Hagiosoritissa and Paraklesis – image types associated with the Virgin's role as intercessor for the faithful; "Decorative Arts," 50. See also *The Oxford Dictionary of Byzantium*, s.vv. "Virgin Hagiosoritissa" and "Virgin Paraklesis" (Nancy Patterson Ševčenko).

29 The representation of rocks as visual referents to specific Holy Land sites was a characteristic feature of a group of ivories produced around the same time as the Protaton reliquary; Henry Maguire, "Ivories as Pilgrimage Art: A New Frame for the 'Frame Group'," *Dumbarton Oaks Papers* 63 (2009): 117–146.

30 Frolow, *La relique*, nos. 126, 473, 662. For more recent literature on these reliquaries, see *BEIÜ* 2: Me16, Me29, Me96. For further discussion, see Hostetler, "The Function of Text," 91.

31 For a list of their grandsons named Alexios, see *BEIÜ* 2: 177. Alexios identifies himself as Doukas, rather than Komnenos, because the former had greater social prestige; Demetrios I. Polemis, *The Doukai: A Contribution to Byzantine Prosopography* (London: Athlone Press, 1968), 134.

32 It is not known how Amalric acquired the reliquary; Frolow, *La relique*, no. 365.

33 Ogier, *Inscription antique*; and Téxier, *Dictionnaire d'orfévrerie*, cols. 833–836.

34 Migne, *Dictionnaire d'épigraphie*, col. 673; Klein, *'wahre' Kreuz*, 220, and *BEIÜ* 2: 174, record that the reliquary measured 12 × 10 cm. Téxier, *Dictionnaire d'orfévrerie*, cols.

833–834 states that the reliquary was 15 × 10 cm and that the height of the cross relic, inside the reliquary, was 12 cm. Texier's measurements are supported by Frolow, *La relique*, 320. Based on the aspect ratio of Ogier's drawings (10.5:15), Texier's dimensions are correct.

35 The description of the lid is recorded in Téxier, *Dictionnaire d'orfévrerie*, col. 835.

36 Téxier, *Dictionnaire d'orfévrerie*, col. 835, notes "et aux pieds une petite figure d'un home avec les mains jointes." Although Texier's description does not explicitly identify the figure at the crucifixion, scholars suggest that, based on the internal evidence of the epigram, it was Alexios; Frolow, *La relique*, 320; Klein, *'wahre' Kreuz*, 220; and *BEIÜ* 2: 175.

37 A similar composition is found on the eleventh–twelfth-century reliquary of the True Cross at Poitiers; Lynn Jones, "Perceptions of Byzantium: Radegund of Poitiers and the Relics of the True Cross," in *Byzantine Images and Their Afterlives: Essays in Honor of Annemarie Weyl Carr*, ed. Lynn Jones (Farnham: Ashgate, 2014), 105–124, Figure 5.7.

38 I know of only two other Middle Byzantine reliquaries with longer epigrams; both were commissioned by members of the Komnenian family. The now-lost reliquary of the True Cross of Emperor Manuel Komnenos (r. 1143–1180) had a 24-verse epigram; Frolow, *La relique*, no. 367; and Brad Hostetler, "Epigrams on Relics and Reliquaries," in *Byzantine Texts on Art and Aesthetics, vol. 3: From Alexios I to the Rise of Hesychasm (1081 – ca. 1330)*, eds. Charles Barber and Foteini Spingou (Cambridge University Press, forthcoming). The cross of Alexios's grandmother, Irene Doukaina, now in the Tesoro di San Marco (Santuario 57), features a 17-verse epigram; *BEIÜ* 2: Me90; and Hostetler, "The Function of Text," 87–89. An eleventh- or twelfth-century reliquary of Saint Demetrios, no longer extant, was also inscribed with a 16-verse epigram; Spyridon Lampros, "Ὁ Μαρκιανὸς κῶδιξ 524," *Νέος Ἑλληνομνήμων* 8 (1911), no. 79; and Hostetler, "The Function of Text," 198–199.

39 Βραχὺν ὑπνώσας ὕπνον ἐν τριδενδ[ρί]ᾳ
ὁ παμβασιλεὺς καὶ θεάν(θρωπ)ος Λόγος
πολλὴν ἐπεβράβευσε τῷ δένδρῳ χάριν·
ἐμψύχεται γὰρ πᾶς πυρούμενος νόσοις
ὁ προσπεφευγὼς τοῖς τριδενδρίας κλάδοις
ἀλλὰ φλογωθεὶς ἐν μέσῃ μεσημβρίᾳ
ἔδραμον, ἦλθον, τοῖς κλάδοις ὑπεισέδυν·
καὶ σῇ σκιᾷ δέχου με καὶ καλῶς σκέπε,
ὦ συσκιάζον δένδρον ἅπασαν χθόνα,
καὶ τὴν Ἀερμὼν ἐνστάλαξόν μοι δρόσον
ἐκ Δουκικ(ῆς) φυέντι καλλιδενδρίας,
ἧς ῥιζόπρεμνον ἡ βασιλὶς Εἰρήνη,
ἡ μητρομάμμη, τῶν ἀνάκτων τὸ κλέος,
Ἀλεξίου κρατοῦντος Αὐσόνων δάμαρ·
ναί, ναί, δυσωπῶ τὸν μόν(ον) φύλακά μου
σὸς δοῦλος Ἀλέξιος ἐ[κ] γένους Δούκας.
Ed. *BEIÜ* 2: 175

40 Frolow, "Deux inscriptions," 233–239. *BEIÜ* 2: 177, cautions against Frolow's thesis because there are characteristics of the epigram's meter that suggest that it was not authored by Kallikles. For Kallikles's cross reliquary epigrams, see the two composed for Irene and Eudokia; Frolow, *La relique*, nos. 241 and 312; and *Nicola Callicle Carmi*, ed. Roberto Romano (Naples: Bibliopolis, 1980), nos. 6 and 27. These reliquaries are no longer extant. It has been argued that Kallikles authored epigrams for two other reliquaries of the True Cross. One was owned by Irene and is now in the Tesoro di San Marco (see note 38); Frolow, *La relique*, no. 308; idem, "Reliquie orientali e reliquiari bizantini," in *Il Tesoro di San Marco*, vol. 2, ed. Hans Hahnloser (Florence: Sansoni, 1971), no. 25; *Nicola Callicle Carmi*, ed. Romano, no. 35; and *BEIÜ* 2: Me90. The other reliquary was owned by Irene's daughter Maria and is now in the Church of St-Eloi in Eine; Frolow, *La relique*, no. 249; Edmond Voordeckers, and Ludo Milis, "La croix byzantine d'Eine," *Byzantion* 39 (1969): 456–488; *Nicola Callicle Carmi*, ed. Romano, no. 33; and *BEIÜ* 2: Me3.

41 For epigrams with the word *stauros*, see Frolow, *La relique*, nos. 146, 212, 273, 276, 528, 661. For those that use the word *xylon*, see nos. 95, 134, 135, 157, 205, 241, 249, 308, 312, 587. For those that use both, see nos. 275, 367, 405, 465, 729. For the epigrams that do not include either word (excluding the Protaton and Grandmont reliquaries), see nos. 126, 338, 407, 427, 467, 473, 662. For further discussion, see Hostetler, "The Function of Text," 33, 91.

42 For the typology of the True Cross as the Tree of Life, see Anatole Frolow, *Les reliquaires de la Vraie Croix*, Archives de l'Orient chrétien 8 (Paris: Institut Français d'Études Byzantines, 1965), 178–186, esp. 183–184.

43 The reliquary is no longer extant, but the epigram is recorded in his corpus of poems; *Nicola Callicle Carmi*, ed. Romano, no. 7. See also Frolow, *La relique*, no. 338.

44 *Tridendria* refers to the three species of wood that supposedly made up the True Cross. Isaiah 60:13: "καὶ ἡ δόξα τοῦ Λιβάνου πρὸς σὲ ἥξει ἐν κυπαρίσσῳ καὶ πεύκῃ καὶ κέδρῳ ἅμα, δοξάσαι τὸν τόπον τὸν ἅγιόν μου." (And the glory of Lebanon shall come to you, with cypress and pine and cedar together, to glorify my holy place.) *Septuaginta*, Editio altera, eds. Alfred Rahlfs and Robert Hanhart (Stuttgart: Deutsche Bibelgesellschaft, 2006); translation from *A New English Translation of the Septuagint*, hereafter cited as NETS, eds. Albert Pietersma and Benjamin Wright (Oxford: Oxford University Press, 2007).

45 John 19:14: "ἦν δὲ παρασκευὴ τοῦ πάσχα, ὥρα ἦν ὡς ἕκτη. καὶ λέγει τοῖς Ἰουδαίοις, Ἴδε ὁ βασιλεὺς ὑμῶν." (Now it was the day of preparation of the Passover; it was about the sixth hour. He said to the Jews, "Behold your King!"), *The Greek New Testament*, 4th rev. ed., eds. Kurt Aland et al. (Stuttgart: Deutsche Bibelgesellschaft, 1983); translation from RSV. The phrase "right at the peak of midday" is also found in the panegyric for the feast day of the Elevation of the Cross (September 14) by the twelfth- and thirteenth-century Cypriot monk and saint Neophytos Enkleistos. *BEIÜ* 2: 175, notes that the wording of this sermon is similar to verses 1 and 6 of the Grandmont epigram: "Ζητοῦσα γὰρ ἐν τῷ ὄρει Γολγοθᾷ ἐν μέσῃ μεσημβρίᾳ εὑρήσω πάντως ἀφυπνοῦντά σε τῇ τριδενδρίᾳ." (Searching on Mount Golgotha, right at the peak of midday, I would, of course, find you sleeping on the three-part tree.), "Πανηγυρικὴ Α," Th. Giagkou and N. Papatriantafyllou-Theodoridi, eds., in *Ἁγίου Νεοφύτου τοῦ Ἐγκλείστου Συγγράμματα*, eds. Ioannes Karabidopoulos et al., vol. 3 (Paphos: Ἱερὰ Βασιλικὴ καὶ Σταυροπηγιακὴ Μονὴ Ἁγίου Νεοφύτου, 1999), 8.72–8.74.

46 The symbolism of the tree's shade is discussed by Gregory of Nyssa in his fourth homily on the Song of Songs: Hermann Langerbeck, ed., *Gregorii Nysseni in Canticum Canticorum*, Gregorii Nysseni opera 6, eds. Werner Jaeger et al. (Leiden: Brill, 1960), 118–119.

47 The use of healing oil from the Tree of Life appears in Rémi Gounelle ed., *Les recensions byzantines de l'Évangile de Nicodème* (Turnhout: Brepols, 2008), 300–302.

48 Psalm 132:3: "ὡς δρόσος Αερμων ἡ καταβαίνουσα ἐπὶ τὰ ὄρη Σιων· ὅτι ἐκεῖ ἐνετείλατο κύριος τὴν εὐλογίαν καὶ ζωὴν ἕως τοῦ αἰῶνος." *Septuaginta*, Editio altera; translation from NETS. See also Silvio Giuseppe Mercati, "Osservazione a CIG 8785," *Bessarione* 27 (1923): 72–73.

49 John Wilkinson, *Jerusalem Pilgrims Before the Crusades* (Warminster: Aris & Phillips, 1977), 81.

50 The healing nature of the True Cross and its association with *anargyroi* saints is reflected in the choice of imagery on the twelfth-century Philotheos Staurotheke, now at the Moscow Kremlin (1141); *BEIÜ* 2: Me97. The *anargyroi* Kosmas, Damianos, Kyros, and Panteleemon are the only figures featured on this reliquary.

51 *The Synaxarion of the Monastery of the Theotokos Evergetis*, 3 vols, trans. Robert H. Jordan (Belfast: Belfast Byzantine Enterprises, the Institution of Byzantine Studies, The Queen's University of Belfast, 2000), vol. 1, 167.

52 Frolow, "Deux inscriptions," 239. A characteristic feature of Kallikles's poetry is his metaphors of nature; *Nicola Callicle Carmi*, ed. Romano, 30. It was also a common theme for Manuel Philes; Andreas Rhoby, "Metaphors of Nature in the Poetry of Manuel Philes (XIVth c.)," in *Le lierre et la statue: La nature et son espace littéraire dans l'épigramme gréco-latine tardive*, eds. Florence Garambois-Vasquez and Daniel Vallat (Saint-Étienne: Publications de l'Université, 2013), 263–273.

53 Brad Hostetler, "The Limburg Staurotheke: A Reassessment," *Athanor* 30 (2012): 7–13.

54 Frolow, *La relique*, 73–94; and Jones, "Perceptions of Byzantium," 105–106.

55 Frolow, *La relique*, nos. 245 and 256.

56 For these reliquaries, see note 40. Irene also has three reliquaries of the True Cross listed in the typikon for her convent in Constantinople dedicated to the Theotokos Kecharitomene; Paul Gautier, "Le typikon de la Théotokos Kécharitôménè," *Revue des études byzantines* 43 (1985): 152.

57 Rhoby states that the word *mētromammē* (grandmother) is also a *hapax legomenon*. There has been some disagreement on the interpretation of this term as either grandmother or great-grandmother; *BEIÜ* 2: 177–178.

58 *Greek-English Lexicon* hereafter cited as *LSJ*, 9th rev. ed., eds Henry George Liddell, Robert Scott, and Henry Stuart Jones (Oxford: Clarendon Press, 1996), s.vv. "κάλλος" and "τρία."

59 *LSJ*, s.v. "ῥίζα."

60 *LSJ*, s.v. "πρέμνον."

61 For the protective function of enkolpia, see Brigitte Pitarakis, "Byzantine Enkolpia," in *Enkolpia: The Holy and Great Monastery of Vatopaidi* (Mount Athos: The Holy and Great Monastery of Vatopaidi, 2001), 13–14; and idem, *Les croix-reliquaires pectorales byzantines en bronze*, Bibliothèque des cahiers archéologiques 16 (Paris: Picard, 2006).

62 Peter Brown, *The Cult of the Saints: Its Rise and Function in Latin Christianity* (Chicago: University of Chicago Press, 1981), 86–87.

63 On the ways in which images validate the authenticity of relics, and sacred matter gives presence to the images they accompany, see Holger Klein, "Eastern Objects and Western Desires: Relics and Reliquaries between Byzantium and the West," *Dumbarton Oaks Papers* 58 (2004): 283–314, esp. 299–300; and Maguire, "Ivories as Pilgrimage," 142–143.

11 Place and surface

Golgotha in late medieval Bruges

Nadine Mai

> *Sed modo non quemamodum quantum olim in venustate atque palliciorum domorumque et templorum ornatur pollet, sanctitate tamen magis propter sacrata in ea loca magis quam olim choruscat. Est enim urbs urbium, sanctarum sanctior, gentium omnium domina, salutis nostre locus in centro mundi, in medio terre posita, a Deo preelecta et sanctificata.*[1]

Taken from a report of a late-fifteenth-century pilgrimage written by the Bruges patrician Jan Adornes, this passage conveys the desires, hopes, and allusions associated with Jerusalem in medieval times. When Jan left Bruges to go on a pilgrimage to the Holy Land with his father, Anselm, in February 1470, the Adornes family had already allocated an area in their hometown for intense veneration of Jerusalem. The site was dominated by a Jerusalem chapel, donated by Anselm's father, Pieter Adornes II, and consecrated in 1429.[2] What this first chapel looked like is uncertain, although we know that it included one altar and two bells and was decorated with some instruments of the Passion and had a Holy Sepulchre within an annexed crypt.[3] It appears that by 1469 it was already in poor condition, and Anselm had assumed responsibility for the whole foundation, the buildings and the charitable as well as liturgical donations.[4] Thus, his pilgrimage was important not only for renewing the family's engagement with the Holy Land, but also for the spiritual and material renovation of their Jerusalem in Bruges. Upon his return in spring 1471, inspired by the original sites, Anselm went beyond his vow to renovate the paternal chapel and completely rebuilt it (Figure 11.1).[5]

Evidently, Anselm planned a broader representation of Jerusalem, trying to combine the particulars of the first chapel, founded by his father, with the knowledge he had acquired while traveling. Completed in 1483/1485,[6] Anselm's chapel was not a copy of a particular site but a conscious mixture of architectural features related to Jerusalem.[7] Its two-part structure consists of a rectangular nave with a high gabled roof and a large octagonal tower with a multilevel one. At its upper part this tower is flanked by two small octagonal towers crowned by a sun and a moon and terminates in a – formerly gilt – copper ball, suggesting Eastern influences as well as some features of Christian iconography. Many panoramas of Jerusalem painted in fifteenth-century Flanders depict similar multilevel towers with arcades and golden domes characteristic of Eastern cities.[8] For this reason, the exterior shape of the Bruges Jerusalem Chapel may be considered an architectural anthology of the Holy Land rather than a reproduction of one specific architectural prototype. Not surprisingly, records related to the

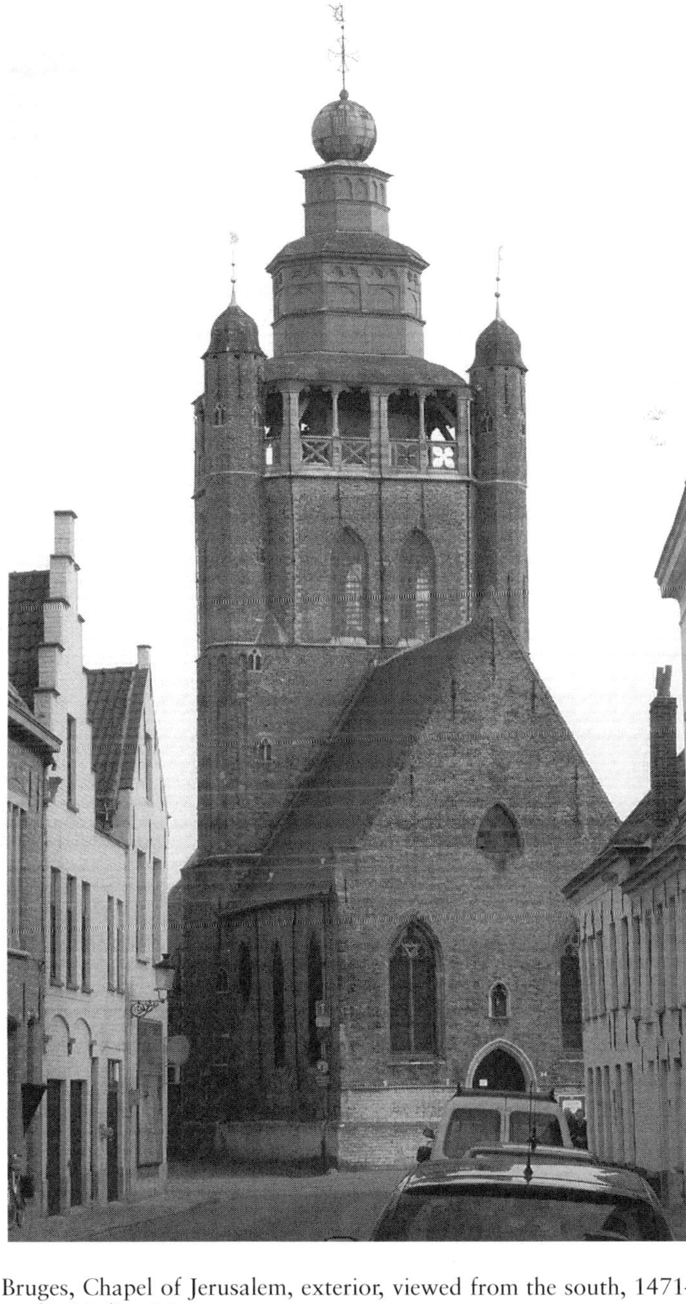

Figure 11.1 Bruges, Chapel of Jerusalem, exterior, viewed from the south, 1471–1485. Photo credit: Nadine Mai.

Bruges chapel from around 1500 reveal how its architectural shape was interpreted by various contemporaries. Described as "*honorem domini sepulcri et illis similtudine*," "*gerentem effigiem pro parte templi Salomonis*," and "*edificatum instar ed ad verum similtudinem templum sancti sepulchrum de Jherusalem*," the building was associated with a wide range of Holy Land edifices.[9]

In contrast, the interior of the chapel evokes more concrete architectural, spatial, and material features of the holy sites in Jerusalem, particularly those within the Church of the Holy Sepulchre. Consisting of a rectangular nave and a tower divided into a low crypt with a high choir above it, the Bruges chapel combines three different spaces. Nave, choir, and crypt are clearly distinguished by their proportions and lighting and are related to each other principally by an altar placed prominently in front of the choir wall and embedded within an impressive stone relief of Mount Calvary (Figs. 11.2 and 11.3).

Some *Arma Christi* displayed on the sculptured rock, including tools as well as horses' hoofprints, alluding to the Roman soldiers who led Christ to Golgotha, evoke a narrative dimension. A large rectangular frame with five, now empty, grated niches indicates that the Calvary monument was also meant as a relic shrine, and, encasing the altar table, it was obviously central to the chapel's liturgy. Crowned by three large crosses this figurative ensemble rises up to the triumphal arch (Figs. 11.2 and 11.4).

As I discuss later, in alluding to the passion and forming a visual junction in the architectural space, this sculptured Calvary seems to reflect the materiality of the holy

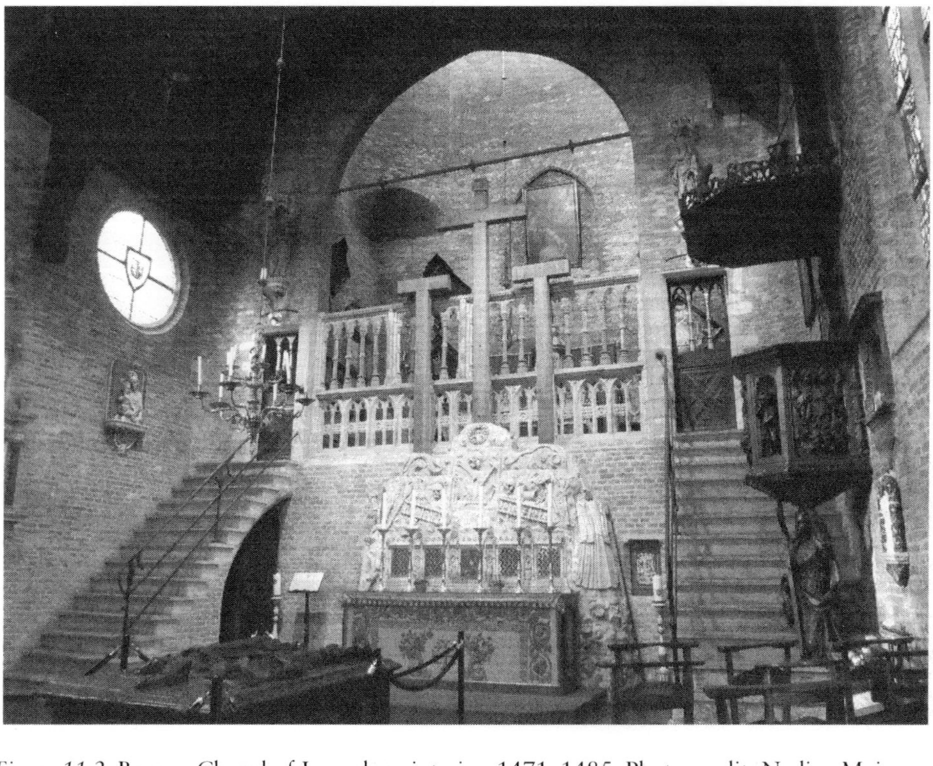

Figure 11.2 Bruges, Chapel of Jerusalem, interior, 1471–1485. Photo credit: Nadine Mai.

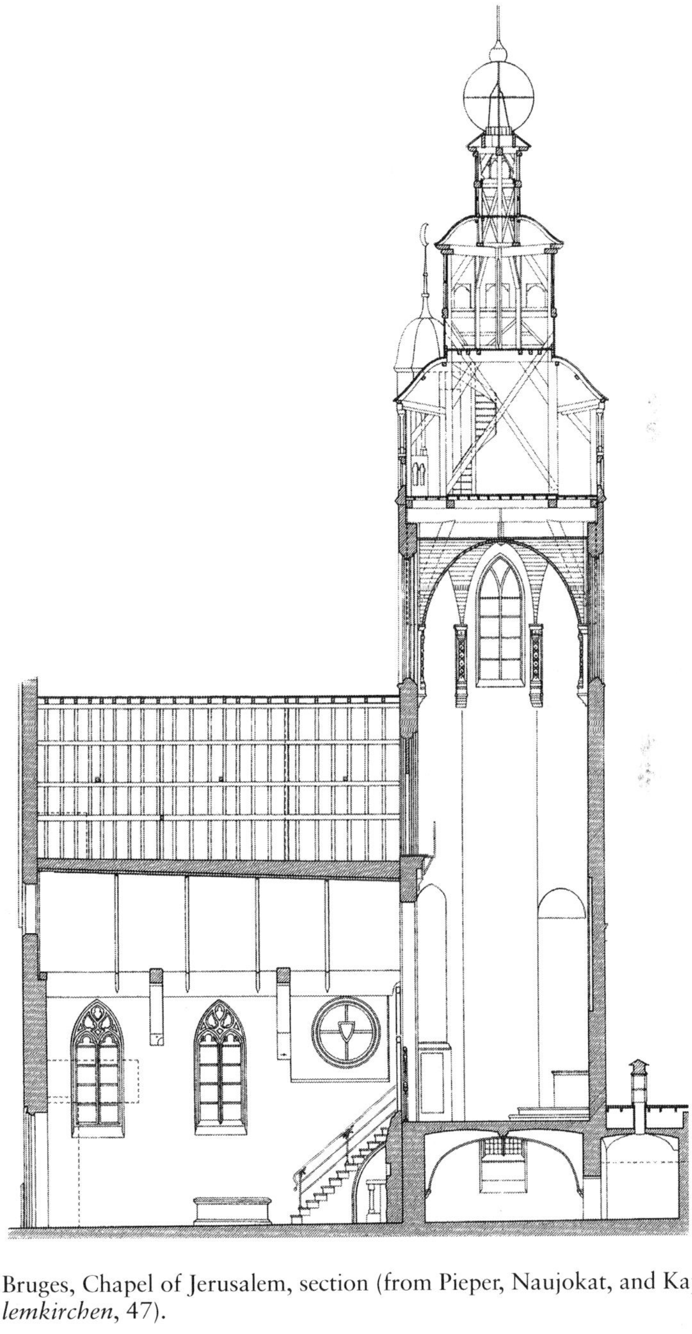

Figure 11.3 Bruges, Chapel of Jerusalem, section (from Pieper, Naujokat, and Kappler, *Jerusalemkirchen*, 47).

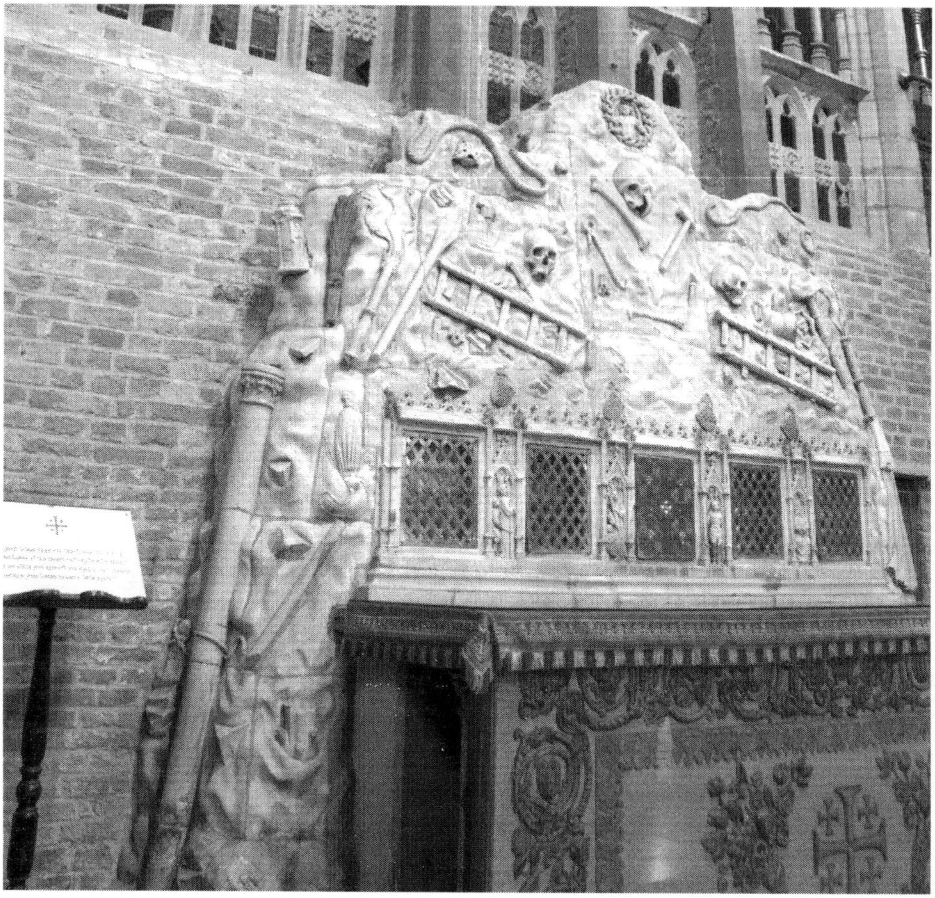

Figure 11.4 Bruges, Chapel of Jerusalem, Calvary, ca. 1475. Photo credit: Nadine Mai.

sites in Jerusalem through certain strategies of approximation and authentication. The way that this centerpiece corresponded to the architectural frame and other furnishings in the chapel also elucidates the practice of translating and re-creating Jerusalem in medieval times.

Building Golgotha

Christian culture imagined the Holy Land as a multimedia embodiment of the Old and New Testaments, or a "Memnotop"[10] of salvation history.[11] With the erection of architectural monuments around the holy sites in late antiquity and the Crusader periods, the biblical narrative was not only fixed to concrete locations but framed and accentuated by art and liturgy.[12] Moreover, the stories were tied to the appearance of the buildings, entangled in specific formal, spatial, material, topographic, and architectural features. Once home, pilgrims such as Jan and Anselm Adornes, who initiated monumental reproductions, sought not only to re-evoke and to preserve their pilgrimage experience, but to transmit the passion story inherent in the original sites. These replicas were designed to inspire an authentic feeling of personal contact and used many media to evoke sacred power.[13] Rather than reproducing exact dimensions, the Bruges chapel *re-enacted* the holy sites and their composition, addressing topographic, spatial, and material aspects to establish a discursive place of veneration, not a substitute.

With a monumental sculptured Calvary marking the chapel's central altar, enlarged with crosses and other passion tools, and Mary and John positioned on top of the choir doors to the left and right, the Bruges installation obviously referred to Christ's martyrdom on Golgotha. In addition to these iconographic references, the two architectural levels of the crypt and the choir apparently reproduced the original crucifixion site in Jerusalem, where the natural rock was lodged in a tiny ground-level room called the Chapel of Adam, which was connected by staircases to an upper chapel dedicated to the Holy Cross.[14] Behind the altar in the Adam Chapel in Jerusalem an opening showed the riven rock of the crucifixion, its large crack traditionally attributed to an earthquake that occurred when Jesus died (Mt. 27,51). In contrast, the Golgotha Chapel upstairs was wide and splendid, decorated with precious mosaics. The floor of this upper chapel was almost entirely covered by marble slabs, exposing only the beginning of the crack and a hole in the rock that indicated where the cross was driven.[15]

The Bruges arrangement, also containing a small, dark space below and a wider space above reached by staircases, was strongly influenced by the original crucifixion site. This connection is also elaborated within the Calvary relief set in front of the crypt. Emerging from the floor and the wall behind and interleaved by architecture, it approximates a natural hillside (Figure 11.4). However, the relief cannot be seen from inside the two level construction, so the inner spaces of both sections may evoke other associations (Figure 11.3). According to Pieper, the contrast of the dark narrow crypt with the high bright choir may allude to another part of the Church of the Holy Sepulchre in Jerusalem: the Anastasis Rotunda, also polygonally arranged and illuminated from above.[16] Following this interpretation, Pieper refers to a replica of Christ's sepulchre that was included in the Bruges crypt from the outset.[17] This iconographic attribution allowed him to describe the Bruges chapel as a "*topische*" perception of the Church of the Holy Sepulchre, illustrating the dialectic of death and resurrection.[18]

Unfortunately, the original Holy Sepulchre of Bruges, which was apparently rearranged in the crypt after the chapel was rebuilt in the 1470s, no longer exists, and nothing about its appearance was ever recorded.[19] However, the renovated larger chapel with its lighted tower and rock retable might have led to a change in the way this sepulchre was perceived within the Bruges chapel. Now hidden behind a monumental rock, the crypt would have been more reminiscent of the biblical tomb cut out of rock. During the Middle Ages, pilgrims were aware that the Church of the Holy Sepulchre in Jerusalem was situated on and around the stony hill of Calvary, as Wilbrand of Oldenburg noted, "in truth the hill and the tomb are very close to one another."[20] This connection is also apparent in Bruges by a small depiction of Christ's tomb on the Calvary relief immediately under the left cross (Figure 11.4). Assembled in the center of the chapel and connecting the crypt and the choir, the sculptured rock of Bruges seems to underscore this topographic coexistence of the crucifixion, entombment, and Resurrection on Golgotha in Jerusalem. The crypt doors, positioned under the stairways to the choir, create mysterious dark caverns flanking the Calvary scene.

This illusion of walking through the rock was probably also enhanced by coloring that highlighted the stones and plots of grass.[21] Although we cannot trace the exact features of this coloring, it may have extended to the choir wall and the flanking staircases, creating a hillside panorama within the architectural setting. In a drawing from the first quarter of the nineteenth century, J.J. Gailliard illustrated the extant parts of a mural that placed the Calvary scene within a large landscape (Figure 11.5). In the natural environment some buildings appear like stations of pilgrimage, and we can see the Bruges Jerusalem chapel on the left.[22] The ensemble thus constitutes a virtual movement through a landscape, which may also have been linked to the donor's pilgrimage, as were other contemporary works, for example, William Wey's large map of the Holy Land or the Gotha panel memorializing the pilgrimage of Duke Frederick the Wise in 1493.[23] Gailliard's drawing conveys the engaging effect of the multilayered arrangement of architecture, sculpture, and mural, which supported imagination and interaction. Thus, for the devoted in Bruges, a visit to the Holy Sepulchre hidden in the crypt provided a strikingly immersive experience. Moving toward that replica and walking under the rocky structure, worshippers would traverse a space pervaded by memories of the passion, as in Jerusalem. According to Lidov's definition of hierotopy – "topographical material concreteness stimulated the power and miraculous efficacy of a spatial image" – the animating concept of the Bruges architecture and its Calvary is to connect the faithful to Jerusalem.[24] Referring directly to the holy sites as the setting

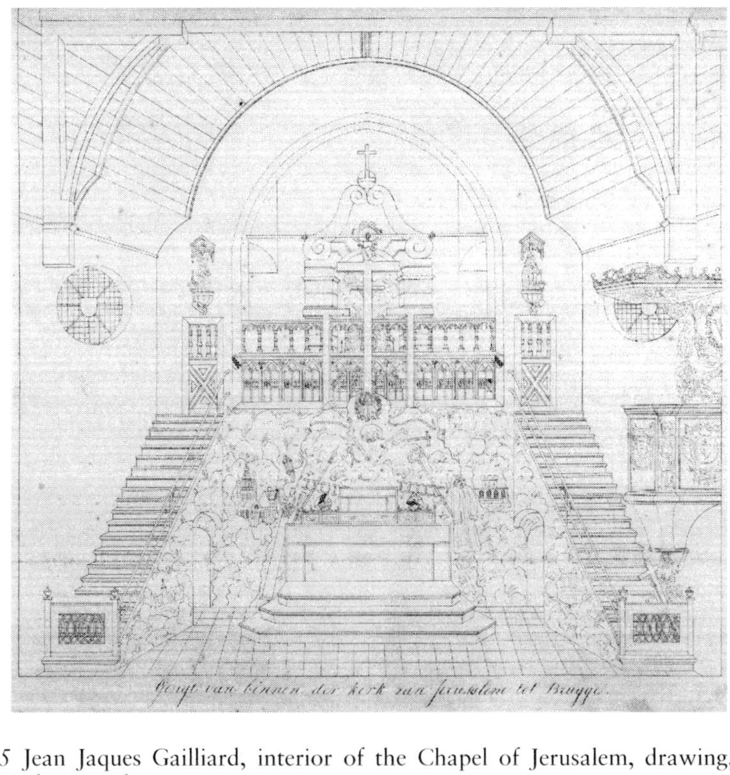

Figure 11.5 Jean Jaques Gailliard, interior of the Chapel of Jerusalem, drawing, ca. 1825. Photo credit: Musea Brugge.

of the biblical story, the ensemble combined features of the original topographic and architectural structures with significant imagery to achieve a spiritually and sensually stimulating holy site abroad.[25]

Toward material visuality

When describing his visit to the site of the crucifixion in Jerusalem, Jan Adornes referred to the original Mount of Golgotha as "a small hill of living stone of white color mixed with red veins, naturally elevated."[26] Such qualities probably influenced the deliberate decision to forgo bright painting – the common altar decoration at that time in Bruges – in favor of an enormous artificial rock in a new arrangement in the family chapel. The Calvary's special impact raises further questions about the contribution of the pilgrim experience to re-creations of the holy sites.[27]

Apparently, no effort was spared to arrange the massive blocks of local sandstone to resemble one rock (Figure 11.4). With its undulating surface, stones strewn about, and protruding grass and tree stumps, the sculpture takes on a compelling resemblance to a natural surface.[28] The *Arma Christi* sculptured onto the artificial rock substantiates this impression decisively.[29] Rather than mere "representations of material objects"[30] or "*signa*"[31] connected to Christ's passion, these life-sized *Arma* evoke a naturalistic, tangible physicality. By not depicting the crucified or dead body of Christ, this passion panorama focuses on material traces, visually compensating for his torment. The relatively haphazard composition forces the viewer to pass and circle the Calvary in order to experience the corporeal presence of each object and to envision its use.[32] Some wooden elements, such as the dice lying on Christ's garments or the scourges on the left side of the relief, increase the visual claim to material reality. Moreover, mnemonic motifs of the passion that appear in some other depictions of the *Arma* – picking hands, a spying tongue, heads of shouting Jews – are totally missing.[33] This Calvary is meant as a *locus sanctus*, where the tools of torture were cast aside after the cruel event, turning the *Arma* from emblems to actual implements of Christ's suffering.[34]

Other tools – broken or destroyed – support this interpretation. In the center a ladder has several rungs hanging awry or snapped off and lying next to it; their rough ends add to the sense of rupture and recall the disruption of the world at the moment of Christ's death, linked to the cleft in the crucifixion rock in Jerusalem.[35] For additional effect in Bruges a fictive corrosion is eroding the tongs, while the lance on the left side is broken into two pieces. Although these destroyed items may refer to the biblical disruption, in my opinion, they also raise the visual "paradox"[36] of sacred materials as humble matter on the one hand and relics touched by Christ on the other.

In terms of the devotional system of pilgrimage, most of the events of the passion were linked to materials, architectural traces, and negligible topographic realities, which became "part and parcel of the ritual experience."[37] As Rachman-Schrire has shown regarding the Church of the Holy Sepulchre, the natural stones displayed at the Jerusalem sites alternated between "categories of *place* and *relic*" owing to the immediate authenticity invoked by their visual and tactile qualities.[38] In addition, material characteristics and purely superficial features, such as color, size, texture, and topographic setting, were inseparable from the story they memorialized, lending the passion a particular narrative and "plot." This strategy was especially evident in the processional routes followed by the faithful to the Church of the Holy Sepulchre, which over the centuries incorporated more and more stones, ruins, and architectural elements.[39] Here, the pilgrim was forced "to take material objects and constructions seriously, to understand them not as mute things but as actors with a persistent past."[40] Consider Burchard of Mount Zion's description of Christ's flagellation column:

> Twenty-four feet east of Calvary is an altar, below which is part of the column at which the Lord was scourged. It has been brought here from the house of Pilate and is covered by the altar stone in such a way that it may be touched, seen, and kissed by the faithful. It is a piece of blackish porphyritic stone, containing natural red spots, which the common people believe to be colorings of Christ's blood.[41]

Burchard lavished attention on the column's properties, totally subverting the legend and enabling the surface to transmit a narrative layer. He also demonstrated that the "common people" took material aspects for reliable evidence. In other words, connected to a certain biblical story, the characteristics of the holy sites not only defined the location but were evaluated as historical witness of the spilling of Christ's blood, becoming a material diorama of his suffering.

As much as materials participated in forming the syntax and experience of pilgrimage to Jerusalem, they became even more significant in the imitation of holy sites at home.[42] A striking example is the so-called flagellation column preserved in the large complex of Santo Stefano in Bologna, designed to reproduce the Church of the Holy Sepulchre in Jerusalem (Figure 11.6).[43] Next to other rebuilt Holy Sites Santo Stefano also preserves a so-called flagellation column, alluding to the already mentioned original in Jerusalem (Figure 11.6). This large monolith of a gray, at some points nearly black, marble is fixed between a Romanesque capital and an antique rectangular pedestal with a floral decoration, clearly marking it as a meaningful *spolia*. An inscription above, added later, identifies it as a monument to the flagellation.[44] Its contextualization as such can be recognized at the lower end, where the column has a lighter colored, abraded area with little crags and indentions. This disjuncture along with its size, aged appearance, and puzzling dark color translates the column visually as a "relic" articulating a sense of history.[45] Described as "a column of marble similar to that where our Lord, Jesus Christ, was beaten and flagellated" in the *Vita* of San Petronio (Bologna, twelfth century), its material appearance was explicitly designed to create a relationship with Jerusalem.[46] Note that the link to Jerusalem was not based on authentic provenance from the Holy Land; it did not derive from the concrete location of the flagellation, but was connected to the event only by its "similarity" to the original site in Jerusalem. The connection to an original by means of a *similtudine* is a well-known way of translating Jerusalem, especially in architectural replicas, and obviously influenced relic devotion as well.[47] Here the sacred bond stems from a mimesis, creating a certain mystical analogy between the arrangement of similar materials in the "New Jerusalems" and the stones truly touched by Christ. Its "relic" character thus resulted not from material legitimacy but from a specific visual quality. Allowing believers to touch, to kiss, to feel the material and also to circle the Bologna installation offered another physical connection to Christ's pain through a piece of natural stone as in Jerusalem. The past was imitated by physical condition and legend by surface resemblance, and matter became act.[48]

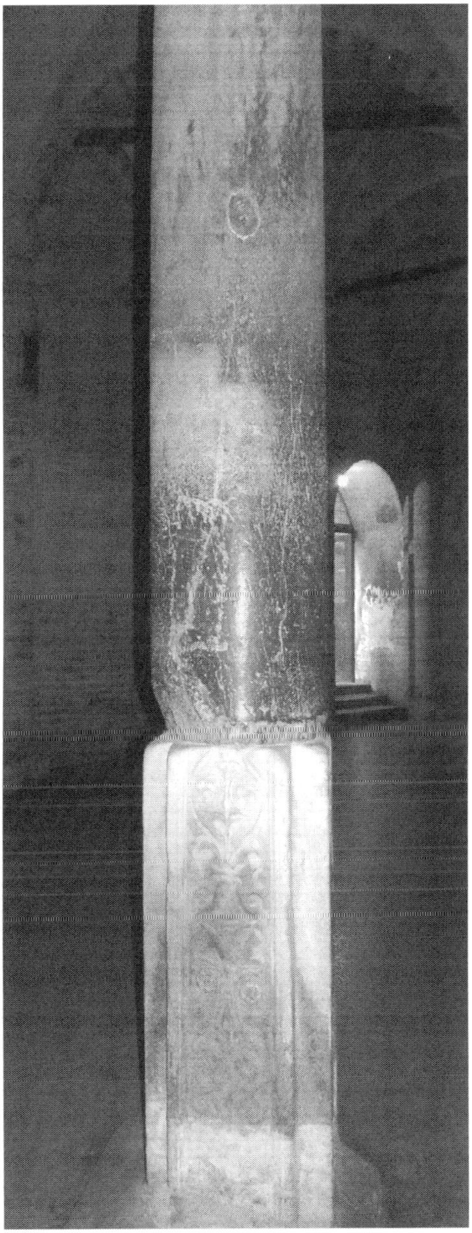

Figure 11.6 Bologna, Santo Stefano, Column of the Flagellation, black and white marble, antique descent, installed in the twelfth century. Photo credit: Nadine Mai.

The Bruges Calvary created similar modes of "metonymic rather than metaphoric"[49] perception by staging material as a trace of the passion, mediating the power of its place. Although not a monolith, it tried to generate visual and haptic correspondences to the rock of Golgotha in Jerusalem and assigned a narrative layer to the material's surface. Similar to the Bologna column offering a reliable illusion of recent torture in the black marble's rough and battered condition, the Bruges Calvary makes Christ's suffering resonant and vibrant.

Apart from the unusual composition of the *Arma*, the horses' hoofprints appearing at the bottom of the relief support this argument. As Jerusalem evidenced numerous *vestigia Christi* within matter, the Bruges example similarly seems to refer to the concept of the Holy Land as contact relic.[50] While the imprints represent a more somatic process, carving the passion into the sculptured hillside, the *Arma* achieve an exalted, subjective tangibility. The rise and fall of the carving might have constituted a kind of devotional instruction comparable to the pilgrimage experience, invoking memory with the mind and physical contact with the body. After all, the particular aesthetic of the imprint, conveying the dialectic interplay of grasp and lose, life and death, present and past is the sustaining principle of sacred places.[51] The Bruges chapel's visual claims joined together with the material authenticity of Jerusalem causes these imprints to make a denotative plea for interlacing material place and biblical event, and by doing so, the Calvary sculpture creates the vivid presence of the material ground, evoking the redemptive permanence of Jerusalem.

Translated materials: the Bruges Calvary as reliquary

As a main paradigm of Holy Land veneration, stones and other materials were thought to have "absorbed" the power of the holy sites, and collecting them was a primary way to maintain the grace of physical connection.[52] Matter deemed holy could be considered an "equivalence" of the *loca sancta*; both originated from the idea of Jerusalem as the real site of the passion and from the *pars pro toto* principle, which holds that the smallest parts provide direct access to the divine.[53] In all likelihood, the grated compartments in the Bruges Calvary exhibited such souvenirs brought home from the Holy Land by Anselm and Jan Adornes. The chapel's treasury and inventories provide insight into the objects they may have housed.[54] While Anselm surely planned to present his holy land relics within the Calvary retabel, his son Jan modified this framing. In a testament addition of 1493, he mentioned that he wanted to donate precious reliquaries for the "*reliquas capelle et etiam pro aliis reliquiis qui portaminus ex t[er]ra s[an]cta.*"[55] This bequest was fulfilled after his death in 1511 when three reliquaries were fashioned, presumably still preserved in the chapel's treasury: a reliquary of the True Cross, a silver resurrection reliquary, and a reliquary showing a cross made of cypress wood (Figure 11.7).[56]

The last two items deserve special attention because they indicate, first, a connection with the pilgrimage of Anselm Adornes and, second, the range of notions about material evidence of the Holy Land.[57] Moreover, they are both registered at the beginning of a 1521 inventory, whereas other relics of famous saints such as St. Kathryn and St. Servatius are listed much further on.[58]

The resurrection reliquary, executed as a tomb of Christ, alludes perfectly to the Holy Sepulchre in Jerusalem and features two small white stones possibly broken off from it. The wooden cross shows scenes from the lives of Christ and Mary and was possibly made in late medieval Byzantium.[59] Although not directly connected to a

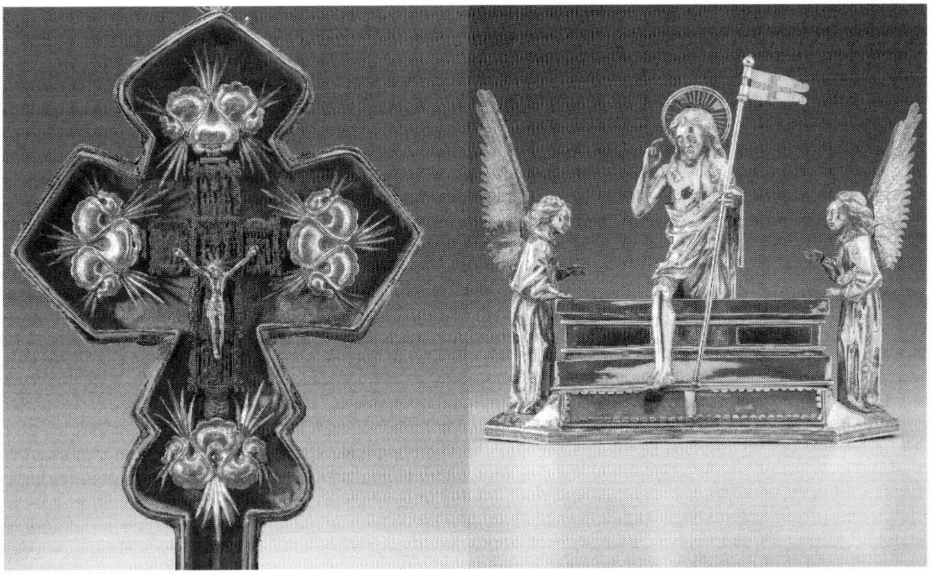

Figure 11.7 Resurrection reliquary, silver, 1511; reliquary with a Byzantine cross, palm wood and silver, 1511; the Christ is an early modern addition. Photo credit: Studio Philippe de Formanoir.

holy site, this reliquary may have had particular status because it was brought from the Holy Land and because of its physical material, cypress wood, duly mentioned in the treasury register of 1521. Such exotic materials carried legendary meaning along with verifiable characteristics of the Holy Land and were thought to transmit the grace of the biblical events.[60] According to the *Legenda Aurea*, the True Cross was made from four different trees: palm, olive, cedar, and cypress.[61] The later addition of the Adornes' silver clouds also visually transformed the Byzantine cross into a relic. The precious decoration was a visual reminder of the donors' pilgrimage, legitimizing the physical artifact's provenance in the Holy Land.

In trying to reconnect the chapel to the city of Jerusalem, these relics were perfect additions to the Bruges foundation. Sources indicate that they were presented to the public during Mass, so the Calvary altar provided an encounter between material evidence and liturgical memory of the passion.[62] These relics, viewed with the tactile surface of the Calvary and the elevated host, the mystical *Corpus Christi*, alluded to different layers of the presence of Christ and enlarged the realm of passion imagination.[63]

Indeed, the place of the passion as a reservoir of relics was not fictional, but rather was rooted in the veneration of the holy sites, especially Golgotha. In the wake of the discovery of the Holy Cross in late antiquity, it was thought that other objects had been found close to the site of the crucifixion, including some original *Arma*, such as the *titulus crucis*, the lance, and the nails.[64] Anselm and Jan Adornes commemorated this legend by explicitly referring to its topography in the subterranean chapel of Saint Helena in Jerusalem, which is in a cave close to the rock of Golgotha, "the place where the crown, the lance, and the nails were discovered, situated below Mount Calvary and carved into it."[65] This passage sheds more light on the juxtaposition of place and materials in the extraordinary iconography of the Bruges Calvary retable.[66] Conveying the ambiguity of Golgotha as both setting and storehouse of the event, the associative topography as well as the naturalistic surface of the Calvary sculpture, especially the tangibility of the *Arma Christi* and their rendering of transiency, establish a further analogy with the holy sites and express the particular visual argument of its material reality, as argued above. The life-sized representation of the passion tools proposes additional ideas of translating the sacred; for example, the veneration of holy measures such as the length of the body of Christ or the crucifixion nails.[67] In any case, the relic culture of Jerusalem may have played a considerable part in creating an "authentic" setting of Golgotha in Bruges, confirming that salvific history could still be traced in Jerusalem.

Conclusion

Having considered the significance of matter – both its provenance and physical qualities – in the creation of devotional experience, I have shown that the Bruges Calvary elicits questions of materiality in a very broad context. The arbitrarily distributed *Arma Christi*, the way the passion tools are almost assimilated into the rocky surface and the not-yet vanished horses' hoofprints express a very individual, if not unique, visualization of the passion via its place, Golgotha. Additionally, converted into a reliquary the sculpture was staged as a material witness. In this regard, the monumental Calvary constituted a multisensory collection and recollection of the sacred, a statement of Christ's mundane permanence, which could be likewise replicated by walking through the surrounding architecture. Visiting the Holy Sepulchre in Bruges intentionally combined passing under the ground and through the rock, while steps leading up to the choir evoked climbing Golgotha. The whole structure required physical movement to support its narrative effects. Simultaneously preserving tools, sites, and relics and allowing liturgical re-enactment, the Bruges installation creates an inspiring link to the original sites of the passion and transformed the chapel into a multifaceted synopsis of what a holy place might be.

Notes

1 "Today, this city does not shine because of its beautiful temples, palaces, and houses, but it is resplendent from the sanctity of the holy sites it embraces. It is the city of cities, the holiest of all sacred cities, the leader of all nations, a place of salvation set in the middle of all territories, in the center of the world, the city elected and blessed by God." My translation, after the French version published in *Itinéraire d'Anselme Adorno en terre sainte (1470–1471)*, eds. Jacques Heers and Georgette de Groer (Paris: Editions du Centre National de la Recherche Scientifique, 1978), 254–256.

2 The document was copied in the late fifteenth century; Stadsarchief Brugge (SAB), *Adornes en Jeruzalemstichting*, 712, fol. 16v.

3 This information can be traced to the consecration document of 1429 (see note 2.) and a papal bull of 1435, which refers to "*sepulcrum nonnullaque alia insignia passionis domini nostri Jhs XPI*" and announces the building of a widows' *beguinage*, a living palace, and a bell tower (inv. 716).

4 SAB, *Adornes en Jeruzalemstichting*, 39, fol. 90r–95r.

5 According to Jean-Pierre Esther, "Monumentenbeschrijving en bouwgeschiedenis van de Jeruzalemkapel," in *Adornes en Jeruzalem: Internationaal leven in het 15de- en 16de-eeuwse Brugge*, eds. Noel Geirnaert and André Vandewalle (Bruges: Stad Brugge gemeentebestuur, 1983), 50–81. The connection between the pilgrimage and the rebuilding of the

chapel is also noted in Mitzi Kirkland-Ives, "Capell nuncapato Jherusalem noviter Brugis: The Adornes Family of Bruges and Holy Land Devotion," *Sixteenth Century Journal: The Journal of Early Modern Studies* 39, no. 4 (2008): 1041–1064; and Jan Pieper, Anke Naujokat, and Anke Kappler, eds., *Jerusalemskirchen: Mittelalterliche Kleinarchitekturen nach dem Modell des Heiligen Grabes* (Aachen: Geymüller, 2003), 17–18.

6 The choir and tower were finished after 1483 under the patronage of Jan Adornes. See SAB, *Adornes en Jeruzalemstichting*, inv. 969, fol. 9r–10.

7 This approach is defined by Richard Krautheimer, "Introduction to an 'Iconography' of Mediaeval Architecture," *Journal of the Warburg and Courtauld Institutes* 5 (1942): 1–33. Overviews of these monuments reproducing the holy sites appear in Geneviève Bresc-Bautier, "Les imitations du Saint-Sépulchre de Jérusalem (IXe-XVe siècles)," *Revue d'histoire de la spiritualité* 50 (1974): 319–342; and Barbara Dietrich, "Anastasis-Rotunde und Heiliges Grab in Jerusalem: Überlegungen zur architektonischen Rezeption im Mittelalter," *Georges-Bloch-Jahrbuch des Kunsthistorischen Instituts der Universität Zürich* 11, no. 12 (2004): 7–29. For a critical assessment of the methods of architectural copying, see Sarah Blick, Rita W. Tekippe, and Vibeke Olson, eds., *Copying in Medieval Art (Visual Resources 20)* (London and New York: Routledge, 2004).

8 See *Avis pour faire le passage d'outre mer* for Phillip the Good, Paris, Bibl. Nat., Ms. Fr. 9087, fol. 85v. Pieper makes the connection to early Netherlandish paintings in Pieper, Naujokat, and Kappler, *Jerusalemskirchen*, 45. Although I agree with his analysis, the tower also recalls some central planned depictions of Jerusalem that were emerging in the same century (*Liber Chronicarum* by Hartman Schedel Nürnberg [Koberger] 1493, folio XVII).

9 "built in honour of the Sepulchre of Christ and similar to it," Episcopal Bull of 1518; SAB, *Adornes en Jeruzalemstichting* inv. 1247; "partly designed as an effigy of the temple of Solomon," letter from Marten Adornes, son of Anselm, in 1494; inv. 39, fol. 194; "built with all similarity after the 'temple' of the Holy Sepulchre in Jerusalem," letter from Jean de la Coste Adornes, c. 1520; inv. 1155.

10 Bruno Reudenbach, "Loca sancta. Zur materiellen Übertragung der heiligen Stätten," in *Jerusalem du Schöne (Vestigia Bibliae 28)*, ed. Bruno Reudenbach (Bern: Peter Lang, 2015), 9–32, esp. 14.

11 The development of a memorial topography in Christian culture is elaborated in Maurice Halbwachs, *La topographie légendaire des Évangiles en Terre Sainte: Etude de mémoire collective* (Paris: Presses universitaires de France, 1941), 149–206; and Glenn Bowmann, "Christian Ideology and the Image of a Holy Land: The Place of Jerusalem Pilgrimage in the Various Christianities," in *Contesting the Sacred: The Anthropology of Christian Pilgrimage*, eds. Michael Sallnow and John Eade (London and New York: Routledge, 1991), 98–121. On the topic of a physical connection, see, for example, Robert A. Markus, "How on Earth Could Places Become Holy? Origins of the Idea of Holy Places," *Journal of Early Christian Studies* 2 (1994): 257–271.

12 See Reudenbach, *Loca sancta*, 12–17; Bruno Reudenbach, "Golgatha – Etablierung, Transfer und Transformation: Der Kreuzigungsort im frühen Christentum und im Mittelalter," in *Räume der Passion: Raumvisionen, Erinnerungsorte und Topographien des Leidens Christi in Mittelalter und Früher Neuzeit (Vestigia Bibliae 32)*, eds. Hans Aurenhammer and Daniela Bohde (Bern: Peter Lang, 2015), 13–28, esp. 14–16.

13 See Bianca Kühnel, "Virtual Pilgrimages to Real Places: The Holy Landscapes," in *Imagining Jerusalem in the Medieval West*, eds. Lucy Donkin and Hanna Vorholt (Oxford: Oxford University Press, 2012), 243–264.The multiple associations are also discussed in Robert G. Ousterhout, "Flexible Geography and Transportable Topography," in *Real and the Ideal Jerusalem in Jewish, Christian, and Islamic Art: Studies in Honor of Bezalel Narkiss on the Occasion of His Seventieth Birthday* (Special Issue of *Jewish Art*, vol. 23), ed. Bianca Kühnel (Jerusalem: Centre for Jewish Studies, 1997), 393–404, esp. 402; Nikolaus Jaspert, "Vergegenwärtigungen Jerusalems in Architektur und Reliquienkult," in *Jerusalem im Hoch- und Spätmittelalter*, eds. Dieter Bauer, Klaus Herbers, and Nikolaus Jaspert (Frankfurt: Campus, 2001), 219–270.

14 Jürgen Krüger, *Die Grabeskirche zu Jerusalem: Geschichte – Gestalt – Bedeutung* (Regensburg: Schnell und Steiner, 2000), 123–133. For the iconographic tradition of the Adam Chapel, see Gustav Kühnel, "Architectural Mise-En-Scène and Pictorial Turns in Jerusalem," in *Jerusalem as Narrative Space: Erzählraum Jerusalem*, eds. Annette Hoffmann and Gerhard Wolff (Leiden: Brill, 2012), 21–31; Reudenbach, *Golgotha*, 20.

15 See, for example, Burchard of Mount Zion, ca.1280: "The floor of this chapel is paved completely in marble," cited in *Pilgrimage to Jerusalem and the Holy Land, 1187–1291*, ed. and trans. Denys Pringle (Farnham: Ashgate, 2012), 295. The building changed over time: See Yamit Rachman-Schrire, "The Rock of Golgotha in Jerusalem and Western Imagination," in *Räume der Passion*, eds. Bohde and Aurenhammer, 32–36.
16 Pieper, Naujokat and Kappler, *Jerusalemskirchen*, 46–47.
17 For documentary evidence regarding the former Holy Sepulchre in Bruges, see note 3.
18 For a definition and examples of the topical approach to Jerusalem, see Pieper, Naujokat, and Kappler, *Jerusalemskirchen*, 8–11.
19 In 1494, the sepulchre replica was still in use; SAB, *Adornes en Jeruzalemstichting*, inv. 39, fol. 194r. In all likelihood it was destroyed in 1522 when the still-preserved Holy Sepulchre chamber was built on the north side of the crypt on behalf of the confraternity of Jerusalem (inv. 787, fol. 6v–7r).
20 Cited in Pringle, *Pilgrimage*, 89.
21 A few remnants of color are still detectable. In all likelihood, the Calvary was painted from the beginning as the first repainting was done in 1529. SAB, *Adornes en Jeruzalemstichting*, inv. 787, fol. 7v.
22 Musea Brugge, Prentenkabinet, inv. 0.3115. The silhouette of the Bruges chapel also appears in the crucifixion mural in the well-preserved oratory.
23 See Pnina Arad, "Pilgrimage, Cartography, and Devotion: William Wey's Map of the Holy Land," *Viator* 43, no. 1 (2012): 301–322, esp. 310–313. "The Gotha panel (Schlossmuseum, Schloss Friedensstein, inv. SG 77)," in *Alltag und Frömmigkeit am Vorabend der Reformation in Mitteldeutschland: Katalog zur Ausstellung 'Umsonst ist der Tod,'* eds. Hartmut Kühne, Enno Bünz, and Thomas Müller (Petersberg: Michael Imhof, 2013), 171–174.
24 Alexei Lidov, "Hierotopy: The Creation of Sacred Spaces as a Form of Creativity and Subject of Cultural History," in *Hierotopy: Creation of Sacred Spaces in Byzantium and Medieval Russia*, ed. Alexei Lidov (Moscow: Progress-Tradition, 2006), 38.
25 For a description of the spatial and tactile evocations of the Bruges chapel revealing a sense of naturalism, see also Laura D. Gelfand, "Illusionism and Interactivity: Medieval Installation Art, Architecture and Devotional Response," in *Push Me, Pull You*, vol. 2: *Physicality and Devotional Practices in Late Medieval and Renaissance Art*, eds. Sarah Blick, Laura Gelfand, and Margaret L. Goehring (Leiden and Boston: Brill, 2011), 87–116, esp. 95–108; see also Kirkland-Ives, *Capell nuncapato*, 1058–1060. General notions on the sensuality and materiality of the Holy Sites in Jerusalem in Bacci, "Remarks on the Visual Experience of Holy Sites in the Middle Ages," in *Mobile Eyes: Peripatetisches Sehen in den Bildkulturen der Vormoderne*, eds. David Ganz and Stefan Neuner (Paderborn: Wilhelm Fink, 2013), 175–197.
26 "*monticulus viva petra albi coloris rubeo qudammodo immixtus, naturaliter elevates*," cited in Heers et al., *Itinèraire*, 265–267.
27 This approach is reflected in Robert Ousterhout, "Loca Sancta and the Architectonical Response to Pilgrimage," *The Blessings of Pilgrimage*, vol. 2: *The Souvenirs and Blessings of Pilgrimage* (Urbana: University of Illinois Press, 1990), 108–124.
28 Devotional aspects of this naturalism are noted in Gelfand, *Illusionism*, 101–104.
29 While representing a life-sized Christ was apparently never intended, four miniature scenes of the passion appear between the niches. Further visual analysis of these reliefs will be provided in my PhD dissertation.
30 Lisa Hyatt Cooper and Andrea Denny-Brown, eds., *The Arma Christi in Medieval and Early Modern Material Culture: With a Critical Edition of O Vernicle* (Farnham: Ashgate, 2014), 3.
31 Robert Suckale, "Arma Christi: Überlegungen zur Zeichenhaftigkeit mittelalterlicher Andachtsbilder," in *Stil und Funktion: Ausgewählte Schriften zur Kunst des Mittelalters*, eds. Peter Schmidt and Gregor Wedekind (Berlin: Deutscher Kunstverlag, 2003), 30–32.
32 On visual representations of tactile qualities, see Iris Wenderholm, *Bild und Berührung: Skulptur und Malerei auf dem Altar der italienischen Frührenaissance* (Berlin and Munich: Deutscher Kunstverlag, 2006), 109–114.
33 On the origin of the *Arma Christi* and their development in medieval art, see Rudolf Berliner, "Arma Christi," *Münchner Jahrbuch der Bildenden Kunst* 3, no. 4 (1955): 35–152.
34 See Berliner, *Arma Christi*, 37–38; different compilations are discussed in Robert Suckale, *Arma Christi*, 15–58.

35 Most pilgrims mentioned it, and it is depicted in other ensembles, most notably in Görlitz (Pieper et al., *Jerusalemskirchen*, 53) and San Vivaldo (Riccardo Pacciani and Guido Vannini, *La* Gerusalemme *di S. Vivaldo in Valdelsa* (San Miniato, 1998), 43.

36 Caroline Walker Bynum, *Christian Materiality: An Essay on Religion in Late Medieval Europe* (New York and Cambridge: Zone books, 2011), 34–36.

37 Robert Ousterhout, "Architecture as Relic and the Construction of Sanctity: The Stones of the Holy Sepulchre," *Journal of the Society of Architectural Historians* 62 (2003): 4–23, esp. 4.

38 Rachman-Schrire: *The Rock*, 31.

39 This development can be traced within pilgrimage reports; for example, Titus Tobler on the *Katholikon*. See his *Golgotha: Seine Kirchen und Klöster nach Quellen und Anschau* (St. Gallen, 1851; rpt. Boston: Adamant, 2007), 318–404.

40 Thomas Coomans, Herman De Dijn, and Jan De Maeyer, eds., *Loci Sacri: Understanding Sacred Places* (Leuven: Leuven University Press, 2012), 9.

41 Cited in Pringle, *Pilgrimage*, 295.

42 On the question of reproduction based on different modes of perception in Jerusalem, see Colin Morris, *The Sepulchre of Christ and the Medieval West: From the Beginning to 1600* (Oxford: Oxford University Press, 2005), 328–333; Laura D. Gelfand, "Sense and Simulacra: Manipulation of the Senses in Medieval 'Copies' of Jerusalem," in *The Intimate Senses: Taste, Touch and Smell*, eds. Lara Farina and Holly Duggan, an issue of *Postmedieval: A Journal of Medieval Cultural Studies* (2012): 407–422.

43 On that complex, see Francesca Bocchi, ed., *Sette Colonne e sette Chiese: La vicenda ultramillenaria del complesso di Santo Stefano in Bologna* (Bologna: Grafis, 1987); on material and liturgical enhancement, see Beatrice Borghi, *In viaggio verso la Terrasanta. La Basilica di Santo Stefano in Bologna* (Bologna: Minerva, 2010).

44 Inscription on the column: "*Questa colonna rappresenta quella alla quale fu flagellato N.S.G.C.E. Si acquistano 200 anni d'indulgenze ogni volta che si visita.*"

45 For the material qualities of relics, see Thomas Raff, *Die Sprache der Materialien: Anleitung zu einer Ikonologie der Werkstoffe* (Münster: Waxmann, 2008), 104–106, 126. Andreas Hartmann recently amassed an extraordinary compendium on the relationships among place, memory, and object also dealing with Jerusalem in *Zwischen Relikt und Reliquie: Objektbezogene Erinnerungspraktiken in antiken Gesellschaften* (Berlin: Verlag Antike, 2010), 57–67, 599–604.

46 "*una colona de marmora a la similitudine de quella là o' lo nostro Signore Jesù Christo fu batudo e flagellado*," in *Vita di San Petronio*, 2nd ed., ed. Maria Corti (Bologna, 2002), 33–35.

47 For example, the stone inserted next to the Holy Sepulchre site in Varallo was "similar" to the one that locked Christ's tomb; see Bram de Klerck, "Jerusalem in Renaissance Italy: The Holy Sepulchre on the Sacro Monte of Varallo," in *The Imagined and Real Jerusalem in Art and Architecture*, eds. Jeroen Goudeau, Mariette Verhoeven, and Wouter Weijers (Leiden and Boston: Brill, 2014), 22–24, 215–236. Mitchell B. Merback presents a very interesting case of relic-making in Passau in *Pilgrimage and Program: Violence, Memory, and Visual Culture at the Host-miracle Shrines of Germany and Austria* (Chicago: University of Chicago Press, 2012), 125–131.

48 On the aesthetics of material language, see Dieter Mersch, "Erscheinung des 'Un-Scheinbaren,'" in *Das Zusammenspiel der Materialien in den Künsten*, eds. Thomas Strässle, Christoph Kleinschmidt, and Johanne Mohs (Bielefeld: Transkript Verlag, 2013), 27–44, esp. 30.

49 Christian Kiening, "Mediale Gegenwärtigkeit: Paradigmen – Semantiken – Effekte," in *Mediale Gegenwärtigkeit*, ed. Christian Kiening (Zürich: Chronos, 2007), 9–70, esp. 17.

50 For such imprints described, for example, by Felix Fabri, see Rachman-Schrire, "Evagatorium in Terrae Sanctae [. . .]: Stones Telling the Story of Jerusalem," in *Jerusalem as Narrative Space*, eds. Hoffmann and Wolff, 353–366, esp. 363–364; Susanne Lehmann-Brauns, *Jerusalem Sehen: Reiseberichte des 12: Bis 15: Jahrhunderts als empirische Anleitung zur geistigen Pilgerfahrt* (Freiburg im Breisgau: Rombach, 2010), 213–242; Christoph Geismar-Brandi and Eleonora Louis, eds., *Glaube – Hoffnung – Liebe – Tod* (Vienna and New York: Springer, 1995), 454–456, 460; Andrea Worm, "Steine und Fußspuren Christi auf dem Ölberg: Zu zwei ungewöhnlichen Motiven bei Darstellungen der Himmelfahrt Christi," *Zeitschrift für Kunstgeschichte* 66 (2003): 297–320, esp. 307–308; and Reudenbach, *Loca sancta*, 16–17.

51 Georges Didi-Huberman, *Ähnlichkeit und Berührung: Archäologie, Anachronismus und Modernität des Abdrucks* (Cologne: DuMont, 1999), 46–48.
52 The impact of the Holy Land as materially storing the passion, for example, is described in Francesco Suriano, *Trattato di Terra Santa e dell'Oriente*, ed. Girolamo Golubovich (Milan: Artigianelli, 1900), 11; on the cult and transfer of relics, see Ousterhout, *Souvenirs*.
53 Reudenbach, *Loca sancta*, 23.
54 See Jos Koldewij, ed., *Geloof en Geluk: Sieraad en Devotie in Middeleeuws Vlaanderen* (Arnhem: Terra, 2006), 186–187.
55 "For the relics of the chapel and for all the relics we brought home from the Holy Land," Rijksarchief Brugge (RAB), *Archief Adornes*, inv. 52, fol. 9r.
56 SAB, *Adornes en Jeruzalemstichting*, inv. 781, fol. 17r. The reliquaries are noted in Geirnaert et al., *Adornes en Jeruzalem*, cat. no. 14, 54; also Koldewij, *Geloof*, 190–192. A large collection of relics, primarily bones, possibly referred to in the inventory of 1522, is preserved in the main altar retable upstairs.
57 Although it evidently bears a very strong connection to the Holy City, the reliquary of the True Cross is very controversial and cannot be thoroughly investigated here. See Nadine Mai, "Creation of the Sacred. Notes on the Jerusalem Chapel in Bruges and its relic of the True Cross," *Handelingen van het Genootschap voor Geschiedenis de Brugge* 153.2 (2016): 266–284.
58 SAB, *Adornes en Jeruzalemstichting*, inv. 969, fol. 2r.
59 A similar cross but from a later date now in the Belgrade Museum of Applied Art is featured in the exhibition catalog *Athos: Monastic Life on a Holy Mountain* (Helsinki: Helsinki City Art Museum, 2006), cat. no. 4.21. The wooden door of Saint Nicholas in Ohrid, ca. 1400, has similar decorations; see Louis Bréhier, *La Sculpture et les Arts Mineurs Byzantins* (Paris: Editions d'Art et d'Histoire, 1936), tab. XLIII.
60 Reudenbach, *Loca Sancta*, 22–24; On foreign and exotic materials as proof see Raff, *Die Sprache*, 114.
61 Jacobus de Voragine, *Legenda Aurea*, ed. and trans. Richard Benz (Darmstadt: Wiss. Buchges, 1997), 350–351. For the tradition of the legend and its relation to art history, see Barbara Beart, *A Heritage of Holy Wood: The Legend of the True Cross in Text and Image* (Leiden and Boston: Brill, 2004).
62 From a document written in 1492 by Jan Adornes we know that these relics should have been exposed during the Eucharist; see SAB, *Adornes en Jeruzalemstichting*, inv. 39, flyleaf on fol. 183r. Liturgical and material memories are linked in Reudenbach, *Loca Sancta*, 28–31.
63 See Robert W. Scribner, "Das Visuelle in der Volksfrömmigkeit," in *Bilder und Bildersturm im Spätmittelalter und in der frühen Neuzeit*, ed. Robert W. Scribner (Wiesbaden: Harrassowitz, 1990), 9–20. Scribner describes the multisensual perception of liturgical contexts. See also Jeffrey Hamburger, "Seeing and Believing: The Suspicion of Sight and the Authentication of Vision in Late Medieval Art and Devotion," in *Imagination und Wirklichkeit: Zum Verhältnis von mentalen und realen Bildern in der Kunst der frühen Neuzeit*, eds. Klaus Krüger and Alessandro Nova (Mainz: Zabern, 2000), 47–70, esp. 55. For a recent overview of *Corpus Christi* concepts, see Ulrike Surmann and Johannes Schröer, eds., *Trotz Natur und Augenschein. Eucharistie – Wandlung und Weltsicht* (Cologne: Greven, 2013).
64 See Tobler, *Golgotha*, 71–73; Beart, A heritage, 27–29.
65 "*loco ubi corona, lancea atque clavi invente fuerunt est sub monte Calvarie et in ipso concavatus*"; cited in Heers et al., *Itinèraire*, 265.
66 The *Besloten Hofjes* are a similar, very distinct form of presenting materials; see Kathryn M. Rudy, *Virtual Pilgrimages in the Convent: Imagining Jerusalem in the Late Middle Ages*, Disciplina monastica 8 (Turnhout: Brepols, 2011), 112–118; Koldewij, *Geloof en geluk*, 231–234.
67 See Geismar-Brandi et al., *Glaube*, 144–148; David S. Areford, "The Passion Measured: A Late-Medieval Diagram of the Body of Christ," in *The Broken Body: Passion Devotion in Late-Medieval Culture*, eds. Alasdair A. MacDonald, Bernhard Ridderbos, and Rita Schlusemann (Groningen: Egbert Forsten, 1998), 211–238. For the Arma Christi in Jerusalem devotion see Rudy, Jerusalem in the Convent, 97–107.

12 Moving stones

On the columns of the Dome of the Rock, their history and meaning

Lawrence Nees

The Dome of the Rock is one of the most famous structures in the world, appearing in every textbook on the history of art and often standing at the beginning of any discussion of Islamic art, of which it is generally reckoned the earliest well-preserved building.[1] Many books focus on the building and especially on its magnificent interior, the section that retains the largest portion of its original decoration from the end of the seventh century. Most of the scholarly attention has been devoted to the remarkable inscription that winds twice around the building's interior, which names its builder, Caliph 'Abd al-Malik, and gives the date of 73 AH, corresponding to 691–692 CE, and to the remarkable mosaic decoration of the interior.[2] The building consists of a tall round central core rising to a wide and high wooden dome, surrounded by two ambulatories, whose outer edges are octagonal. The interior is dominated by twenty-eight massive columns, sixteen supporting the roof in the outer colonnade between the two ambulatories, arranged in eight pairs, each pair set between eight piers, and twelve supporting the great dome in the inner colonnade, arranged in four sets of three, between four piers.[3] They are all apparently in their original locations and have their twenty-eight original capitals.[4] To my knowledge no scholar has suggested that their position has ever been changed, and no evidence has been presented concerning possible alterations. That they stand in their original positions from the seventh-century building campaign has always been an assumption, and I proceed on that basis in what follows.

The Dome of the Rock has often been referred to as if miraculously preserved through its long history, implying that what we see today is its original appearance.[5] It is a bit of a shock to read in the second paragraph of Oleg Grabar's recent and fundamental essay on the building that "Nearly everything one sees in this marvelous building, both inside and outside, was put there in the second half of the twentieth century."[6] The famous golden dome on the exterior is quite new, erected as recently as 1999, imitating the dome built in 1960–1962,[7] and there have been renovations throughout, certainly including at least some of the famous mosaics, likely some of the marble paneling, and the floors and ceilings. The best preserved original feature of the building is its set of columns, which, with their capitals, all appear to be from the original seventh-century building campaign. The capitals have been cataloged by John Wilkinson,[8] but it is remarkable that to date there has not been an extended study of the column shafts, arguably the best preserved part of the structure, essentially unchanged from when the building was first erected. Their different colors and origins and their arrangement inside the Dome of the Rock have never been studied; nothing like a complete description, much less a catalog, exists. Unless and until a detailed study can be made, all statements must be regarded as conditional and tentative. All of the column shafts appear to be Roman *spolia*, removed from one or more Roman buildings, whose identity and location are unknown, but whose superb quality is beyond doubt. An investigation of the column shafts and their possible significance, along the lines of the important work by Dale Kinney with the spoliated columns at Santa Maria in Trastevere or most recently by Lex Bosman in connection with Old Saint Peter's Basilica in Rome would be highly desirable, and likely very revealing.[9] The colors of the Dome of the Rock's columns vary dramatically (Plate 5), as was common in late antiquity, when color and *varietas* in color as well as other features were so important, something that we have lost sight of for most ancient and medieval buildings.[10]

For example, Bosman's study of the columns of Old Saint Peter's showed that many different colors were used, organized in pairs on opposite sides of the nave, with the black *africano* shafts greeting the visitor just inside the entrance.[11] The recent restoration through a team led by Elizabeth Bolman of the Red Monastery at Sohag in Upper Egypt, dating probably from the fifth to the seventh century and thus chronologically and geographically much closer to the Dome of the Rock,[12] gives a sense of the rich color that was sought; in that case it is with paint imitating what a building like the Dome of the Rock achieved with colored stones. After the sixth century most of the quarries that supplied these colored marbles were closed,[13] and the reuse of older column shafts was a necessity. Whether and to what extent there were ideological issues involved in the use of Roman *spolia* is a complex and fascinating issue that I do not address here.[14]

To the best of my knowledge, we still await a study that will identify the kind of stone in each column shaft of the Dome of the Rock, which would allow scholars to address the issue of arrangement on a secure basis. Is there a pattern to the arrangement, as was clearly the case in other early buildings such as Old Saint Peter's that have been studied, or is the arrangement random, even haphazard? Such deliberate ordering of colored column shafts has been suggested for at least one other early Islamic building. Christian Ewert and Jens-Peter Wisshak studied the layout of thirteen differently-colored columns in the Great Mosque of Kairouan and found what they saw as an implicit "building within a building," as the colored shafts appear to form an octagonal shape in the middle of the prayer hall, which they compared to the Dome of the Rock.[15] As previously noted, the remarks here are necessarily preliminary in the absence of any scholarly catalog and study, and are based on a very incomplete published photographic record and a brief few minutes inside the building in January 2010; the visit was not long enough to have permitted complete notes to be made about the disposition of the columns, much less to have examined any of them in detail.

The most important observation, it seems to me, at least for the present study, has to do with the arrangement of the columns in the inner circular arcade. There are twelve columns there, the largest in the building, arranged in four groups of three, separated by piers at the northwest, northeast, southeast, and southwest, so that there are three columns on the north, south, east, and west. It is striking that the arrangement of the three columns is the same on the north, east, and south sides (Plate 5), but different on the west.[16] In the north, south, and east, the central column of the set of three is a striking and attention-getting example, boldly patterned with white and a deep red verging almost to purple. I pretend no expertise in the study of marble and am most reluctant to claim that I can properly name this remarkable stone, but for the sake of convenience I call it "pavonazzetto," based on its similarity to an illustration so designated in the book on ancient marble by Raniero Gnoli.[17] As the photograph shown here was taken in poor light, the red color of the pavonazzetto columns is not clearly apparent. It appears better in the splendid professional photographs by Saïd Nuseibeh,[18] although it seems to be the same type of material used for the two outermost columns on the exterior north porch of the Dome of the Rock (Plate 6), photographed in better light, and with the red color of the pavonazzetto far more evident.[19]

At least to my eyes, this stone is distinct from a second, more uniformly red stone used for the columns in the interior of the Dome of the Rock,[20] and also in the exterior, for example, the central columns flanking the entrance on the south porch (Plate 7).[21]

Such stone was mentioned (as Phrygian marble) by many Roman authors (Ovid, Tibullus, Martial, Juvenal),[22] and was used in buildings of the highest prestige in Rome, apparently for the twenty-four columns of the Basilica of San Paolo fuori le mura taken from the Mausoleum of Hadrian and destroyed by fire in 1823,[23] and for twelve great columns still in San Lorenzo fuori le mura.[24]

Whence came these remarkable column shafts? Obviously, we cannot say with certainty, but perhaps because their color has not heretofore been cataloged or even specially remarked, there has been no consideration of this issue. It is striking that in his long description of Justinian's new church of the Virgin Mary, commonly known as the Nea Ekklesia (new church), Procopius made a major point of the columns needed for the huge church not being available until "God revealed a natural supply of stone perfectly suited to this purpose in the nearby hill, one which had either lain there in concealment previously, or was created at that moment."[25] Procopius goes on to say that "the church is supported on all sides by a great number of huge columns from that place, which in color resemble flames of fire [τῷ χρώματι πυρός τινα φλόγα]," and that "two of these columns stand before the door of the church, exceptionally large and probably second to no column in the whole world."[26]

The Nea's history is complex and need not be thoroughly reviewed here.[27] Briefly, built by Justinian in the mid-sixth century on an enormous scale and at great expense, this huge basilica is now known to have stood to the southwest of the Haram, where extensive remains have been found in excavations conducted after 1967, some inside and some continuing outside the Ottoman city wall. In his useful summary discussion of those excavations, Meir Ben-Dov discussed the columns at length, and suggested, plausibly, that the story of the miraculous discovery of a hidden supply of such magnificent stone in a quarry near Jerusalem is highly unlikely, and a common literary trope, and then suggested that the "flame-colored" columns might have been taken from the Temple Mount/Haram al-Sharif, where they had formed part of Herod's great stoa along the southern side of the platform.[28] He did not suggest, and as far as I can tell no one else has suggested, a possible connection between these "flame-colored columns" of the Nea Ekklesia and those boldly patterned red pavonazzetto columns featured so prominently in the inner arcade of the Dome of the Rock. Nor are the columns of the Dome of the Rock mentioned in a more recent study by Yoram Tsafrir on the Nea Ekklesia, in which he suggested that there might indeed have been a quarry of red-veined stone near Jerusalem, one still containing a cracked monolith twelve meters tall, so that Procopius's "miracle" might be a true story, even if not one requiring divine intervention.[29] In Tsafrir's opinion, the "flame-colored columns" of the Nea Ekklesia may be the hard and difficult-to-quarry *mizzi ahmar* ("red stone") used in Jerusalem, taken from beds north of the Old City, "commonly used in Jerusalem since the late nineteenth century, when modern quarrying technology based on the use of heavy machinery and explosives became common."[30] The cracked monolith found in the 1860s in the quarry northwest of the Old City is an indication that the stone was used, or at least that there was an attempt to use it, much earlier, whether for the Herodian or Justinianic constructions in the city being uncertain. Whether this *mizzi ahmar* is the same as that used so prominently in the Dome of the Rock remains to be established, and seems to me a question well worth posing.[31]

I hasten to say that Procopius's "flame-colored" could be variously interpreted, assuming it is to be taken literally at all.[32] Presumably it requires us to think of a stone with red in it, and owing to their bolder pattern and the play of the extensive amount of white against the red, the pavonazzetto shafts prominently displayed in the Dome of the Rock seem to me to fit the "flame" term better than the other red columns appearing elsewhere in the building. What about the size of these columns? Are they such as might have been deployed in the Nea Ekklesia? It is not clear when that church was destroyed; some scholars have suggested that it was in the earthquake of 749, but it may well have been demolished during the Persian sack of the city in 614,[33] although the most recent study of Jerusalem in the latter half of the first millennium of the common era suggests that the pervasive and extensive destruction claimed by some textual sources, and most modern scholarship, is not supported by the archeological evidence.[34] Certainly most of its material must have been removed and at least some of it reused, or the huge church would not have been so difficult to find, its traces now almost entirely limited to foundations. Indeed, the extensive excavations by Nahman Avigad, as published by Oren Gutfeld, indicate that no columns or capitals of the Justinianic building were discovered, even though columns from the Crusader-period structure abutting it were found,[35] and a number of column shafts and capitals were unearthed during the excavation of the neighboring Cardo area.[36] The many columns described by Procopius and envisaged in the various reconstructed plans of the basilica were evidently all entirely removed at some point, the only questions being when they were removed and for what purpose.

It seems that a portion of the lintel of the church, including a large cross-in-wreath with rosettes and the alpha and omega in the quadrants, was reused for secondary construction in the eighth-century Umayyad "palace" area at the southwest corner of the Haram,[37] which would have been very near the site of the Nea, and it seems to me that this fact lends support to the view that the destruction of the Nea is more likely to have taken place in the seventh than in the eighth century, since that early Islamic structure is now dated to the early eighth century, decades before the earthquake of 749. Is it possible that these prominent pavonazzetto columns axially aligned with the north, east, and south doors, and, forming a pair at the west door, probably always, as now, the principal entrance to the Dome of the Rock,[38] were either taken from the Nea Ekklesia, or if not literally taken from the building, might be thought to have referred to its striking "flame-colored" columns? That the new Islamic buildings on the Haram al-Sharif were in some sense competing with the two great Christian structures immediately to their west, the Church of the Holy Sepulchre and the Nea Ekklesia, has been suggested by many scholars, so such a reference would not be altogether unexpected.[39]

One other feature of Procopius's description of the Nea Ekklesia that may deserve mention is his contention that "two of these [flame-colored] columns stand before the door of the church, exceptionally large and probably second to no column in the whole world."[40] Singling out two of the columns, those flanking the entrance, for special mention, in a huge church in sight of the Temple Mount might well be thought to be a reference to the two columns flanking the entrance to the porch of Solomon's Temple, according to the account in 1 Kings 7:21, columns named Jachin and Boaz.[41] Is it a coincidence that the east, north, and south doors of the Dome of the Rock have a single "flame-colored" column shaft, whereas the entrance door on the west has a pair? Striking also is the description by the eleventh-century Persian traveler Nasir-i Khusraw indicating that there were "two marble pillars the color of red carnelian" flanking the mihrab of the al-Aqsa Mosque; whether this description accords better with the "flame-colored" columns inside the Dome of the Rock and on the north porch (Plates 5 and 6) or with the darker red columns on the south porch (Plate 7) is uncertain.[42]

It is hardly likely that this arrangement in the Dome of the Rock or perhaps also in the al-Aqsa Mosque was any kind of reference to the Nea Ekklesia, very possibly still standing when the Dome of the Rock was created. However, as in Procopius's description if not in the intention of the builders, it might well have been a reference to Solomon's Temple, which many today believe stood on the same site. We do not know if the builders of the Dome of the Rock believed that they were building on the site of Solomon's Temple, although numerous Solomonic references make this seem entirely possible and perhaps likely.[43] Nasir-i Khusraw certainly thought that this was the case – that the Rock was revealed to Moses as the direction of prayer, and that Solomon built the temple (he calls it a mosque) "around the rock, with the rock in the middle."[44] Moreover, if Ben-Dov's suggestion that the flame-colored columns in the Nea Ekklesia were taken from the ruins of Herod's enormous stoa on the southern edge of the Temple Mount is correct, is it not possible that the flame-colored columns in the Dome of the Rock also came from that extensive ruin?[45] Was some of this natural material translated from the Temple Mount down to Justinian's Nea church just beneath it and some translated from one portion of the ruined Temple Mount to a different location on the Haram al-Sharif?

Much more work has to be done on these columns before such questions can be answered, but it is important to pose them and to urge that a thorough photographic documentation and archeological study of the columns of the Dome of the Rock be undertaken and made available to scholars for study. Here it must suffice to say that the columns appear to have been arranged in a deliberate, not haphazard, fashion. Whether or not this arrangement carried meaning or had a particular function remains to be investigated.

Notes

1 The Dome of the Chain, which stands immediately beside the Dome of the Rock on the Haram al-Sharif in Jerusalem, most likely dates to an earlier period. See Lawrence Nees, "Muslim, Jewish and Christian Traditions in the Art of Seventh-Century Jerusalem," in *Age of Transition: Byzantine Culture in the Islamic World*, ed. Helen Evans (New York: Metropolitan Museum of Art, 2015), 2–19, and Lawrence Nees, *Perspectives on Early Islamic Art in Jerusalem*, Art and Architecture of the Islamic World (Leiden: Brill, 2015).

2 Bibliography on the Dome of the Rock is vast and growing so rapidly that there is little doubt that any list offered here would be out of date by the time it was published. For the most convenient and concise overview, and a recent one, with a survey of the earlier literature, including his fundamental earlier studies, see Oleg Grabar, *The Dome of the Rock* (Cambridge, MA and London: The Belknap Press of Harvard University Press, 2006), and

Andreas Kaplony, "635/638–1099: The Mosque of Jerusalem (Masjid Bayt al-Maqdis)," in *Where Heaven and Earth Meet: Jerusalem's Sacred Esplanade*, eds. Oleg Grabar and Benjamin Z. Kedar (Jerusalem: Yad Ben-Zvi and Austin, TX: University of Texas Press), 100–131, esp. 105–125.

3 For the best published photographs, see Saïd Nuseibeh and Oleg Grabar, *The Dome of the Rock* (New York: Rizzoli, 1996).

4 John Wilkinson, *Column Capitals in al Haram al Sharif (from 138 A.D. to 1118 A.D.)* (Jerusalem: The Adm. of Wakfs and Islamic Affairs, and the Islamic Museum al-Haram al-Sharif, printed in Jerusalem by Dar al Aitam, the Islamic Industrial Orphanage, 1987).

5 General books suggesting that the Dome of the Rock is well preserved include Markus Hattstein and Peter Delius, *Islam: Art and Architecture*, trans. George Ansell *et alia* (Cologne: Könemann, 2000), 64: "The interior decoration . . . has been largely maintained in its original state." Barbara Bernd, *Islamic Art* (London: British Museum and Cambridge, MA: Harvard University Press, 1991), 24: "The earliest Islamic building to survive in its original form is the Dome of the Rock. Though it has often been restored down the ages, much of the seventh-century work is still visible." That the building is well preserved in its original form is literally textbook wisdom, e.g., Fred S. Kleiner, *Gardner's Art through the Ages: A Global History*, 13th ed. (Boston: Thomson Wadsworth, 2009), 344: in contrast to the much restored exterior, "The interior's rich mosaic ornament has been preserved and suggests the original appearance of the exterior walls."

6 Grabar, *Dome of the Rock*, 1. For Ottoman and early modern restorations, see Beatrice St. Laurent and András Riedlmayer, "Restorations of Jerusalem and the Dome of the Rock and Their Political Significance, 1537–1928," *Muqarnas* 19 (1993): 76–84.

7 Grabar, *Dome of the Rock*, 72.

8 Wilkinson, *Column Capitals*, reproduces each capital, gives its location and type, and also gives a concise description and historical analysis, especially concerning its time and place of origin.

9 Dale Kinney, " 'Spolia' from the Baths of Caracalla in S. Maria in Trastevere," *Art Bulletin* 68 (1986): 379–397, and more recently on related topics, Dale Kinney, "Bearers of Meaning," *Jahrbuch für Antike und Christentum* 59 (2007): 139–153; Lex Bosman, *The Power of Tradition: Spolia in the Architecture of Saint Peter's in the Vatican* (Hilversum: Uitgiverij Verloren, 2004), esp. 29–56, which also discusses how many of the columns from the fourth-century basilica, many of them *spolia*, were respoliated in the sixteenth-century rebuilding, especially around altars, where they are still to be seen, and Lex Bosman, " 'Spolia' in the Fourth-Century Basilica," in *Old St. Peter's, Rome*, eds. Rosamond McKitterick, John Osborne, Carol M. Richardson, and Joanna Story (Cambridge: Cambridge University Press, 2013), 65–80.

10 On this issue with specific reference to columns, see Beat Brenk, "Spolien und ihre Wirkung auf die Ästhetik der varietas: Zum Problem alternierender Kapitelltypen," in *Antike Spolien in der Architektur des Mittelalters und der Renaissance*, ed. Joachim Poeschke (Munich: Hirmer, 1996), 49–92, and more recently the discussion and bibliography in Kinney, "Bearers of Meaning," 15.

11 Bosman, " 'Spolia'," 68–69, figs. 3.1 and 3.3, and pl. 2.

12 Elizabeth S. Bolman, "Late Antique Aesthetics, 'Chromophobia', and the Red Monastery, Sohag, Egypt," *Eastern Christian Art* 3 (2006): 1–24.

13 See Alfred Michael Hirt, *Imperial Mines and Quarries in the Roman World: Organizational Aspects 27 BC–AD 235* (Oxford: Oxford University Press, 2010). This fact was noted in reference to the marble columns and capitals on the Haram by Wilkinson, *Column Capitals*, 4, whose view is that the very last to close did so ca. 550 CE, and that the others had closed at an earlier date. Bosman, " 'Spolia'," 73, noted that the use of colored marbles in Roman architecture was increasingly common in the first and second centuries CE, probably peaking in the third century, and declining thereafter, but he found evidence that at least some of the columns in the fourth-century Vatican basilica were newly made rather than spoliated.

14 For discussion of *spolia* in general, see Friedrich Wilhelm Deichmann, *Die Spolien in der spätantiken Architektur* (Munich: Bayerische Akademie der Wissenschaften, 1975), Beat Brenk, " 'Spolia' from Constantine to Charlemagne: Aesthetics versus Ideology," *Dumbarton Oaks Papers* 41 (1987): 103–109, Dale Kinney, "Roman Architectural 'Spolia',"

Proceedings of the American Philosophical Society 145 (2001): 138–161, Dale Kinney, "The Concept of 'Spolia'," in *A Companion to Medieval Art: Romanesque and Gothic in Northern Europe*, ed. Conrad Rudolph (Oxford: Blackwell, 2006), 233–252, and Richard Brilliant and Dale Kinney, eds., *Reuse Value: Spolia and Appropriation in Art and Architecture, from Constantine to Sherrie Levine* (Farnham: Ashgate, 2011).

15 Christian Ewert and Jens-Peter Wisshak, *Forschungen zur Almohadischen Moscheen* 1, Madrider Beiträge 9 (Mainz, 1981), 15–20 and especially 31–54, Figure 20.

16 Owing to the temporary partitions now erected inside the western door, the only entrance when I visited the building in 2012, and the scaffolding in the central space under the dome, it was not possible to see the western column shafts clearly. The only published view I know of, which shows the different colored shafts clearly, is found on the "Virtual Walking Tour" of the Haram al-Sharif led by Oleg Grabar, on the website of *Saudi Aramco World*. Digital facsimile: http://www.saudiaramcoworld.com/issue/200901/al-haram/tour.htm, at number 8 (accessed September 25, 2014).

17 Raniero Gnoli, *Marmora Romana*, rev. ed. (Rome: Edizioni dell'Elefante, 1988), 169–171, Figure 125. The stone is also called Phrygian marble, *synnadicu*, or *docimenium*, and comes from a quarry in Anatolia, modern Turkey. See also Raniero Gnoli, Maria Cristina Marchei, and Attilia Sironi, "Repertorio," in *Marmi antichi*, ed. Gabriele Borghini, rev. ed. (Rome: Edizioni de Luca, 2001), 264–265, no. 109, figs. 109a and b. A striking red-veined marble that appears to me similar, if not necessarily the same, is used for the facing of the main piers in San Vitale, Ravenna, and in the most extensive publication is termed Carian marble, or *cipollino rosso*, by Isotta Fiorentini and Piero Orioli, *I marmi antichi di San Vitale* (Faenza: Edit Faenza, 2003), 75, figs. 10–17 and 22.

18 Saïd Nuseibeh and Oleg Grabar, *The Dome of the Rock* (New York: Rizzoli, 1996), 63–64, and a slightly different view showing all three columns more clearly on the volume's cover. The photograph in Grabar, *Dome of the Rock*, Figure 3, also shows all three columns, but the colors are less clear.

19 Unfortunately, time did not permit me to investigate the columns in the outer, octagonal arcade, and I cannot say one way or the other whether the same stone was used for any of the columns there. Some published views suggest that it does occur, although variations in lighting make it difficult to say for certain; see Nuseibeh and Grabar, *Dome of the Rock*, 62, 65.

20 For a good example, see Nuseibeh and Grabar, *Dome of the Rock*, 69, in the southeast octagonal arcade.

21 The tops of these columns are visible in Nuseibeh and Grabar, *Dome of the Rock*, 23.

22 See Gnoli, *Marmora romana*, 169.

23 Borghini, *Marmi antichi*, 265.

24 Gnoli, *Marmora romana*, 169.

25 Procopius, *De Aedificiis* V.vi., 22; see H.B. Dewing, trans., *Procopius*, vol. 7: *Buildings*, Loeb Classical Library, vol. 343 (Cambridge, MA: Harvard University Press and London: Heinemann, 2000), 342–349, esp. 346. The passage was quoted in full in Meir Ben-Dov, *In the Shadow of the Temple: The Discovery of Ancient Jerusalem*, trans. [from Hebrew] Ina Friedman (New York: Harper & Row, 1982), 233–235. On this text, see Charlotte Roueché, Jean-Michel Carrié, and Noël Duval, eds., *De Aedificiis: le texte de Procope et les réalités*, in *L'antiquité tardive* 8 (2000): 7–180, including the essay by Yoram Tsafrir, "Procopius and the Nea Church in Jerusalem," 149–164. The portion of the text bearing upon the Nea is presented by Ben-Dov in Dewing's English translation in *Jewish Quarter Excavations in the Old City of Jerusalem Conducted by Nahman Avigad 1969–1982*, vol. 5: *The Cardo (Area X) and the Nea Church (Areas D and T, Final Report)*, ed. Oren Gutfeld (Jerusalem: Israel Exploration Society and Institute of Archaeology, Hebrew University of Jerusalem, 2012), 246–247.

26 Procopius, *loc. cit.*, quoted in Ben-Dov, *In the Shadow of the Temple*, 235.

27 For the most recent archeological study of the Nea Ekklesia, see Gutfeld, *Jewish Quarter Excavations*, esp. 141–267, cited by Gideon Avni, *The Byzantine-Islamic Transition in Palestine: An Archaeological Approach*, Oxford Studies in Byzantium (Oxford: Oxford University Press, 2014), 109.

28 Ben-Dov, *In the Shadow of the Temple*, 233–241, with remains, plan, and reconstruction.

29 Tsafrir, "Procopius and the Nea Church in Jerusalem," 162–164, cited by Michael Greenhalgh, *Marble Past, Monumental Present: Building with Antiquities in the Mediaeval Mediterranean*, The Medieval Mediterranean: Peoples, Economies and Cultures, 400–1500, vol. 80 (Leiden and Boston: Brill, 2009), 281.
30 Tsafrir, "Procopius and the New Church," 163, note 5 and Figure 13.
31 Tsafrir cites two articles regarding this stone, L. A. Picard "Geology of Jerusalem," in *Sefer Yerushalayim [Book of Jerusalem]: Jerusalem, Its Natural Conditions, History and Development from the Origins to the Present Day*, vol. 1, *The Natural Conditions and the History of the City from Its Origins to the Destruction of the Second Temple*, ed. Mikael Avi-Yonah (Jerusalem and Tel Aviv: Bialik Institute, 1956), 5–44 (in Hebrew), and D. Gill, "The Geology of the City of David and Its Ancient Subterranean Waterworks," in *Excavations at the City of David 1978–1985*, vol. 4 (*QEDEM* 35), eds. Donald T. Ariel, Alon De Groot, Elisheva Kamaiski, Raz Kletter, and Aryeh E. Shimron (Jerusalem, 1996), 3–28. Neither shows the stone in question in color or describes it in detail. Gill discusses this stone at some length as the lowermost unit beneath Jerusalem (p. 9) and lists places where it is exposed today, as well as several where it was found during excavations for new buildings, but he says nothing about whether or not it has been used in the modern period for building of any kind, much less for tall monolithic columns. An electronic article in Wikipedia says that *mizzi ahmar* was used by the Mamluks in Jerusalem for multi-colored *ablaq* masonry, but does not specify where; see Digital facsimile: http://en.wikipedia.org/wiki/Jerusalem_stone (assessed October 20, 2014). The only color illustration I have found of what is identified as *mizzi ahmar* is on a commercial site, and the detail shown is too small to be compared to the "flame-colored" column shafts of the Dome of the Rock; see Digital facsimile: http://www.alibaba.com/product-detail/Jerusalem-Red-Mizzi-Ahmar-7800L_157094176.html (accessed October 20, 2014). The red stone alternating with white in the voussoirs of the arcade may or may not be *mizzi ahmar*, but has scant resemblance to the column shafts.
32 Greenhalgh, *Marble Past, Monumental Present*, 281, cited a recent study finding that Procopius's descriptions are usually reliable, Denis Roques, "Les 'constructions de Justinien' de Procope de Césarée: document ou monument?" *Académie des inscriptions et belles-lettres, Comptes rendus* (November 1998): 989–1001, although sometimes overemphasizing the personal role of Justinian in connection with structures begun or even completed by his predecessors Justin I and Anastasius (in 997) or claiming some supernatural intervention as at the Nea Ekklesia in Jerusalem but also at Sinai and in Constantinople in 993–994. Andrew Palmer, "Procopius and Edessa," in *De Aedificiis*, eds. Roueché et al., 127–136 at 136, is significantly more critical about relying on Procopius, whom he sees (at least insofar as building activities at Edessa are concerned) as "intentionally misleading," warning that "archaeologists should not take this book as a guide."
33 This is the view advanced by Ben-Dov, *In the Shadow of the Temple*, 241.
34 Avni, *The Byzantine-Islamic Transition, passim.*
35 Gutfeld, *Jewish Quarter Excavations*, 197 and photo 5.58.
36 Gutfeld, *Jewish Quarter Excavations*, esp. 103–110 and many photographs.
37 Ben-Dov, *In the Shadow of the Temple*, 241 and Figure p. 237. It may be worth noting that the equal-armed cross-in wreath appearing on this lintel also appeared on some of the capitals installed in the Dome of the Rock, for which see Wilkinson, *Column Capitals*, 75 and 104–105, nos. 40–41 and 58–59. The capitals were mentioned, but not illustrated, by Rudolf Kautzsch, *Kapitellstudien: Beiträge zu einer Geschichte des spätantiken Kapitells im Osten vom vierten bis ins siebenten Jahrhundert*, Studien zur spätantiken Kunstgeschichte 9 (Berlin and Leipzig: Walter De Gruyter, 1936), 103 and nos. 23, 24 and 25.
38 The mid-eleventh-century description of Jerusalem by Nasir-i Khusraw notes that the main entrance to the mosque, by which he clearly means the Haram al-Sharif and not the structure commonly known today as the al-Aqsa Mosque, was from the gate on the western side; Jonathan M. Bloom, "Nasir Khusraw's Description of Jerusalem," in *No Tapping around Philology: A Festschrift in Honor of Wheeler McIntosh Thackston Jr.'s 70th Birthday*, eds. Alireza Korangy and Daniel J. Sheffield (Wiesbaden: Otto Harrassowitz Verlag, 2014), 395–406, esp. 396–398. For the text see Wheeler M. Thackston, trans., *Nasir-i Khusraw*, Persian Heritage Series, 36 (Albany: State University of New York Press, 1986). The populated area of Jerusalem when the Dome of the Rock was constructed was primarily to the

west of the Haram; see Yoram Tsafrir, 70–638, "The Temple-less Mountain," in *Where Heaven and Earth Meet*, eds. Grabar and Kedar, 72–99 at 95, Figure 57; the largest number of gates leading to the Haram are on the western side.

39 Most recently, see Avni, *The Byzantine-Islamic Transition*, 113–114, referring especially to Oleg Grabar, *The Shape of the Holy: Early Islamic Jerusalem* (Princeton: Princeton University Press, 1996), 104–110, with earlier literature, and Grabar, *Dome of the Rock*, 24–25.

40 Procopius, *loc. cit.*, quoted in Ben-Dov, *In the Shadow of the Temple*, 235.

41 Gutfeld, *Jewish Quarter Excavations*, 243 and Figure 5.23, suggests that the location of two huge columns at the entrance to the Nea Ekklesia, as described by Procopius, does not seem feasible, and that the huge columnar pair might have flanked the altar. This may be the case, but it seems to me that the essential point being made by the author here is not so much an accurate description of the building as the evocation of Solomon's Temple, and to make the point the two special columns had to be put at the entrance, whether they were, in fact, there or not. Nasser Rabbat, "The Meaning of the Umayyad Dome of the Rock," *Muqarnas* 6 (1989): 12–21, esp. 13–14, emphasizes the Solomonic and Davidic references in the building, expressed especially in al-Wasiti's early eleventh-century *fada'il*, which claims to be transmitting earlier traditions.

42 Wheeler M. Thackston, Jr, ed. and trans., *Nasir-i Khusraw's Book of Travels [Safarnama]: A Parallel Persian-English Text*, Bibliotheca Iranica Intellectual Traditions Series 6 (Costa Mesa, CA: Mazda, 2001), 34. I am grateful to an anonymous reader for bringing this text to my attention.

43 See Grabar, *Shape of the Holy*, 110–114, and the discussion and references in my forthcoming book.

44 Thackston, *Nasir-i Khusraw*, 28–30.

45 Grabar, *Shape of the Holy*, 124, notes that "in Umayyad times it is most likely that the stoa of Herod was still largely standing"; for further discussion he refers to his earlier study "A Note on the Chludoff Psalter," *Okeanos [Harvard Ukrainian Studies]* 7 (1983), but notes that his suggestion has not been supported by other scholars. In fact, in the earlier article (pp. 264–265) he says that Herod's stoa was "destroyed by the Romans and left fallow in Byzantine times"; his hypothesis is not that it survived, but that the Chludoff Psalter is an "evocation" of the Islamic structures built during the seventh to ninth centuries. I was very much interested in the announcement of a lecture by Orit Peleg-Barkat (Hebrew University of Jerusalem) on "Herodian Artistic and Architectural Influences on the Umayyad Haram al-Sharif," which was to have been presented on August 12, 2014, at a conference on "The Religious Institutions of Late Antique and Early Medieval Jerusalem and Its Hinterland" at the Kenyon Institute in Jerusalem. The conference was, sadly, postponed for security reasons. It is clearly a topic of much interest.

13 Christ's unction and the material realization of a stone in Jerusalem*

Yamit Rachman-Schrire

The Stone of Unction in Jerusalem came to be associated with the object upon which Christ's body was laid and anointed after he was taken down from the cross and prior to his entombment. By 1335, the stone was being shown to pilgrims at the entrance hall of the Church of the Holy Sepulchre in Jerusalem. As fourteenth-century pilgrims' accounts suggest, it was a flat slab of porphyry or marble at floor height, framed by a black-and-white checkered pattern. An illustration that accompanied the account of Niccolò of Poggibonsi, who made a pilgrimage to Jerusalem in 1347, gives a basic notion of what the stone looked like (Figure 13.1).[1] The green stone, *la pietra verde* as Poggibonsi termed it, is depicted from above surrounded by a schematic checkered frame. A woodcut taken from the sixteenth-century *Viaggio da Venetia*, often attributed to Noè Bianchi,[2] provides more detail, showing the checkered pattern, arranged as an alternating set of black and white squares (Figure 13.2).The stone resides today within the southern arm of the transept of the Crusader Church of the Holy Sepulchre, near the entrance. As it appears now, the fourteenth-century slab is no longer visible but is covered by a reddish stone lid elevated above floor height.[3] The whole structure is framed by a stone barrier inscribed with Greek letters and illuminated from above by eight oil lamps (Plate 8).

Parts of the checkered frame of the medieval stone are still visible between the modern stone lid and the barrier. Behind the stone, a modern mosaic on the nineteenth-century wall shows the removal of Christ's body after the crucifixion, the anointing of his body upon the Stone of Unction, and the entombment.[4]

Unlike the Rock of Golgotha and the Tomb of Christ, which were integrated into the church complex in the fourth century, the Stone of Unction was introduced into the church at a later stage, sometime between 1327 and 1335. Prior to that, the site of the unction was marked in the choir of the church, but no stone was on display there.

How and why did a *stone* in the Church of the Holy Sepulchre come to be associated with the unction of Christ? What was the role of its materiality? This chapter discusses the migration of the Stone of Unction as a relatively new component in the space of the church. I show that the Stone of Unction in Jerusalem was placed in dialogue with a different slab that was presented as the Stone of Unction in the Pantokrator Church in Constantinople as early as in 1169 and with the visual renderings of the stone in Franciscan imagery of Mary's lamentation over her dead son.

Jerusalem: place of the unction

The unction of Christ's body is narrated in John 19:40. After the crucifixion, Joseph of Arimathea asked Pilate for the body; then he and Nicodemus took it "and bound it in

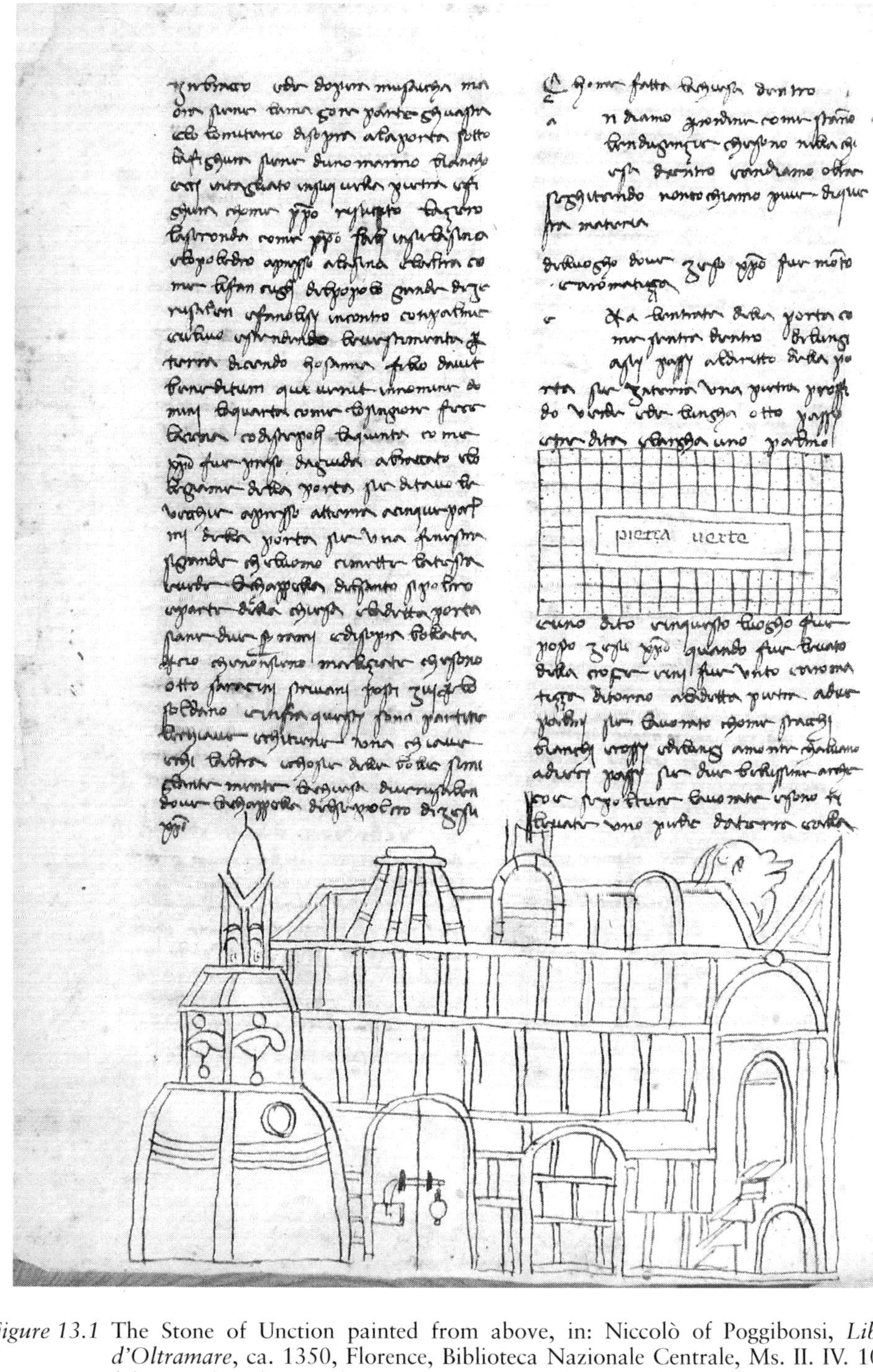

Figure 13.1 The Stone of Unction painted from above, in: Niccolò of Poggibonsi, *Libro d'Oltramare*, ca. 1350, Florence, Biblioteca Nazionale Centrale, Ms. II. IV. 101, fol. 6v (right column). Photo credit: *Ministero dei beni e delle attività culturali e del turismo/Biblioteca Nazionale Centrale di Firenze*.

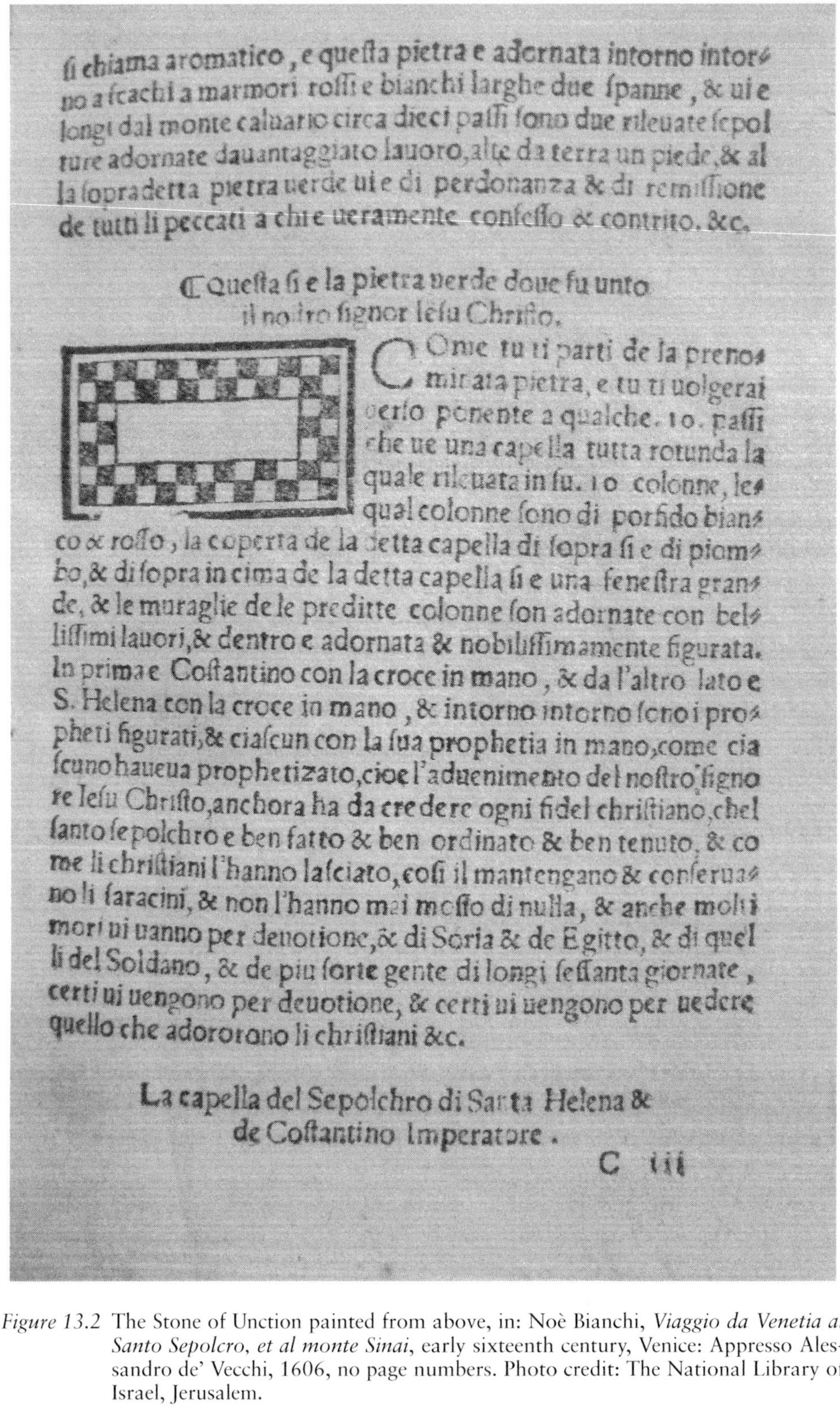

ſi chiama aromatico, e queſta pietra e adornata intorno intor-
no a ſcachi a marmori roſſi e bianchi larghe due ſpanne, & ui e
longi dal monte caluario circa dieci paſſi ſono due rileuate ſepol
ture adornate dauantaggiato lauoro, alte da terra un piede, & al
la ſopradetta pietra uerde ui e di perdonanza & di remiſſione
de tutti li peccati a chi e ueramente confeſſo & contrito. &c.

¶ Queſta ſi e la pietra uerde doue fu unto
il noſtro ſignor Ieſu Chriſto.

COme tu ti parti de la preno-
minata pietra, e tu ti uolgerai
uerſo ponente a qualche. 10. paſſi
che ue una capella tutta rotunda la
quale rileuata in ſu. 10 colonne, le-
qual colonne ſono di porfido bian-
co & roſſo, la coperta de la detta capella di ſopra ſi e di piom-
bo, & di ſopra in cima de la detta capella ſi e una feneſtra gran-
de, & le muraglie de le predicte colonne ſon adornate con bel-
liſſimi lauori, & dentro e adornata & nobiliſſimamente figurata.
In prima e Coſtantino con la croce in mano, & da l'altro lato e
S. Helena con la croce in mano, & intorno intorno ſono i pro-
pheti figurati, & ciaſcun con la ſua prophetia in mano, come cia
ſcuno haueua prophetizato, cioe l'aduenimento del noſtro ſigno
re Ieſu Chriſto, anchora ha da credere ogni fidel chriſtiano, chel
ſanto ſepolchro e ben fatto & ben ordinato & ben tenuto, & co
me li chriſtiani l'hanno laſciato, coſi il mantengano & conſeruā-
no li ſaracini, & non l'hanno mai meſſo di nulla, & anche molti
mori ui uanno per deuotione, & di Soria & de Egitto, & di quel
li del Soldano, & de piu ſorte gente di longi ſeſſanta giornate,
certi ui uengono per deuotione, & certi ui uengono per uedere
quello che adororono li chriſtiani &c.

La capella del Sepolchro di Santa Helena &
de Coſtantino Imperatore.

C iii

Figure 13.2 The Stone of Unction painted from above, in: Noè Bianchi, *Viaggio da Venetia al Santo Sepolcro, et al monte Sinai*, early sixteenth century, Venice: Appresso Alessandro de' Vecchi, 1606, no page numbers. Photo credit: The National Library of Israel, Jerusalem.

linen clothes with the spices, as is the burial custom of the Jews." Thus, the anointing of Christ's body occurred after the crucifixion and before the burial; however, there is no mention in John's narrative of a stone in connection with the unction. The Synoptic Gospels allude to the practice of anointing: three days *after* Christ's entombment, the women, carrying ointments in their hands, found the tomb empty, and the angel told them that Christ was resurrected. No *act* of anointing is narrated, only the intention to do so (Mark, 16:1; Luke, 24:1).

During the Crusader period (1099–1187), the newly built choir of the Church of the Holy Sepulchre was gradually established as the place of Christ's unction, the same site where the center of the earth (*medium terrae*) was marked; no stone was shown there.

John of Würzburg, who visited Jerusalem ca. 1170, mentioned seeing an elevated marble table (*tabula de marmore*) and a network of iron chains that looked like an altar; under it, stone slabs inscribed with small circles marked the *medium terrae*; and a lamp hung there. At the same place, he continued, Christ appeared to Mary Magdalene after he was resurrected, and "some people say (*quidam asserunt*) that Joseph begged the body of Jesus from Pilate . . . he took it down from the cross, reverently washed it, anointing it with costly liquids and spices, and wrapped it in clean linen."[5] In Theoderich's account of his visit to Jerusalem sometime between 1169 and 1174, we read that he saw a small cross inscribed within the circle on the floor marking the exact spot where Joseph and Nicodemus laid Christ's body for anointing.[6] These accounts indicate that the place of the unction was identified with the same spot where the center of the earth was shown. The marking of the site as circles inscribed in the floor specifically alluded to the notion of the *medium terrae*. This, alongside John of Würzburg's phrase "*quidam asserunt*" concerning the identification of the place of the unction, suggests that Christ's anointing was only a secondary tradition there.

At the same time that the unction was associated with the center of the earth in the choir of the Church of the Holy Sepulchre, it was also linked with another site in the church: a bedlike altar (*lectus*) within the tomb aedicule, said to be the object upon which Christ's body was laid.[7] Behind this altar in the Chapel of the Angel, the Crusaders affixed a mosaic showing Christ placed in the sepulchre by Nicodemus and Joseph with the Virgin Mary and the three Marys offering pots of perfumes, and the angel sitting above the sepulchre, rolling the stone away, saying, "behold the place where they have laid him." An inscription refers to Christ's resurrection.[8] We see that in the Crusader church, two different sites were associated with Christ's unction: the *lectus* within the tomb, alluding to the narrative of the Synoptic Gospels, and the place in the middle of the choir, according to the narrative of John. After the Crusaders' reconstruction of the Church of the Holy Sepulchre as a Romanesque pilgrimage church, it could hold thousands of worshippers. The site of the unction in the middle of the choir was suitable for public celebrations and could accommodate the Crusaders' liturgical feasts, which followed the narrative sequence: Golgotha (crucifixion) through the choir (unction) to the tomb.[9]

Following the Crusaders' loss of Jerusalem in 1187, the unction site was associated with both the center of the earth in the choir and the *lectus* in the tomb aedicule. Notably, during the Crusaders' subsequent brief return to Jerusalem (1229–1244), it was more firmly associated with the *medium terrae*,[10] an indication of its central role in the Crusaders' liturgy.

Jerusalem: from the place of the unction to the Stone of Unction

After the Crusaders left Jerusalem for good, the site of the unction was gradually separated from the spot marking the *medium terrae*. However, accounts vary, and determining its exact location is difficult. According to an anonymous mid-thirteenth-century Greek account, the site was at the entrance to the church, marked by a lamp.[11] Another anonymous Greek account, this one from the end of the thirteenth century, locates it directly in front of Christ's tomb and mentions a holy table, next to which Nicodemus and Joseph laid and anointed Christ's body.[12] According to this source, the place was marked by four marble columns supporting a dome in the form of a cross and eight lamps hung there day and night. As no other record mentions this structure, this pilgrim may have mistaken the outer part of the aedicule, by then partially open, for the site of the unction.[13]

It was in that same period that a stone associated with the unction was first mentioned in Jerusalem. The Dominican pilgrim Riccoldo of Monte Croce, who spent several months in Jerusalem in 1288/1289, referred to a stone next to and behind the place where the Virgin Mary stood during the crucifixion: "[T]hey showed there the stone where they placed the body, bound it with linen cloth and preserved it with spices for burial."[14] He noted that the center of the earth was marked by a column,[15] but his testimony must be taken with caution; the indefinite language he used to describe the stone's location makes an exact identification impossible. Moreover, as there was no mention of the stone in pilgrim accounts in the following decades, it is probable that what Riccoldo saw was one of the altars in the area of the choir.

Pilgrims' accounts from the 1320s indicate that the site of the unction was imputed to the southern transept at the entrance of the Church of the Holy Sepulchre, as noted, for example, by the Franciscan pilgrims Symon Semeonis and Odoric of Pordenone, who associated the unction with this location, but made no mention of a stone there.[16] It is only in the account of Antoninus de Cremona, dated to 1327–1330, that a stone at the entrance hall of the church was mentioned for the first time. According to Antoninus' description, a stone was placed next to the site of the unction, which was said to be made according to the measurements of Christ's body, as well as according to the measurements of the stone within the tomb aedicule.[17] He did not specify the form or the position of the stone in the entrance hall (horizontal or vertical) or its material, yet the link between the body of Christ and the Stone of Unction was made concrete through the stone's measurements. It is tempting to consider Antoninus' account a description of the early stages of fixing a stone at the site of the unction.

From 1335 onward, the Stone of Unction is mentioned by almost all the pilgrims who visited the Church of the Holy Sepulchre. Jacopo da Verona, who visited the church in that year, described a black stone close to floor height.[18] According to an anonymous English pilgrim who visited Jerusalem in 1344–1345, it had a bronze hue ("*lapis preciosus aërie coloris*") and was also believed to be the stone upon which Jacob slept in Bethel.[19] The Franciscan pilgrim and priest Niccolò of Poggibonsi, who voyaged to the Holy Land between 1346 and 1350, wrote:

> There is on the ground, inside the door, as you enter, and six steps from the door, a stone of green porphyry, eight palms three fingers long and two palms and one finger wide: and in this place was laid Christ when he was taken down from the Cross, and here he was anointed and embalmed. Around the said stone there is

a chequer border two palms wide, and near to the choir wall, two steps away, are two beautiful sarcophagi, that is, tombs, carved and raised a foot above the ground; and at the above mentioned stone there is a plenary indulgence.[20]

Here, we see the process by which the site of the unction was established at the southern arm of the transept near the entrance to the church, sometime between 1327 and 1335. The stone's attraction as noted in pilgrims' accounts demonstrates that it became one of the most important objects in the church, where indulgences were granted. In the following decades, the stone was consistently described as a flat porphyry or marble slab; its color was variously described as green, black, black spotted with red dots, and rust. [21]

The Franciscan connection

The introduction of the Stone of Unction into the Church of the Holy Sepulchre is a unique example of a major transition in the inner space of the church carried out after the time of the Crusaders and before the nineteenth-century restorations.[22] Notably, the stone was introduced into the entrance hall of the church during the period that the Franciscan order began establishing itself in Jerusalem and the Holy Land.[23] In 1309, a *firmān* of Sultan al-Nasir Muhammad gave the Franciscans the exclusive right among the Latins to reside in the Church of the Holy Sepulchre.[24] In 1332–1333, in the name of the kings of Naples, Friar Roger Guérin obtained permission from the sultan for the Franciscans to occupy the church and to take precedence over other denominations.[25] As the church custodians, the Franciscans had to ensure that the Latin liturgical services were maintained and be responsible for pilgrims arriving from Europe.[26] Although not given absolute ownership of the sites, they were given priority where limited space prevented all clergy from celebrating together.[27] As pilgrims' accounts dated to the 1330s and 1340s testify, the Eastern Christians were required to comply with Western practice.[28] The Franciscan friars and Eastern clergy were supposed to care for the sanctuaries and to conduct the rituals together.[29]

During the 1330s, the Franciscans initiated and developed the Via Dolorosa, into which the Stone of Unction was eventually integrated as a separate station on the devotional route.[30] Moreover, Franciscan pilgrims were the first to report on the stone.[31] It is most likely then that it was the Franciscans who introduced the Stone of the Unction into the Church of the Holy Sepulchre.

How can we explain the choice of marking the unction of Christ specifically with a stone? What could have been behind the stone's physical shape, materiality, and framing? In what follows I trace two iconographical sources for the identification of the unction of Christ with a slab of stone in the Church of the Holy Sepulchre in Jerusalem: first, in Constantinople, where another slab of stone was marked as the Stone of Unction as early as in 1169, and second, in Franciscan iconography of Mary's Lamentation over her dead son, where the Stone of Unction appears as the object upon which Christ's body was laid.

The Stone of Unction in Constantinople

More than a hundred years before the Stone of Unction was introduced into the Church of the Holy Sepulchre in Jerusalem, a different slab of stone, also identified by pilgrims as the Stone of Unction was on display at the Pantokrator Church in Constantinople. The story of its arrival in Constantinople was told by the twelfth-century Byzantine chroniclers John Kinnamos (ca. 1143–1200) and Nicetas Choniates (1150–1215).[32] They contended that the Stone of Unction was brought from Ephesus to Constantinople in 1169 and that Emperor Manuel Comnenus I carried it on his shoulders up the hill from Boukoleon harbor to the *Pharos* Church.[33] After the emperor's death in 1180, his widow, Eirene, had the stone moved to his burial chapel in the Pantokrator Church; its transfer from Ephesus to Constantinople was commemorated *in situ* with an inscription.[34]

Choniates wrote that it was "a slab of red marble the length of a man which received veneration . . . and was commonly reported to be that on which Christ was washed with myrrh and wrapped in burial linen clothes after he had been taken down from the cross."[35] Kinnamos remarked on the unique feature of the stone: traces of the Virgin's tears on its surface.[36] Choniates explained that after Christ was crucified his mother "laid Him prone, as was customary, on this stone; falling down, she lamented deeply, as was reasonable, and the tears from her weeping reached the stone and still remain there, unexpunged, something rather miraculous."[37] The stone was conceptualized not only as the object upon which Christ's body was laid but also as the visual manifestation of the Virgin's tears.[38] The tears in the stone had visible, physical form; they were at once presented and represented in the matter. In a sense, the stone became the medium through which the tears themselves turned into petrified objects, and it thus embodied the sanctifying narrative materially and visibly. This special attribution – spots on the surface identified as the Virgin's tears – was repeated in later descriptions.

In April 1204, the city of Constantinople fell to the Crusaders. It was then, at the latest, that the Stone of Unction became known in the Latin West. Robert of Clari, a French knight from the Picardy region who participated in the conquest, chronicled the events and noted seeing the stone in the Pantokrator Monastery: "In this abbey . . . was the marble table (*le tavle de marbre*) whereon Our Lord was laid out after He had been taken down from the cross; and there, too, were seen the tears that Our Lady had shed over Him."[39]

The Stone of Unction continued to occupy the Western imagination after 1261, at which time the Crusaders lost Constantinople. In 1431–1432, a knight, Bertrandon de La Brocquière, described a stone or table of "diverse colors" (*une pierre ou table de diverses couleurs*). He added that at first glance he mistook the tears for drops of wax, but after a closer look, horizontally and against the light, they seemed like frozen drops of water.[40]

Unlike the many passion relics plundered and taken to the West during Latin rule, the Stone of Unction was on display in the Pantokrator Monastery at least until 1453.[41] That means that from 1335 onward, *two* Stones of Unction were on display simultaneously: one in Jerusalem and one in Constantinople.[42]

The Stone of Unction was one of the most important relics in Constantinople; its importance in both Eastern Christendom and the Latin West is apparent from the introduction of the stone as a new iconographical feature into visual images: In the East, the figure of the stone was added to the iconography of the Threnos (i.e., the lamentation of the Virgin Mary over the dead body of Christ), following the arrival of the stone to the Pantokrator in 1169.[43] In the West, the Stone of Unction was introduced as a new feature in visual imagery of the lamentation/entombment. Of special importance for our discussion is the integration of the figure of the stone into early Franciscan imagery,[44] to which I now turn.

The unction of Christ in early Franciscan imagery

From the beginning of the twelfth century, with the growing devotion to the life and suffering of Christ and the emphasis on the stages following the crucifixion, the scene of the unction gained prominence in Western imagery. Christ's unction was depicted either as a separate scene, with Joseph and Nicodemus anointing Christ's body, or combined with the lamentation of the Virgin or the entombment.[45] In such images Christ's body was most often laid upon a sarcophagus-like form, sometimes with three *oculi* on its frontal face, alluding to Christ's tomb in Jerusalem.[46]

Several early Franciscan images diverge from this common scheme to position Christ's body horizontally on the Stone of Unction rather than on a sarcophagus. Two such images, attributed to the *Master of Saint Francis*, attest to the centrality of the stone in Franciscan imagery as both a Marian and a Christological relic. The first is one panel from a double-sided altar piece from San Francesco al Prato in Perugia, dated to ca. 1272 (Plate 9).[47]

This panel shows the body of Christ lying on a red rectangular stone with a grainy surface. Mary is sitting on the stone with Christ's upper body in her lap and his legs stretched out horizontally on the surface of the stone. The iconography of Mary sitting on the Stone of Unction holding Christ on her lap is found in Eastern images.[48] Except for Christ, the Virgin Mary is the only figure on the surface of the stone; the others – Joseph of Arimathea, Nicodemus, John the disciple, and the women – are behind it, and in the foreground Mary Magdalene kneeling on a green swath is embracing, slightly lifting, Christ's legs. Owing to the exceptional seating of the Virgin Mary upon the stone and Mary Magdalene in front of it, lifting Christ's legs so that his body remains horizontal, the body seems to be floating. The figure of the stone dominates the composition: central and relatively large, its reddish color and unique grainy texture strikingly recall the descriptions of the Stone of Unction in Constantinople with the Virgin's tears frozen on its surface. In accordance with Byzantine tradition, the figure of the stone in this altar-piece panel is corporally related both to Christ, whose body is laid upon its surface, and to the Virgin Mary, who sits on it, her tears imprinted on the stone.[49] Christ's wounds are visible and his red blood is still flowing, saturating the surface of the stone.

The Stone of Unction also appears in a fresco attributed to the *Master of Saint Francis* in the nave of the lower basilica of San Francesco in Assisi. It is one of five frescoes depicting Christ's Passion, which was originally placed next to five frescoes depicting the life of Saint Francis as analogous to that of the Savior.[50] It also shows Mary's lamentation over her dead son (Plate 10).[51]

Christ is shown lying half-naked upon a great slab of stone. To the left, close to her son's head, the fainting Virgin Mary has her left hand upon his head, almost touching his halo. Three women are supporting her collapsing body. Of the two male figures behind the rectangular stone, the one on the left may be John the disciple, wearing an orange dress, his palms raised in opposite directions in a gesture of deep sorrow. The figure beside him is severely damaged and cannot be identified exactly, but he might be Joseph of Arimathea or Nicodemus. As in the panel from Perugia, the stone is very prominent owing to its sharp outlines and remarkable whiteness. Although the right side of the fresco has been virtually destroyed, the scene is clearly the Virgin's lamentation at the Stone of Unction. The accentuated white of the stone's surface is probably related to an early tradition claiming that the relic in Constantinople was a white stone dyed by the Savior's blood.[52] Whereas in the panel from the altar piece in Perugia the Stone of Unction is incorporated with Mary's corporal and material presence (she sits on it and her tears are imprinted on its surface), the fresco from Assisi emphasizes Christ's body as is clear from its vivid contrast against the white surface. The stone is no longer a Marian relic but is rather related to Christ's body. In fact, the Virgin swoons away from it.[53] In other Franciscan images, typically on painted crosses, the figure of the Stone of Unction appears in the scene of the lamentation and, like the Perugia altar piece, represents a Marian relic visually connected to the relic in Constantinople by the surface ingrained with Mary's tears.[54] Notably, in Assisi, where the fresco is part of a grand plan paralleling Franciscus and Christus, the stone is "returned" to Christ's body without reference to Mary's tears.

Over the course of the thirteenth century, the Franciscans, the most important Latin order involved in pilgrimage to the Holy Land, became central actors in maintaining the relationship between Western Europe and Byzantium, including the Mediterranean Basin and the Holy Land.[55] As Ann Derbes contends, their constant movement between Italy and the Levant after 1261 and the restoration of Byzantine rule in Constantinople left a mark on the visual culture of both regions.[56] Consequently, the Franciscans were principal agents in mediating traditions and iconographical schemes among Constantinople, Italy, and Jerusalem. Nevertheless, I must stress that the Stone of Unction entered Franciscan visual imagery before an actual stone was placed in Jerusalem. The Stone of Unction in Constantinople and its visual imagery in Duecento Italy should therefore be seen as the context in which the scene of the unction in Jerusalem was concretized in a stone. Similarly to the way in which the figure of the Stone of Unction entered Franciscan visual images of Mary's Lamentation, the *object* of the Stone of Unction was introduced into the space of the Church of the Holy Sepulchre in Jerusalem.

Visual and material translation: Constantinople and Jerusalem

Juxtaposing the two Stones of Unction – the one in Constantinople and the one in Jerusalem – provides another layer in the interpretation of the emergence of the stone in Jerusalem. The evidence suggests that the Stone of Unction in Constantinople was a marble slab set on an elevated platform; it was either red marble or a slab of "diverse colors" and was believed to carry the marks of the Virgin's tears, making manifest its sanctifying narrative.[57] Thus, the stone stood as both a Christological and a Marian relic.

Unlike its counterpart in Constantinople, the Stone of Unction in Jerusalem was set in the floor of the Church of the Holy Sepulchre, apparently from ca.1335 onward, either flush or slightly elevated. Its alignment with floor level is evident from the famous description of Felix Fabri, who accidently stepped on it.[58] In Jerusalem, the stone was reproduced as a two-dimensional surface, its three-dimensional qualities almost entirely concealed. Pilgrims' illustrations of the stone from above emphasize its shallow appearance (Figs. 13.1 and 13.2). The translation of the three-dimensional relic from Constantinople into a two-dimensional surface in Jerusalem suggests that in Jerusalem the stone itself stood as an amalgamation of relic and image, both a presentation and a representation of a stone.

From a material point of view, the stone in Jerusalem stood in dialogue with the one in Constantinople. As the accounts suggest, the former was made of porphyry or marble. Marble and porphyry share certain visual qualities that gave them a special status in the medieval imagination, as their speckling and colored veins were a metaphor for the sacrificial blood of Christ.[59] Whereas in Constantinople the material of the stone was associated with Mary via the identification of her tears in the substance, in Jerusalem the slab of porphyry or marble was symbolically connected to Christ; its materiality validating and "restoring" the sacred connotations of his body.[60]

The separation of the unction site from the center of the earth and its association with a stone surface in the entrance hall of the church suggested a quality of presence. In this location, the materiality of the stone surface was more prone to invoke devotional practices, inviting ritual and spontaneous gestures. Thus, setting the stone in the entrance hall might have contributed to consolidating the ritualistic nature of the visit to the church under the guidance of the Franciscans. Flat, lying on the same plane as the floor, marked by a checkered frame, the stone invited pilgrims to "participate" in the scene of the unction, imagining the body of Christ from the perspective of the Virgin Mary standing near and above her son. Furthermore, the position of the Stone of Unction in Jerusalem recalls the narrative of the death of Saint Francis, who asked to be laid naked on the ground as he was dying.[61] Thus the Stone of Unction also alludes to the narrative of *Franciscus alter Christus* in the narrative space of Christ in Jerusalem.

Unlike stones taken from the Holy Land and displayed elsewhere, which held the meaning of their place of origin *pars pro toto*, the Stone of Unction in Jerusalem is – perhaps counter-intuitively – the *in situ* realization of iconographical traditions based on the relic in Constantinople and its visual representations in Italy. The transition from an elevated stone in Constantinople into a stone surface at floor level in Jerusalem can be conceptualized as the material concretization of an absent relic *in situ*. In other words, the surface in Jerusalem is the "material shadow" or the "material trace" of the stone of Constantinople. In spatial terms, the stone in Jerusalem reverses the state of affairs between Constantinople (often considered as a "New Jerusalem") and Jerusalem, suggesting a dialogical model for the translation of iconographic schemes and their realization in stone.

Notes

* I thank Anna Gutgarts, Bianca Kühnel, Ora Limor, Iris Shagrir, Gerhard Wolf, and the anonymous reader of this chapter for their helpful comments at different stages of the study. I thank Julie Edelson for her careful editing.

1 This illustration appears in Ms. II. IV. 101 in the Biblioteca Nazionale Centrale of Florence. According to Kathryn Blair Moore, this is the oldest of the newly found illustrated manuscript copies of Niccolò of Poggibonsi and could be either the one made by Niccolò himself after his return to Italy in 1350 or an immediate copy of the lost original. See Kathryn Blair Moore, "The Disappearance of an Author and the Emergence of a Genre: Niccolò da Poggibonsi and Pilgrimage Guidebooks between Manuscript and Print," *Renaissance Quarterly* 66, no. 2 (2013): 357–411.

2 Kathryn Blair Moore, in "The Disappearance of an Author," shows that the attribution is wrong, and traces the origins of the *Viaggio* text and woodcut illustrations to the mid-fourteenth-century *Libro d'Oltramare* by Niccolò da Poggibonsi.

3 The Church of the Holy Sepulchre was extensively restored following the fire that broke out on October 12, 1808. The lid covering the Stone of Unction was fixed there during the renovations carried out in 1809–1810, J. Robert Wright, "An Historical and Ecumenical Survey of the Church of the Holy Sepulchre in Jerusalem, with Notes on Its Significance

for Anglicans," *Anglican and Episcopal History* 64, no. 4 (1995): 494. It is not known if the fourteenth-century porphyry slab is still to be found beneath the modern lid. Fifteenth-century pilgrims' reports of the stone's color are inconsistent, and there is a possibility that the stone had been already covered (perhaps more than once) in the fifteenth century. See note no. 21 below.

4 Martin Biddle, Gideon Avni, Jon Seligman, and Tamar Winter, *The Church of the Holy Sepulchre* (New York: Rizzoli, 2000), 26.

5 R.B. C. Huygens, ed., *Peregrinationes Tres: Saewulf, John of Würzburg, Theodoricus*, Corpus Christianorum Continuatio Mediaevalis 139 (Turnholt: Brepols, 1994), 119–120: "In medio choro dominorum, non longe a loco Calvariae, est quidam locus elevatione tabularum de marmore et reticulorum ferreorum concatenatione in modum altaris designatus, infra quas tabulas in pavimento orbiculis quibusdam factis meditullium terrae dicitur designatum, iuxta illud: *operatus est salutem in medio terrae.*" English translation in: John Wilkinson, Joyce Hill, and William F Ryan, eds., *Jerusalem Pilgrimage, 1099–1185* (London: Hakluyt Society, 1988), 260.

6 Huygens, *Peregrinationes Tres*, 151.

7 See, e.g., Theoderic's description: "altare quidem parvum sed reverendum habetur, ubi corpus dominicum antequam sepulture daretur positum fuisse a Ioseph et Nychodemo narrator," ibid., 147. This altar on which they laid the body of Christ was mentioned as early as in the eighth century in the account of Willibald, who terms it *lectus*: "et in orientali plaga in illa petra sepulchri est ostium factum, per quod intrant homines in sepulchrum orare. Et ibi est intus lectus, ubi corpus Domini jacebat. Et ibi stant in lecto quindecim craterae aureae cum oleo ardentes die noctuque. Ille lectus, in quo corpus Domini iacebat, stat in latere aquilonis intus in petra sepulchri, et homini est in dextra manu, quando intrat sepulchrum orare," see Willibaldus, "S. Willibaldus: Vita seu potius hodoeporicon Sancti Willibaldi," in *Descriptiones Terrae Sanctae ex saeculo VIII. IX. XII. et XV*, ed. Titus Tobler (Leipzig: J.C. Hinrichs'sche Buchhandlung, 1874), 30.

8 This mosaic, which no longer exists, is depicted in the account of Theoderich: Huygens, *Peregrinationes Tres*, 147–148. According to Theoderich, the mosaic was placed over the entry to Christ's sepulchre from the Chapel of the Angel. For a discussion of the inscription and the mosaic, a fusion of the entombment and the Three Marys at the Tomb, see Jaroslav Folda, *The Art of the Crusaders in the Holy Land, 1098–1187* (Cambridge: Cambridge University Press, 1995), 231–232. Since the mosaic was not mentioned in pilgrims' accounts prior to John of Würzburg and Theoderich, Folda concludes that it was part of the Crusaders' renovations, ibid., 231.

9 Here I follow Iris Shagrir's discussion of the adaptation of the Crusaders' liturgical feasts to the architectural plan of the church in "The 'Visitatio Sepulchri' in the Latin Church of the Holy Sepulchre in Jerusalem," *Al – Masāq* 22, no. 1 (2010): 57–77.

10 An anonymous French account (1229–1239) associates the place of the unction with the center of the earth, Anonymous, *Les Sains Pelegrinages que l'en doit Requerre en la Terre Sainte*, 472. On liturgical ceremonies in the Church of the Holy Sepulchre between 1229 and 1244, see Shagrir, "*Visitatio Sepulchri*," 59.

11 English translation of the Greek account: Pringle, *Pilgrimage*, 192.

12 In *Patrologia Graeca*, ed. J.-P. Migne, vol. 132 (Paris: Garnier Fratres, 1884), cols. 981–982. I thank Stefan Schorch for helping me with the Greek source, which is discussed in Denys Pringle, *The Churches of the Crusader Kingdom of Jerusalem: A Corpus* (Cambridge and New York: Cambridge University Press, 1993), 32–33.

13 The oldest illustration of the tomb aedicule, which is to be found in a fourteenth-century manuscript (Biblioteca Apostolica Vaticana, Cod. Urb. Lat. 1362, f. 1), shows it this way. For the illustration, see Jürgen Krüger, *Die Grabeskirche zu Jerusalem: Geschichte, Gestalt, Bedeutung* (Regensburg: Schnell and Steiner, 2000), 151.

14 "Ibi retro et iuxta eam erat locus et ostendebant lapidem, ubi deposuerunt corpus et ligaverunt linteis et condiderunt aromatibus, ut sepelirent," in Johann Karl Mauritz Laurent et al., *Peregrinatores Medii Aevi quatuor: Burchardus de Monte Sion, Ricoldus de Monte Crucis, Odoricus de Foro Julii, Wilbrandus de Oldenborg* (Leipzig: J.C. Hinrichs, Bibliopola, 1864), 112; Pringle, *Pilgrimage*, 373.

15 Laurent et al., *Peregrinatores*, 112. Riccoldo noted that the Stone of Unction was situated behind and next to the place from where Mary witnessed the crucifixion, so it could have

been either in the choir or in the southern transept in front of the church's entrance, depending on where the pilgrim was standing; cf. Shagrir, "'Visitatio Sepulchri'," 70, n. 43.

16 Semeonis, "Itinerarium Symonis Semeonis ab Hybernia ad Terram Sanctam": Odorico da Pordenone, "Odorici de Foro Iulii Liber de Terra Sancta," in *Peregrinatores medii aevi quatuor; Burchardus de Monte Sion, Ricoldus de Monte Crucis, Odoricus de Foro Julii, Wilbrandus de Oldenborg*, 2nd ed., ed. J.C.M. Laurent (Leipzig: J.C. Hinrichs, 1873), 108–109, 149.

17 ". . . item locus ubi Josep et Nicodemus leverunt corpus Yhesu. Ibique eodem loco in titulum unus lapis, quem feriticus secundum mensuram longitudinis Christi mensuravi, et est eiusdem mensurae sicut sepulcrum Christi"; Antoninus de Cremona de Reboldis, "Itinerarium ad Sepulcrum Domini et Ad Montem Sinai," in *Biblioteca Bio-Bibliografica Della Terra Santa E dell'Oriente*, ed. Girolamo Golubovich (Florence: Quaracchi, 1919), vol. 3, 326–342.

18 "In predicta ecclesia sepulchri statim in introitu porte ad dextram ecclesie est unus lapis contiguus pavimento et est niger: super illum lapidem positum fuit corpus Jhesu Cristi quum depositum fuit de cruce et [Maria] aromatis et unguentis preciosis unxit eum et involvit in loco illo eum in lincheo mundissimo et posuit sudarium circa caput suum et de loco illo portavit eum ad monumentum," in *Liber Peregrinationis di Jacopo da Verona*, ed. Ugo Monneret de Villard (Rome: La Libreria dello Stato, 1950), 30. See also Ludolphus de Sudheim, "De Itinere Terre Sancte," in *Archives de l'Orient Latin*, ed. G.A. Neumann (Paris: Ernst Leroux, 1884), vol. 2, 305–377.

19 "Itinerarium cuiusdam Anglici Terram Sanctam et alia loca sancta visitantis," in *Biblioteca Bio-Bibliografica*, ed. Golubovich, vol. 4, 452.

20 English translation: *A Voyage beyond the Seas (1346–1350)*, eds. T. Bellorini and E. Hoade (Jerusalem: Franciscan, 1945), 13. "Tra l'entrata della porta, come s'entra dentro, di lungi da sei passi al diritto della porta, si è in terra una pietra di porfido verde, ed è lunga otto palmi e tre dita, larga uno palmo e uno ditto; e in questo luogo fu posto Cristo, quando fu levato dalla croce, e ivi fu unto e aromatizato. D'intorno alla detta pietra a due palmi si e lavorato come scacchi, et apresso del muro del coro, a due passi, si à due bellissime arche, cioè sepulture, lavorate e sollevate un piè sopra la terra; e alla sopradetta pietra si à indulgenzia di colpa e di pena". In: *Libro d'Oltramare di Fra Niccolò da Poggibonsi*, ed. Alberto Bacchi, (Bologna: Gaetano Romagnoli, 1881), 54–55.

21 The terms used in the accounts to describe the stone are: "*pietra di porfido*," "*lapis niger marmoreus*," "*lapis preciosus aërei*," "*marmelsteyne*," "*marbyl styn.*" In some of the accounts the stone is simply termed "*lapis niger.*"

22 Krüger, *Die Grabeskirche zu Jerusalem*, 200.

23 Sabino de Sandoli, *The Peaceful Liberation of the Holy Places in the XIV Century: The Third Return of the Frankish or Latin Clergy to the Custody and Service of the Holy Places through Official Negotiations in 1333* (Cairo: Franciscan Centre of Christian Oriental Studies, 1990), 72; Pringle, *Pilgrimage*, 32–33.

24 Golubovich, ed., *Biblioteca Bio-Bibliografica*, vol. 3, 128–130; Pringle, *Churches*, 32–33.

25 Pringle, *Churches*, 33.

26 Jerome Murphy-O'Connor, *Keys to Jerusalem: Collected Essays* (Oxford: Oxford University Press, 2012), 110; Sandoli, *Peaceful Liberation*, 72.

27 Andrew Jotischky, "The Mendicants as Missionaries and Travelers in the Near East in the Thirteenth and Fourteenth Centuries," in *Eastward Bound: Travel and Travellers, 1050–1550*, ed. Rosamund Allen (Manchester: Manchester University Press, 2004), 96–97.

28 Ibid.

29 The anonymous English pilgrim who arrived in Jerusalem in 1344–1345 told how pilgrims were received by four Friars Minor and four Greeks, chanting, who led them into the church, see "Itinerarium Cuiusdam Anglici," 451–452. Sandoli argues that when the Franciscans entered the Church of the Holy Sepulchre in 1333 they lived in "harmony" with the Eastern Christian clergy. Even if this position is a radical one, accounts do attest to a certain degree of cooperation: see *Peaceful Liberation*, 67–69.

30 Niccolò of Poggibonsi describes how the pilgrims were led through the church by the Friar Minor, following the path of the indulgences assigned to each holy place rather than according to a chronological narrative: from the place of the unction to the Anastasis Rotunda, the Holy Column, the site of the appearance to Mary Magdalene, the prison, Calvary, and the Chapel of Golgotha.

31 I could not find sources concerning the possession of the place in the fourteenth century. Yet, by the beginning of the seventeenth century the southern transept of the Church of the Holy Sepulchre, where the stone was situated, was under Franciscan control. See Oded Peri, *Christianity under Islam in Jerusalem: The Question of the Holy Sites in Early Ottoman Times* (Leiden: Brill, 2001), 107.
32 John Kinnamos wrote a history of the reigns of John II (1118–1143) and Manuel I (1143–1180) in *Deeds of John and Manuel Comnenus*, trans. Charles M. Brand (New York: Columbia University Press, 1976). Nicetas Choniates chronicled Byzantine history from the death of Emperor Alexios I Komnenos on August 15/16, 1118, to autumn 1207 in *O City of Byzantium: Annals of Niketas Choniatēs*, Byzantine Texts in Translation (Detroit: Wayne State University Press, 1984), ix. The story of the arrival of the Stone of Unction in Constantinople based on these two Greek sources is discussed in Cyril Mango, "Notes on Byzantine Monuments," *Dumbarton Oaks Papers* 23/24 (1969–1970): 369–375, and in Mary Ann Graeve, "The Stone of Unction in Caravaggio's Painting for the Chiesa Nuova," *Art Bulletin* 40, no. 3 (1958): 223–238; Robert Ousterhout, "Architecture, Art and Komnenian Ideology at the Pantokrator Monastery," in *Byzantine Constantinople: Monuments, Topography, and Everyday Life*, The Medieval Mediterranean 33, ed. Nevra Necipoğlu (Leiden: Brill, 2001), 133–150.
33 Choniates, *O City of Byzantium*, 125.
34 The inscription on or near the stone told the story of its transfer, evoking a comparison between the Emperor Manuel Comnenus I and Christ and between Eirene and the Virgin Mary. The inscription is recorded in the seventeenth-century *Geography* of Meletios of Ioannina (d. 1714). For the Greek inscription and an English translation, see Mango, "Notes," 372–373. See also Nancy P. Ševčenko, "The Service of the Virgin's Lament Is Revisited," in *The Cult of the Mother of God in Byzantium: Texts and Images*, eds. Leslie Brubaker and Mary B. Cunningham (Burlington, VT: Ashgate, 2011), 257. For the building phases of the Pantokrator and a discussion of the *opus sectile* floor as a reflection of early Christian themes as well as an expression of Komnenian ideology see Ousterhout, "Architecture."
35 Choniates, *O City of Byzantium*, 125.
36 Kinnamos, *Deeds*, 207.
37 Choniates, *O City of Byzantium*, 125.
38 Cf. Graeve's discussion of the iconographical sources of Caravaggio's *Christo in Scurto* in "Stone of Unction." She notes that it was simultaneously a Marian and a Christological relic. See also Fabio Barry, *Painting in Stone: The Symbolism of Colored Marbles in the Visual Arts and Literature from Antiquity until the Enlightenment* (PhD dissertation, Columbia University), 2011, 432–434.
39 Robert de Clari, *The Conquest of Constantinople*, trans. Edgar Holmes McNeal (New York: Columbia University Press, 1936), 112–113. "En chel abeïe si estoit le tavle de marbre ou Nostre Sires fu estendus, quant il fu despendus de le crois, et si I paroient encore les lermes que Nostre Dame avoit plouré deseure," de Clari, *La Conquête de Constantinople: Édition Bilingue*, trans. Jean Dufournet (Paris: Champion Classiques, 2004), 184.
40 Charles Schefer, ed., *Le voyage d'Outremer de Bertrandon de la Broquière* (Paris: Ernest Leroux, 1892), 160–161.
41 George P. Majeska, "The Relics of Constantinople after 1204," in *Byzance et les Reliques du Christ*, eds. Jannic Durand and Bernard Flusin (Paris: Centre d'histoire et civilisation de Byzance, 2004), 183–190.
42 An explicit reference to this duplication is found, for example, in the fifteenth-century account of William Wey, see *The Itineraries of William Wey* (Oxford: Bodleian Library, 2010), 130.
43 On the Stone of Unction in Eastern visual imagery, see Ioannis Spatharakis, "The Influence of the Lithos in the Development of the Iconography of the Threnos," in *Byzantine East, Latin West: Art: Historical Studies in Honor of Kurt Weitzmann*, eds. Christopher Moss and Katherine Kiefer (Princeton: Princeton University Press, 1995), 435–441.
44 Anne Derbes, *Byzantine Art and the Dugento: Iconographic Sources of the Passion Scenes in Italian Painted Crosses* (unpublished PhD dissertation, University of Virginia, 1980), 320–324, 333–339.
45 Gertrud Schiller, *Iconography of Christian Art* (London: Lund Humphries, 1971), 172–174. Derbes relates the proliferation of images of the anointing of Christ in the mid-twelfth

century in the West to the Second Crusade (1145–1149) and the preaching of Bernard of Clairvaux in whose sermons the scene of the unction prefigures the Crusaders' "anointing" of the holy places, "Byzantine Art," 300–303.

46 For the three *oculi* on Christ's sarcophagus and Christ's tomb in Jerusalem, see Derbes, *Byzantine Art*, 299–300, and William H. Forsyth, *The Entombment of Christ: French Sculptures of the Fifteenth and Sixteenth Centuries* (Cambridge, MA: Harvard University Press, 1970), 9.

47 Dillian Gordon, "A Perugian Provenance for the Franciscan Double-Sided Altar-Piece by the Maestro Di S. Francesco," *Burlington Magazine* 124, no. 947 (1982): 70–77.

48 In Spatharakis's typology of images of the Threnos, it is the second type. See "Influence of the Lithos," 440.

49 Compare: Derbes, "Byzantine Art," 323–324.

50 Originally the frescoes flanked the narrow windows in the three bays on both sides of the nave of the lower church. All the paintings of the lower church were designed to accentuate typological parallels that the Franciscans drew between Christ (*passio*) and Saint Francis (*compassio*).

51 For the attribution of the works to the Master of Saint Francis, see Henry Thode, *Franz von Assisi und die Anfänge der Kunst der Renaissance in Italien* (Vienna: Phaidon, 1934), 86–87, 113. For the dating of the frescoes, see Joanna Cannon, "Dating the Frescoes by the Maestro Di S. Francesco at Assisi," *Burlington Magazine* 124, no. 947 (1982): 65–69.

52 For the Greek source, see Barry, *Painting in Stone*, 432. See also Andreas Prater, "Mantegnas Cristo in Scurto," *Zeitschrift für Kunstgeschichte* 48, no. 3 (1985): 283.

53 Joachim Poeschke relates the expressive gestures of the women in the fresco to passion plays that were performed in Italy during the same period. Joachim Poeschke, *Die Kirche San Francesco in Assisi und ihre Wandmalereien* (Munich: Hirmer, 1985), 17–18.

54 Derbes studied the migration of the Stone of Unction into early Franciscan painted crosses as a fusion of the Byzantine Threnos and the Western anointing of Christ in *Byzantine Art*, 320–324, 333–339, and *Picturing the Passion in Late Medieval Italy: Narrative Painting, Franciscan Ideologies, and the Levant* (Cambridge and New York: Cambridge University Press, 1996), 1–11.

55 Derbes, *Picturing the Passion*, 24–27; Heyden B.T. Maginnis, "Places beyond the Seas: Trecento Images of Jerusalem," *Notes in the History of Art* 13, no. 2 (1994): 1–8.

56 Derbes, *Picturing the Passion in Late Medieval Italy*.

57 For the staging of the stone in Constantinople on a platform, see Barry, "Painting in Stone," 435.

58 Konrad Dietrich Hassler, ed., *Fratris Felicis Fabri Evagatorium in Terrae Sanctae, Arabiae et Aegypti Peregrinationem* (Stuttgart: Sumtibus Societatis Literariae Stuttgardiensis, 1843), vol. 2, 283.

59 Georges Didi-Huberman, *Fra Angelico: Dissemblance and Figuration* (Chicago: University of Chicago Press, 1995), 31. For the sacrificial association of porphyry, see Suzanne B. Butters, *The Triumph of Vulcan: Sculptors' Tools, Porphyry, and the Prince in Ducal Florence* (Florence: L.S. Olschki, 1996), vol. 1, 50. For the perception of marble as a "living rock" with "animalistic veins" symbolically associated with Christ's flesh, see Fabio Barry, "Painting in Stone," 3–4.

60 For the principle of overt materiality, see Herbert L. Kessler, *Seeing Medieval Art*, *Rethinking the Middle Ages* (Peterborough, ON: Broadview, 2004), ch. 1; Caroline Walker Bynum, *Christian Materiality: An Essay on Religion in Late Medieval Europe* (Cambridge, MA: Zone Books, 2011), 53ff.

61 This story is told in both biographies of the saint, the *Vita Prima* written by Thomas of Celano in 1228 and the *Legenda maior* written by Bonaventure in 1263.

IV

Contemporary re-enactment

14 Susan Hiller's *Homages to Joseph Beuys*

Mystics, cult, and anthropology

Kobi Ben-Meir

Starting in 1969, the London-based artist Susan Hiller collected water from various locations that are considered sacred, bottling it in hundreds of small glass containers that had been used for medicine during the Victorian and the Edwardian periods. She carefully labeled each vial, noting the time and place of collection. This project was concluded in 2013, with the storing of the vials in felt-lined wooden cabinets, most of which were early twentieth-century first-aid cabinets, each of which held tens of vials. These cabinets were first exhibited in 2011 in the large retrospective of Hiller's work in London's Tate Britain, all subtitled *Homage to Joseph Beuys* (Figs. 14.1 and 14.2).[1]

In what follows I present three models for Hiller's water collection and bottling and for her creation of the *Homages to Joseph Beuys*: Pilgrimage to Holy Places, which includes the collection of holy substances and its preservation in containers; the art

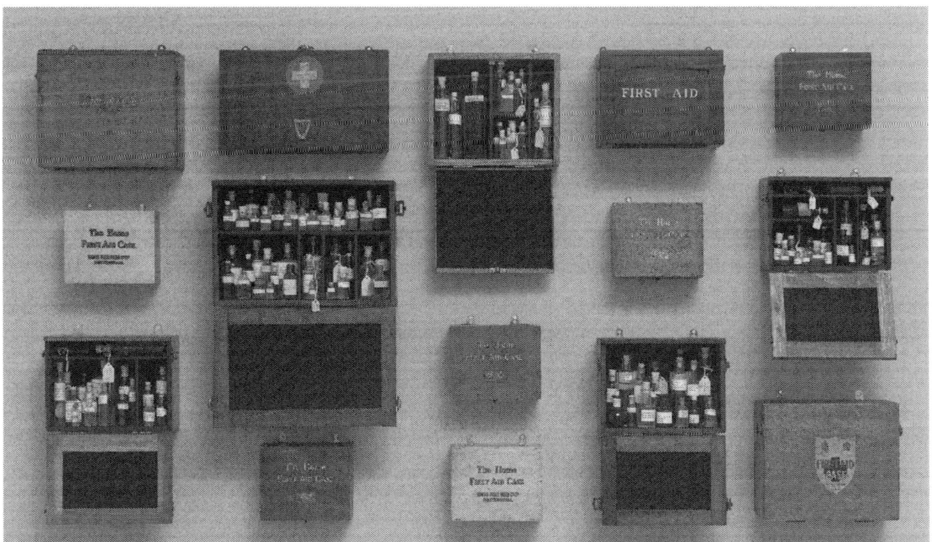

Figure 14.1 Susan Hiller, *Home Nursing: Homage to Joseph Beuys*, 1969–2011, fifteen wooden felt-lined first-aid cabinets, containing 116 bottles of holy water and vintage first-aid supplies, 110 × 191.3 cm. Collection of the artist. Photo Credit: ©Susan Hiller.

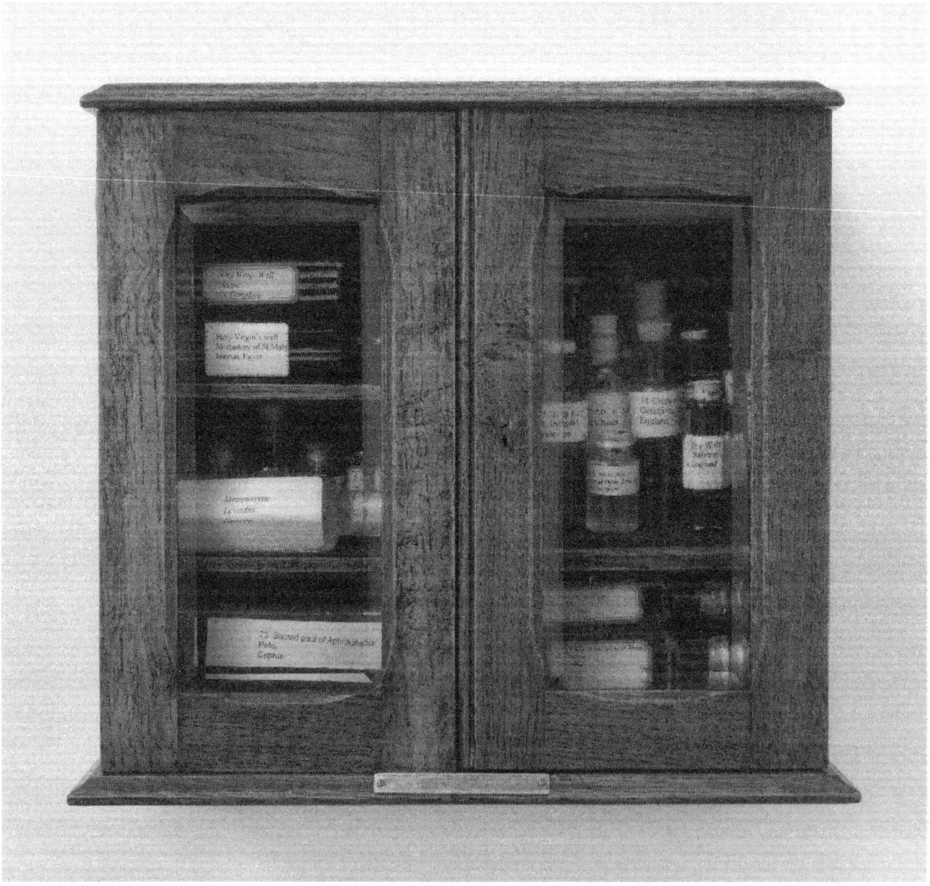

Figure 14.2 Susan Hiller, *Fortes Saintes: Homage to Joseph Beuys*, 1969–2013, double Edwardian oak, felt-lined cabinet with bottles of holy water, 28.6 × 32.5 × 19 cm. Collection of the artist. Photo Credit: ©Susan Hiller.

of Joseph Beuys, which aimed at healing through shamanistic use of materials; and anthropology, the collection of research material for the scientific analysis of culture. These three methodologies – religion, art, and science – may seem to be in opposition to one another, but Hiller's work is all about confrontation of alleged contradictions in pursuit of their reconciliation. Throughout her career, Hiller has undermined the binarism between science and ecstasy, rationality and irrationality, reality and fiction. She says:

> Our culture more than most makes a distinction between the rational and irrational, between empiricism and intuitive ways of apprehending the world. In my experience those kinds of distinctions don't have any validity. In my work, I'm trying to approach a kind of reconciliation of rational and irrational, factors which seem to me a lived truth for many people. For myself, I can only say that this is part of the way I see things.[2]

Homage to pilgrimage

Hiller journeyed all around the world to visit holy places, sacred to different religions, with natural water sources at their center. Water is one of the most basic, ubiquitous, and as the artist puts it – banal – of materials. Nevertheless, owing to its power to purify, it has profound cultural meaning and is associated with an entire array of religious customs and rituals, both ancient and modern. The common denominator of the sites that the artist visited is that their waters are strongly associated with sacred historical events from the mythical past; thus the waters are thought to have magical powers to purify the body and the soul, heal, protect, and provide heavenly blessing. As such, these sites are pilgrimage destinations for believers around the world.

Hiller realized that the cult of holy water is a universal phenomenon, not particular to any religion or specific to any period, but occurring in most cultures. As an example, the Zamzam Well in the compound of Masjid al-Haram in Mecca, the central destination of Haj, has miraculously generated water since Ismail's cry for water in the desert.[3] According to the Hadith, "A person drinking Zamzam water should intend and hope for healing, blessings and whatever is best for him in this life and in the hereafter. The Prophet (peace be upon him) said, 'The water of Zamzam is (good) for whatever it is intended' " (Fiqh-us-Sunnah, Fiqh 5.81). Another example is the Ganges River in Varanasi, India: putting the body of the deceased in the river in Varanasi, the city of Lord Shiva, cleanses the soul of the dead Hindu and guarantees *Moksha*, liberation from the misery of the cycle of death and rebirth.[4]

Holy natural sources of water are very significant in the hagio-topography of the Holy Land. Pseudo-Antoninus of Piacenza, a sixth-century pilgrim to the Holy Land, listed ten different natural water sources there in which believers bathed for blessing. When he wrote about the Siloam in Jerusalem, he made a distinction between the natural fountain and the man-made pools that contained the holy water for human use:

> Descending from that arch down to the fountain of Siloam by many steps, we saw the round church, from beneath which Siloam rises. This church has two baths made by the hands of man, out of marble; between the two baths runs a partition; in the one men, and in the other women, bathe for a blessing. In these waters many cures are effected, and even lepers are cleansed.[5]

The most sacred water source in the Holy Land is the Jordan River, where Christ was baptized. Christian pilgrimage to the Jordan started during the late Roman period[6] and continues to this day. Hiller visited the site and, though sparing the religious blessing, collected and bottled water from there, in a practice not very different from that described by Pseudo-Antoninus:

> We celebrated the Epiphany by the side of the Jordan, and wonders take place on that night in the place where the Lord was baptized. . . . After matins, as day begins to dawn, the deacons begin the holy mysteries and celebrate them in the open air; the priest descends into the river, and at the hour when he begins to bless the water, at once the Jordan, with a mighty noise, rolls back upon itself, and the water stands still until the baptism is completed. And all the men of Alexandria who have ships, with their crews, holding baskets full of spices and balsams, at the hour when the priest blesses the water, before they begin to baptize, throw those baskets into the river, and take thence holy water, with which they sprinkle their ships before they leave port for a voyage.[7]

According to this sixth-century witness, water from the Jordan was collected, for it was expected to perform miracles. The Jordan River water was sanctified because it touched the body of Christ, as explained in the sixth century by Saint Gregory of Tours: "Faith believes that everything that the sacred body touched is holy."[8] The power invested in water from the Jordan is due to the fact that it comes from and thus represents the river in which Christ was baptized.

Collecting liquids from holy places was probably widespread. Numerous glass containers dating to the second half of the sixth century, parallel to the period of the pilgrimage of Pseudo-Antoninus, have been found, and according to their quality and quantity were probably mass-produced for pilgrims.[9] Similar to the well-known Monza and Bobbio flasks, some of these containers include depictions of a cross mounted on two or three horizontal lines, surmounted by an arch (Figure 14.3).

This is probably a schematic representation of Golgotha: the stairs leading upward from the inner atrium of the Constantinian Anastasis compound to the site of the crucifixion, where pilgrims could admire the True Cross displayed underneath a ciborium. This representation marked the vessels as containing holy material from the Holy Sepulchre; whether it was water, oil, or earth is unknown.[10]

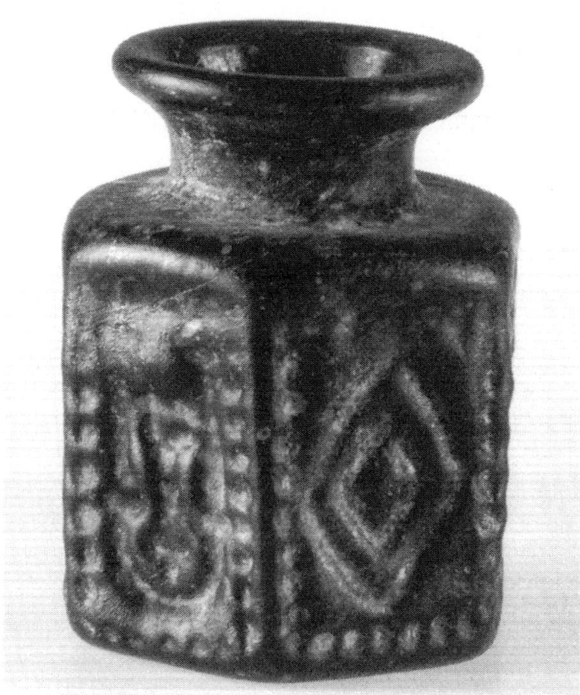

Figure 14.3 *Eulogia Jar with Crosses*, provenance unknown, late sixth to early seventh centuries, glass, height 15 cm. Israel Museum, Jerusalem. Photo Credit: © Israel Museum, Jerusalem.

Customs of collecting and bottling holy water are far from being unique to the past, as most contemporary religious pilgrims to the Jordan River collect its water in bottles as *eulogia*, blessings, as the river still holds the blessing of the Messiah. The same custom is common at other water-related pilgrimage sites such as the Ganges in Varanassi and the Zamzam well in Mecca, mentioned above, and the pools of Lourdes.[11] One can also buy holy water in bottles that carry the blessing of the holy place in tourist shops or even order them online in various designs and shapes.

In Hiller's *Homages*, the water is treated as a holy substance, and even as a relic: each vial is labeled according to its place of origin and the time of its collection, similar to the method of labeling Christian relics. For instance, the sixteenth-century bishop of Milan, Carlo Borromeo, instructed that relics be labeled by the names of the saints, the date of their deposition, and the places from which they had been translated.[12] Similar to many medieval reliquaries, Hiller's reliquary-like cabinets hold different "relics," translated from different places at different times.[13]

Hiller's *Homages* is a modern artistic interpretation of the tradition of collecting and bottling holy water. Judging by the act itself, one might have considered Hiller a modern pilgrim, as she journeyed to pilgrimage sites and collected their holy water. Moreover, judging by her method of collecting, labeling, and storing the holy substance in cabinets, *Homages* appears to be religious-art, but it could fit as easily in a modern art museum. However, the cabinets contain waters that are holy to different religions: holy waters from Christian, Muslim, Hindu, and Buddhist pilgrimage sites are placed side-by-side, attesting to the universality of the custom of collecting water and to the fact that it is not unique to any one religion. While pilgrims go to sites sacred only to their own faith, Hiller did not distinguish among different denominations, neither in her pseudo-pilgrimage nor in the presentation of the pseudo-relics. Rather than focusing, as a pilgrim would, on the location, Hiller translocated the holy water.

Another important point to consider is that of movement. As Coleman and Elsner point out, movement is a key factor in pilgrimage. A pilgrim does not just go to a sacred site; he is steered by official guides or official texts to a succession of places where a specific sequence of rites should be performed. The movement through space is controlled in order to create a narrative, or even a topographical text, from the site and to "extract" the most out of the religious experience and the holy sites.[14] Hiller definitely moved from one place to another as she traveled around the world, but her movements were arbitrary and not determined by any rigid textual or human guidance; often she collected the water in certain countries that she visited for holiday or work, when neither the visits to the holy places nor the collection of the holy water was the primary motivation for the trip. Moreover, the route was not predetermined and there was no inherent sense of progression or narrative, other than the overall theme of the collection of holy water.[15]

In his book *Dramas, Fields and Metaphors*, Victor Turner argued that pilgrimage is a liminal phenomenon.[16] Many times it is not promoted by official religious institutions because the pilgrim lives in a liminal social stratum during his pilgrimage. Leaving one's normative social environment when embarking upon a pilgrimage means leaving behind many of the social norms and creating a new, more flexible, pilgrim society. Throughout her career, Hiller has shown great interest in liminal phenomena, bordering on ecstasy,[17] and thus I contend that this liminal quality is what attracted Hiller toward the performance of her own unique version of pilgrimage – to "flirt" with the idea of a religious cult.

Hence, in the act of making the cabinets, Hiller was not a pilgrim, but simply acted as one, following in the footsteps of past and present pilgrims to distant sacred places. Her work affirms that the custom of collecting and bottling water is not confined to a specific period, place, or religion. Although she deals with the spiritual, the miraculous, and the ecstatic connected with the healing holy water of pilgrimage sites, Hiller does not relate to the power of the water, nor does she judge if that power is actual or fictional; she narrows the representation of the water to the physical acts of collecting and bottling.[18] While transforming the eulogia into an artifact, she transformed the religious-sacral aura of the water into an artistic aura, that is, the unique genial product of the artistic conceptual work. Thus, Hiller is neither projecting supernatural powers upon the water nor totally reducing it into basic archeological and cultural evidence. The bottles are what they are; it is the observer's role to categorize and judge their nature.

Physical and chronological transformation

What attracted Hiller to water in the first place was its most basic quality – being in constant flow – which does not allow it to be fixed. Although chemically stable in room (or cabinet) temperature, the visual appearance of water's shape is determined by the shape of its container. Like water, Hiller's works are not static objects, but are in a constant flux. She worked on the *Homages to Joseph Beuys* from 1969 to 2013. This prolonged creation of the work actually turns it into an action, and as it spanned more than 40 years, it can be regarded as a mirror to the artist's career and even to her life. Likewise, the action is not limited to the space of the studio, but rather, by its nature, broadened to the whole globe.

The water is stored in containers and cabinets that have their own histories, preceding that of the *Homages*. The cabinets are reused storage facilities produced during the Edwardian period. The vials, small glass bottles for medicine, were found by the artist during private excavations in nineteenth- and twentieth-century garbage dump sites. During the Victorian and Edwardian periods in Britain, garbage was buried underground, usually along river banks, and those dump sites are where the artist found the containers. Hiller also bought some old containers on eBay and in antique shops. Hence, the cabinets and the containers have a history, and they bring their own baggage, sets of memories, times, and places to the piece; the inclusion of the old Edwardian bottles within the pieces made by Hiller loads them with a new, unexpected, set of associations. By translocating the bottles and transforming their function, the artist gave them new lives: they were excavated from underground, brought to light and to sight, and placed in a new, unconventional setting. Hence, the symbolism of water bringing life, rejuvenating the body and soul, is paralleled by the exhumation of the containers; moreover, whereas the time, place, and function of the containers and the cabinets are transformed, they transform the artistic works themselves.

Similar to the study of a universal rite of collecting holy water, Hiller's cabinets are not limited to specific chronology or geography. The *Homages* cannot be labeled as a product of only 2010–2013 London – the time and place of assembling the material components of the cabinets; they also connect to 1969–2009, the forty years during which the artist collected the holy water from various locations; and to late-nineteenth and early twentieth-century Britain – the time and place of the manufacture and the burial of the containers. The water itself echoes remote locations and remote times, extending the times and places of reference, thus achieving universality. It resonates with religious legends from distant pasts and landscapes such as the baptism of Christ in the Jordan River, the appearance of the Virgin in the grotto of Massabielle, the descent of the Ganga to Earth, and the creation of the Zamzam well by the angel Jibril's kick on the Earth. As Hiller's cabinets reference periods and locations beyond the "here and now" of the perceived objects, the concepts of process and transformation are also applicable to the outcome and not only to the creative process: the meaning of the *Homages* is in constant dialectic states of evolution and dissolution.

Process and transformation are persistent motifs in Hiller's work. By rejecting objectivity, that is, rejecting a single truth or a single possible appearance of an object, Hiller inquires into the nature of materials, re-makes finished works, and examines potential transformations. Her early works from the 1970s are transformations of her own earlier paintings. In a series of works titled *Painting Blocks* (Figure 14.4), Hiller cut up her older oil paintings and banded the cuts into ten blocks. Characteristically, this work/process took place over a long period: she started it in 1970 and completed it in 1984.

The elements of *Painting Blocks* seem remarkably minimalistic, and, from a formal point of view, they can be compared to other minimalistic works, such as those of Donald Judd. Judd's cubical constructions deal only with form, reduced to the minimum, and represent pure conceptual art that deals only with the art itself. It is in regard to this precise point that the difference between Hiller's work and conceptual minimalism becomes apparent: conceptual minimalism is all about "here and now," the actual appearance of a form in a certain space, whereas Hiller's works are also

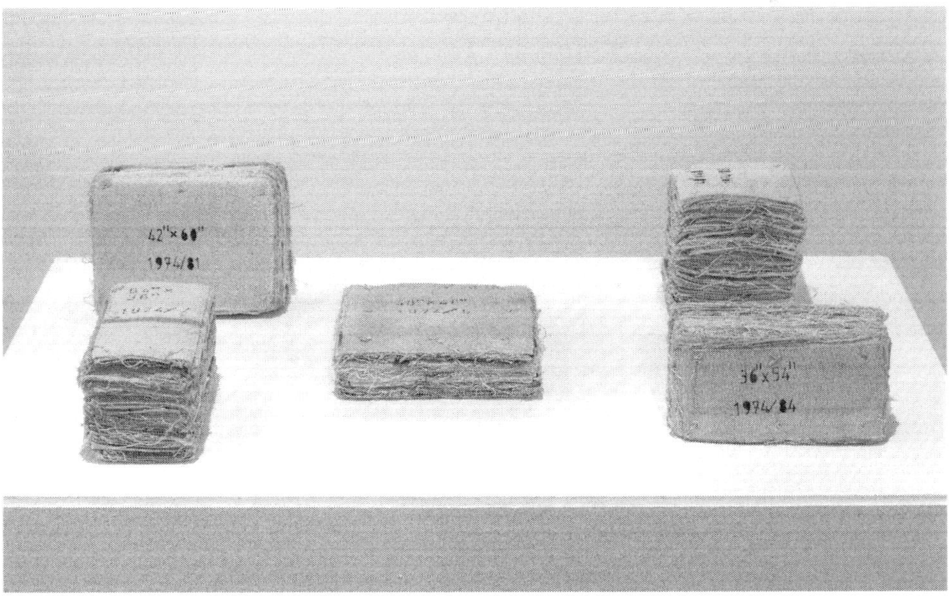

Figure 14.4 Susan Hiller, *Painting Blocks*, 1970–1984, five oil-on-canvas paintings, cut and bound with thread into blocks, dimensions variable. Collection of the artist. Photo Credit: ©Susan Hiller.

about "then and there," and bear the memory of former appearances. In contrast to the American High Modernism of the 1960s, which aspired to purge art of any context, that is, to rid art of non-artistic or non-formalistic associations, Hiller's minimalism is highvly contextual.[19] Furthermore, the elements of *Painting Blocks* hold unseen secrets within them, as the artist banded them in such a way that the paintings themselves remain hidden, and only the back of the canvas can be seen. While employing High-Modernist vocabulary such as the repetition of cubical forms (the block, the cabinet), the inclusion of *objets trouvés* transforms the object into a carrier of memory and context. These objects are receiving new lives through her art, and thus both the objects and the work of art acquire new contexts.

The fluidity of materials and medium was an important factor in Hiller's time-based installation *Work in Progress* in Matt's Gallery, London, in 1980. During the first week of the exhibition, she unwove a canvas of one of her older paintings and formed a three-dimensional sculptural piece with the threads of the former two-dimensional canvas. A week later, she remodeled the groups of threads – some were freely arranged and others were cut into rectangles. The following week, the threads were piled on shelves.[20] Solid forms became flux as liquid, and like water they were in constant circulation, changing their appearance according to their container.

Most artists who have destroyed their works were motivated by the desire to edit their artistic legacy, that is, not to include unqualified works as part of their oeuvre; in contrast, Hiller's destruction of her own works was also constructive, as it was an inherent concept of the construction of new pieces of art.[21] Another example of the process of material transformation is *Hand Grenades* (1972), for which Hiller burned some of her earlier paintings from 1969, collected the ashes, and stored them in glass chemical containers, labeled according to the group of burned works. The paintings were not reduced to ashes, but rather transformed into ashes, bearing the material essence of the canvases and the pigment they once held. The ash itself is treated as a relic and stored in a kind of reliquary. The inclusion of fire in the creation of the work gives the piece a performative, even shamanistic character. Furthermore, the artist continues to burn one of her older paintings every year, making this action into a ritual.[22]

The Beuysian heritage of spiritual transformation

In their title and their materiality, Hiller's cabinets directly reference the art of the German Joseph Beuys. They are paying homage to Beuys on several levels: on the formal level through the inclusion of cabinets, water, and felt; on the issue of transformation; on the issue of religious practice; and on the issue of the social role of art and the artist. Hiller met Joseph Beuys in 1974 in the Institute of Contemporary Arts in London – the only encounter between the young artist and Europe's artistic hero of the time – but nonetheless Beuys's art had a great influence on Hiller, especially in regard to the function of art.

The persona of Beuys as an artist was as influential on post-World War II art as was his art. The foundation of Beuys's art and personal image was the myth of his own rebirth. Beuys claimed that during the war, when serving as an aircraft radio operator in the German Luftwaffe, his plane crashed in the Crimean desert. The Tatars of Crimea saved his life by wrapping his almost frozen body in fat and felt. This healing by the Tatars was Beuys's initiation into shamanistic practice, in which felt was the primary healing instrument. According to his biography, Beuys became a re-born man, a newly transformed individual, whose past was irrelevant to his post-crash life. As the myth goes, the Tatars actually told Beuys "*Du nix njerncky, du Tatar*" ("You are not a German, you are a Tatar").[23] This story corresponds to the mythical initiation of a shaman: Mircea Eliade contended that a near-death experience of a shaman is a call to his new role in the community. The cycle of suffering, death, and rebirth is a principal motif in the shaman initiation stories in various cultures. Once the shaman has triumphed over death and confronted the spirits, he is capable of healing his community and serving as its mediator with the spiritual realm. [24] Christian and animistic motifs were combined in Beuys's projected and perceived messianic persona: like Christ, Beuys was resurrected from the dead and suffered in order to heal his community which he served as a shepherd. [25]

Like all myths, that of Beuys is a product of edited memories and fantasies, adapted to the persona that the artist wished to convey. When Gilles Deleuze analyzed the role of myths in the Platonic philosophical discussions, he argued that a myth is an axiom that defines the boundaries of the narrative.[26] The myth has its axiomatic status because everything stems from it and everything comes back to it: it is the source of all claims, which are logically proven by the way they accord with the myth. Thus, whether the myth of Beuys is true or not is irrelevant; what is relevant is how it motivated Beuys's actions. The myth provides the correlation between art and life: Beuys's life experience constructed his art formalistically and thematically, and one may claim that his whole artistic career is a visual expression and expansion of his self-fabricated autobiography. Just as important, the myth explains the transformation from a wounded German soldier to a healing shaman. Beuys aimed not only at expanding the borders of art and paralleling it to life itself, but specifically at healing wounded post-World War II Germany. Indeed, the resurrected Beuys was perceived as a cultural messiah and the champion of postwar renaissance. [27]

Baptismal water often appears in the art of the priest-shaman Beuys. As early in his career as 1960, Beuys presented the ready-made *Bathtub*, a response to Marcel Duchamp's 1917 *Fountain*. Whereas Duchamp's *Fountain* is a urinal elevated to the status of an artifact and a mockery of art, Beuys's *Bathtub* is an artistic ready-made object elevated to a medium of purification. In *Bathtub* one cannot only cleanse one's body, but also one's spirit. In total awareness of the symbolic time, the performance of the *Celtic (Kinloch Rannoch) Scottish Symphony* during the Easter Holy Week of 1970 started with Beuys washing the feet of seven people, an allusion to Christ's washing of the feet, and concluded with Henning Christiansen pouring water on Beuys from a funnel, an allusion to the baptism of Christ. The baptized artistic messiah also wished to cleanse the bodies and souls of his believers.

Another, less well-known example, is Beuys's 1968 *Evervess II*, a handmade wooden cabinet containing two bottles of Evervess brand tonic water (Figure 14.5). One of the bottles was left untouched as a ready-made object, whereas the two labels on the other are concealed by pieces of felt, marking it as a medium of healing and transformation. There are instructions on the back of the cabinet to drink the contents of the second bottle, that is, to absorb the magical powers of the sacramental water.[28]

Beuys' use of cabinets and display vitrines is ubiquitous. It is possible to separate his art into two main types: the performative and the sculptural. Objects used by the artist during his performances were later rearranged and stored in vitrines; but these vitrines should be regarded not only as sculptural display cases, but also as reliquaries. As

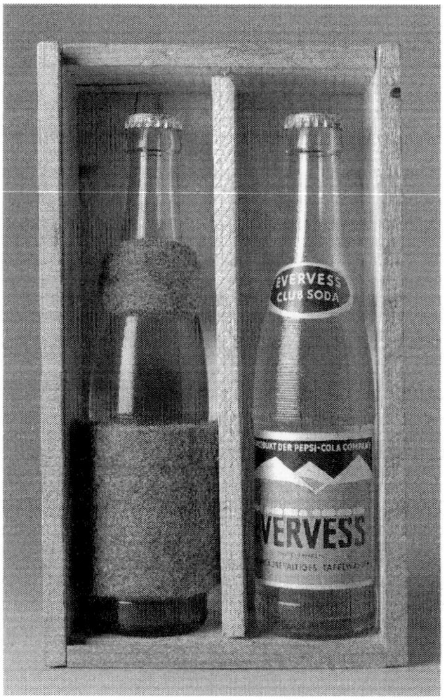

Figure 14.5 Joseph Beuys, *Evervess II*, 1968, two bottles of tonic water, felt application, wooden box, 27 × 16.5 × 9.5 cm. Israel Museum, Jerusalem. Photo Credit: © Israel Museum, Jerusalem/ Licensed by VG Bild-Kunst, Bonn.

with many shamans and spiritual healers, the medium of the Beuysian transformation is based on materials whose transformation – or shall we refer to it as transubstantiation – initiates the spiritual transformation of the observer. These materials are relics, objects touched by the shaman, who impressed his spiritual powers upon them during his performative/shamanistic rituals. Their previous involvement in the master's actions transformed them into containers of spiritual energy, so they do not symbolize past actions, but rather harbor the power of past transubstantiations.

At first glance, Beuys' and Hiller's water cabinets seem quite similar: they both include display cases, felt, and water endowed with some kind of sanctity, and both are outcomes of physical actions associated with spirituality: the pseudo-religious rites of Beuys, and the pseudo-pilgrimage of Hiller. Hiller's felt-lined *Homages* might seem like works by Beuys himself, but the two cabinets actually make the difference between the two artists quite clear. Both Beuys and Hiller deal intensively with transformation, but the German master's transformation is one of the spirit, and is metaphysical and shamanistic by nature, whereas Hiller focuses on the transformation of materials through the most basic concepts of the physics of space and time (a canvas is transformed to ashes, a two-dimensional painting becomes a three-dimensional sculpture), making the study of history, religion, and cult quasi-scientific. By applying Beuysian materials and presentation methods, Hiller is questioning the role of Beuys' artistic persona in the acceptance of performative rituals as high art, and at the same time questioning the role Beuys' shamanistic persona played in the acceptance of his art as objects of *eulogia*.

A second look reveals that Hiller's cabinets are more like medicine cabinets, rather than like the reliquaries of Beuys. Some of them, as I noted, are actually reused first-aid cabinets that used to hold boxes of pills and in their prior lives the vials containing the holy water were mostly used for medicine. Hiller draws an analogy between modern medicine and holy water, between the scientific and the religious ways of healing the body and the spirit. It is noteworthy that this analogy between medical cabinets and reliquaries goes back to the Middle Ages. As demonstrated by Anja Kalinowski, both a reliquary from the Chur Cathedral and one from the Sancta Sanctorum Chapel in the Vatican are repurposed medicine boxes; even some of the subdivisions in the box, in which different kinds of medicines had been stored, were left to store various relics.[29] Beuys drew a similar analogy in several works; for instance, his 1974–1975 *Show Your Wound* included medical instruments, such as dissection tables, test tubes, clinical thermometers, and preserving jars with gauze, alongside four metal containers of fat. In *Show Your Wound*, the medical environment of a clinic and the process of healing physical wounds were metaphors for the exposition of spiritual wounds and the process of healing them by the magical Beuysian material fat.

Do Hiller's cabinets aim at healing, and if they do, do they heal the body, the spirit, or both? They actually do not aim at healing either, as she uses different and allegedly contradictory methods of representation and partially sabotages their proper representation and context. Her cabinets reflect upon healing powers attributed to holy water, upon Beuysian materials, and upon modern medicine, but still function only as a work of art, to be displayed and viewed in an art gallery. They deny any inherent power beyond that of the visible object itself. Hiller does not judge whether medicine or holy water actually works, and while reflecting upon these traditions of healing and visually evoking them, she accords great power to the interpretative gaze of the observer.

Transforming anthropology

Hiller's method of water collection and bottling can easily be regarded as collecting "research material," especially in connection with the meticulous labeling of the containers, according to time and place. Hence, the same method of labeling that I associated earlier with the religious realm of relics could be as easily interpreted as secular and scientific. Furthermore, if I have interpreted Hiller's vitrines as either medicine cabinets or reliquaries, I now suggest that they also resemble showcases in anthropological museums. As a matter of fact, modern art and anthropology are strongly linked. After the end of the Napoleonic wars, many artists, including Eugène Delacroix, Paul Gauguin, and Emil Nolde, journeyed out of Europe in order to enrich European art by learning about non-European cultures. Beuys made several trips to rural areas in England, Scotland, and Ireland, which he perceived as non-Western locations. Though rejecting these artists' colonial appropriative approach, Hiller continues this heritage of the artist as an anthropologist. Henry Moore wrote that the modern artist is similar to the anthropologist, hailing the renewed friendship between the two fields.[30] While directly referring to Henry Moore's concept of sculpture as "truth to material," Hiller insists that her preoccupation with cultural materials is provoked equally by anthropology and modern art history. [31]

Hiller studied for a PhD in anthropology in the United States, but during the mid-1960s, after several field trips to Mexico, Belize, and Guatemala, she felt uncomfortable with anthropology's aspiration for objectivity and decided to dedicate herself to the subjective field of art. Yet, she continued to use the methods of anthropology, such as collecting, comparing, analyzing, and presenting in vitrines, in her artistic work.[32] One may contend that she is still an anthropologist, but one who changed her position from the passive, remote observer to the participant, as she says: "Artists' work does not allow discontinuities between experience and reality, and it eliminates any gap between the investigator and the object or situation investigated."[33] Hiller's 1976–1977 *Fragments* is a good example of an allegedly objective methodological research manifested in subjective art (Figure 14.6).

In this work she assembled broken pieces of pottery made by Pueblo Native American women and, using a traditional archeological method, sorted those fragments and studied the origins of their decorations. According to her "scientific" results, the design and decoration of the pottery reflect tradition and personal innovation, and she concluded that those anonymous women artists function much as she, the well-known artist, does.[34]

Hiller's art puts the works of the repressed artists – socially, nationally, gender-wise, and economically – at the front and sheds light on neglected cultural fragments. In both *Fragments* and the *Homages to Joseph Beuys*, the investigation of material fragments of the past is actually research into the neglected dark holes of material and social history. Beuys's holy water and the holy water collected by ancient and modern

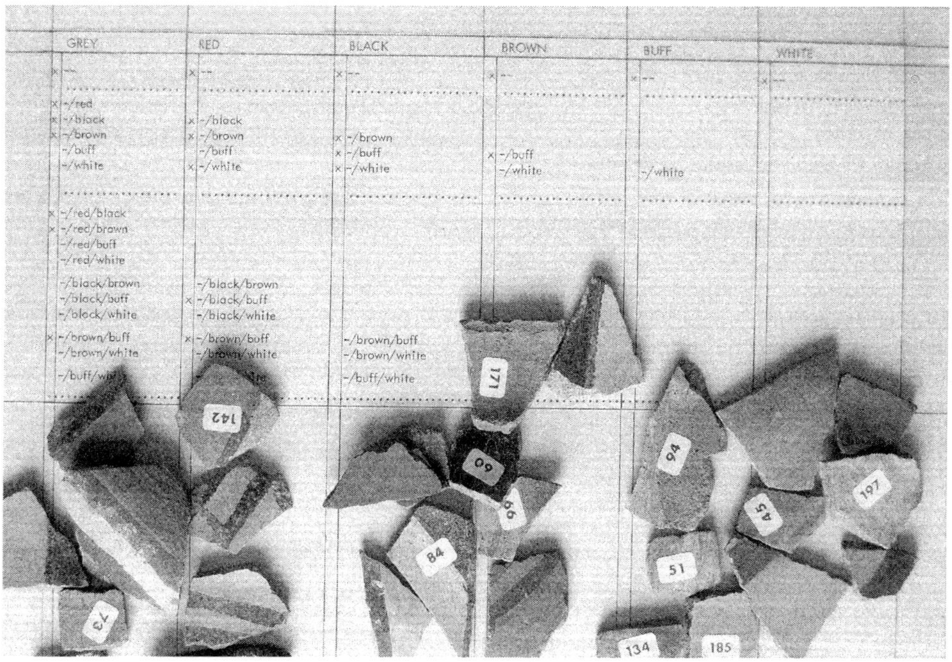

Figure 14.6 Susan Hiller, *Fragments* (detail), 1976–1977, mixed media, dimensions variable. Collection of the artist. Photo Credit: ©Susan Hiller/Boyd Webb.

pilgrims are both *eulogia*, but whereas the first is considered high art to be installed and revered in a museum, the second is an outcome of what is regarded as popular belief, often existing at the boundaries of official religion. Although the eulogia of Beuys is labeled as being the product of the artist-shaman, the more common and widespread containers of water from cult pilgrimage sites, collected by anonymous pilgrims, remain unlabeled. Likewise, whereas Beuys is very much admired as a creator and thus widely shown, most pilgrims remain in the shadows.

In order to collect the water, Hiller followed in the footsteps of anonymous pilgrims, putting their marginal actions, sometimes categorized as superstitious, to the front, labeling their collected materials as art and actually curating a show of these beliefs and places. The concept of "rejuvenation" applies not only to the material of water and to the excavation of containers; the artist rejuvenated the act of collecting water as well, and her artistic interpretation of this traditional custom gives it a new, contemporary level of meaning.

Hiller functioned as both an anthropologist collecting samples of local materials from remote locations and as a pilgrim collecting eulogia from cult pilgrimage sites. Although the two acts were the same thing, the motivations were completely different: the first approach is scientific, removed, and allegedly objective, whereas the second was spiritual and subjective. The bridge between the two is the artistic form: remote and communicative, objective (as in being an object) and allowing subjective interpretation at one and the same time. These water containers become a "piece of art" through their decontextualization from a specific place. Displaying vials that contain water from different places, sacred to different religions, Hiller de-contextualized the action of collecting the water from the specific place and the specific religious group, thus undermining both anthropological and religious methodologies. Rather than making distinctions between places and people – distinctions that are common in both anthropology and religion – Hiller's artistic installation makes a statement about the universal perception of water as holy. Furthermore, undermining the traditional curatorial practice of presenting materials according to their origin and culture and reassembling fragments into a coherent object with a coherent objective meaning, Hiller kept their fragmentary appearance, allowing different subjective motivations, aspirations, and views to reside between their cracks.

Hiller has three different approaches to her cabinets. The first is the scientific approach of anthropology, which gives no real meaning to the water but rather to its significance for the community; the second is the popular fetishism of pilgrims (the object of the anthropological study), conferring sanctity on materials that originate from a sacred physical space; and the third is the genial spirituality of Beuys, whose materials, transformed by the artist-shaman, play a significant role in the healing of the community. Hiller's cabinets form bridges among the three. They are respectful of the traditions of religion, science, and art history, without accepting or rejecting any of them. The transformation in the cabinets is not only of space and time, but also of methodology. Generally speaking, until the Renaissance, science, religion, and the creation of art were tightly connected: the scientific truth was the religious one, and art had to represent it. Hiller's containers of water are located in the modern, wider space, somewhere between data collection, relic adoration, and art appreciation.

Hiller draws inspiration from these three traditions, but at the same time purposely sabotages their proper practice, making the *Homages* not less and – more importantly – not more than a work of art. The vials and the cabinets should not be venerated, and neither should they be studied scientifically. They should only be observed, not for what they can offer (be it redemption or scientific proof), but for what they are – materials presented in space.

As an artist, Hiller deals with materials and their presentation, and the presented glass vials filled with water and the felt-lined wooden cabinets are anything but neutral: they are what the artist refers to as "cultural materials" – objects with piles of subjective associations upon them. As such, the cabinets reflect upon preconceived notions and methodologies, with honor and doubt at the same time. They also allow an ongoing active reflection: as the cabinets are placed in the gallery space, the different visual and contextual appearances of the water and the vials give them different meanings depending on the observer. Hiller, in a reaction against conceptual art, insists that the artist's views on her or his work should not be taken as the last word, but rather as a starting point for a conversation with the observer. As can be seen in most of the references cited herein, Hiller usually discusses her work with a variety of audiences, rather than writing about it; hence, the meaning, if there is any, is the outcome of a tension among the viewer, the artist, and the artifact. Hiller's study of transformation is therefore an inquiry into the elusive nature of the self and of the forces at work in the construction of subjectivity.[35] The starting point is the actual material, which evokes different and sometimes contradictory sentiments and thoughts, thus provoking contemplation. The artist says:

> Most of my work involves cultural representations and artifacts that are as peculiar as the mermaid, although less folkloristic. There is something elusive, uncanny, fascinating beneath the surface of what at first seems easy to understand, or ordinary, or banal. I like to work with materials that have been culturally repressed or misunderstood, what's been relegated to the lunatic fringe or what's so boring we can't even look at it anymore. . . . I particularly like the way that the mundane becomes special as soon as you pay attention to it; I particularly like the way we hide the depth of things from ourselves; I particularly like the way the shape of things shift when you look hard at them.[36]

According to the Platonic binary system, ideas are eternal and static whereas matter is temporary and in constant change. While undermining one of the most rooted concepts of art as eternal, that is, enduring through the ages by being static, Hiller also undermines the concept of art as carrying a specific and total preconceived meaning. Indeed, the art of Susan Hiller exists beneath the surface. Like an archeological mound, Hiller's *Homages to Joseph Beuys* has many layers in different times and different places, which are still active; as these cabinets deny total grasp or understanding, they still acquire new layers of meaning by each subjective interpretative gaze.

Notes

1 Most of the information about the *Homages to Joseph Beuys* in this chapter is based on conversations I conducted with the artist in 2010 and 2013.

2 Susan Hiller, "Dedicated to the Unknown Artists: An Interview with Rozsika Parker," in *Thinking about Art: Conversations with Susan Hiller*, ed. Barbara Einzig (Manchester: Manchester University Press, 1996), 29.

3 Ahmad Ghabin, "The Zamzam Ritual in Islam and Its Jerusalem Connection," in *Sacred Space in Israel and Palestine: Religion and Politics*, eds. Marshall J. Breger, Yitzhak Reiter, and Leonard Hammer (New York: Routledge, 2012), 117–118.

4 Julian Crandall Hollick, *Ganga: A Journey down the Ganges River* (Washington, DC: Island Press, 2007), 119.

5 Aubrey Stuart, trans., *Of the Holy Places Visited by Antoninus Martyr* (London: Palestine Pilgrims' Text Society, 1887), 20–21. See also: Barbara Baert, "The Healing of the Blind Man at Siloam, Jerusalem: A Contribution to the Relationship between Holy Places and the Visual Arts in the Middle Ages," *Arte Cristiana* 95 (2007): 49–69 (Part 1) and 121–130 (Part 2).

6 Kenneth G. Holum, "Hadrian and St. Helena: Imperial Travel and the Origins of Christian Holy Land Pilgrimage," in *The Blessings of Pilgrimage*, ed. Robert Ousterhout (Chicago: University of Illinois Press, 1990), 67–68; E.D. Hunt, *Holy Land Pilgrimage in the Later Roman Empire AD 312–460* (Oxford: Clarendon Press, 1982), 8–9.

7 Stuart, *Of the Holy Places*, 10–11.

8 Gregory of Tours, *Glory of the Martyrs*, trans. Raymond Van Dam (Liverpool: Liverpool University Press, 1991), 27.

9 Dan Barag, "Glass Pilgrim Vessels from Jerusalem: Part 1," *Journal of Glass Studies* 12 (1970): 47–48.

10 Andre Grabar, *Ampoules de Terre Sainte: Monza, Bobbio* (Paris: C. Klincksieck, 1958), 57, 60.

11 Deana L. Weibel, "Of Consciousness Changes and Fortified Faith: Creativist and Catholic Pilgrimage at French Catholic Shrines," in *Pilgrimage and Healing*, eds. Jill Dubisch and Michael Winkelman (Tucson: University of Arizona Press, 2005), 126, 129–130.

12 Alexander Nagel, "The Afterlife of the Reliquary," in *Treasures of Heaven: Saints, Relics, and Devotion in Medieval Europe*, eds. Martina Bagnoli et al. (New Haven and London: Yale University Press, 2010), 211.

13 A comparison to this assemblage of relics could be the fourteenth-century Charlemagne reliquary from the Aachen cathedral treasury. It displays relics from different places and times, such as of the passion, with pieces of the cross, a nail, and a thorn (Jerusalem, first century), dust of Saint John the Evangelist (Ephesus, first century), a tooth of Saint Catherine (Alexandria, fourth century), an arm of Saint Nickolas (Myra, fourth century), and particles of bone and three teeth of Charlemagne (Aachen, ninth century): Herta Schmitz-Cliever-Lepie, *Treasury of the Cathedral of Aachen* (Aachen: Aachen Cathedral, 1986), 60.

14 Simon Coleman and John Elsner, *Pilgrimage Past and Present in the World Religions* (Cambridge: Harvard University Press, 1995), 209–217.

15 Jörg Heiser, "Messages Suppressed by Culture Do Not Cease to Exist," in *Susan Hiller*, ed. Ann Gallagher (London: Tate Britain, 2011), 135–136. Hiller's journeys might resemble the ones of the cultural tourist: a tourist going from place to place out of cultural curiosity, and collecting souvenirs along the way. Continuing the approach of the Turners, who coined the well-known statement that "a tourist is half a pilgrim, if a pilgrim is half a tourist," modern research shows the similarities between pilgrimage and tourism. Both involve leaving one's home in order to visit unknown places and meet with new experiences, both are usually guided by a certain defined route, both provide a detachment from daily routine, and both are associated with the sacred. Tourism might be regarded as a secular, modern version of pilgrimage, but visiting tourist attractions is a result of yearning for a place and involves the expression of awe and admiration. Furthermore, the two allegedly different journeys constantly mix in modernity: embarking upon a pilgrimage also involves visits to secular tourist attractions, as many holidays include visits to holy places, mostly according to the tourist's own religion. Nelson H.H. Garburn, "Tourism: The Sacred Journey," in *Hosts and Guests: The Anthropology of Tourism*, ed. Valene L. Smith (Philadelphia, PA: The University of Pennsylvania Press, 1978), 25–27; David L. Gladstone, *From Pilgrimage to Package Tour: Travel and Tourism in the Third World* (New York: Routledge, 2005), 171–177; Victor Turner and Edith Turner, *Image and Pilgrimage in Christian Culture* (New York: Columbia University Press, 1978), 20.

16 Victor Turner, *Dramas, Fields and Metaphors: Symbolic Action in Human Society* (Ithaca: Cornell University Press, 1974), 166–169; Turner and Turner, *Image and Pilgrimage*, 250.

17 A prominent example is *Witness* from 2000. In this work, Hiller installed hundreds of small speakers suspending from the ceiling, transmitting vocal testimonies in different languages from people all over the world about their encounters with UFOs or aliens. The visitor listened to the testimonies and acted as an interpreter of the texts and the phenomena: Virginia Button and Charles Esche, *Intelligence: New British Art* (London: Tate Gallery, 2000), 41–42; Susan Hiller, "On Surrealism," in *Susan Hiller: The Provisional Texture of Reality: Selected Talks and Texts, 1977–2007*, ed. Alexandra M. Kokoli (Zurich: JRP-Ringler and Dijon: Les Presses de Réel, 2008), 223; Louise Milne, "On the Side of the Angels," in *Susan Hiller: Recall: Selected Works 1969–2004*, ed. James Lingwood (Basel: Kunsthalle Basel, 2004), 141–143, 149. The method in the *Homages to Joseph Beuys* is similar: first, a presentation of material testimonies in space that invites the observer to interpret their meaning; second, a universal action that aims at the elimination of territorial distinctions, attesting to the universality of the phenomenon; and third, dealing with the paranormal and working on the fine line between the rational and the irrational. As people – usually socially considered irrational and even borderline – are telling about their encounters with UFOs in a calm and rational manner, pilgrims collecting holy water might be considered rational and irrational at the same time. Pilgrimage is a rationally well-planned journey whose motivation is the encounter with sacred materials providing spiritual and physical health.

18 Denise Robinson, "Encounters with the Work of Susan Hiller," in *Contemporary Art and Anthropology*, eds. Brand Schneider and Christopher Wright (New York: Berg, 2006), 79.

19 Joseph Kosuth, "Art after Philosophy," in *Art after Philosophy and After: Collected Writings, 1966–1990*, ed. Gabriele Guercio (Cambridge: MIT Press, 1991), 15–17.

20 Stuart Morgan, "Susan Hiller," in *Rites of Passage: Art for the End of the Century*, eds. Stuart Morgan and Frances Morris (London: Tate Gallery, 1995), 112.

21 Ann Landi, "Auto-Destructive Tendencies," *Art News* (December, 2012). Digital facsimile: http://www.artnews.com/2012/12/24/auto-destructive-tendencies/ (accessed July 21, 2013).

22 Rosemary Betterton, "Susan Hiller's Painted Works: Bodies, Aesthetics and Feminism," in *Susan Hiller: Recall: Selected Works 1969–2004*, ed. James Lingwood (Gateshead: Baltic, 2004), 16–18.

23 Benjamin H.D. Buchloh, "Beuys: The Twilight of the Idol," *Artforum* 16/1 (1980): 38.

24 Mircea Eliade, *Shamanism: Archaic Techniques of Ecstasy* (London: Routledge-Kegan Paul, 1964), 33–34; Joan Halifax, *Shaman: The Wounded Healer* (London: Thames and Hudson, 1982), 10.

25 Alain Borer, *The Essential Joseph Beuys* (Cambridge: MIT Press, 1997), 31.

26 Gilles Deleuze, "Plato and the Simulacrum," trans. Rosalind Krauss, *October* 27 (1983): 47.

27 Borer, *Essential*, 12, 21; Buchloh, *Twilight*, 36–38; Stefan Germer, "Haacke, Broodthaers, Beuys," *October* 45 (1989): 67; Mario Kramer, "Joseph Beuys: Auschwitz Demonstration 1956–1964," in *German Art from Beckmann to Richter: Images of a Divided Country*, ed. Eckhart Gillen (New Haven: Yale University Press, 1997), 267.

28 Adina Kamien-Kazhdan, *Modernism in Dialogue: 20th Century Painting and Sculpture in the Israel Museum* (Jerusalem: The Israel Museum, Jerusalem, 2010), 219.

29 Anja Kalinowski, *Frühchristliche Reliquiare im Kontext von Kultstrategien, Heilserwartung und sozialer Selbstdarstellung* (Wiesbaden: Reichert, 2011), 93. I am grateful to Julia M. H. Smith for drawing my attention to this issue.

30 Alan Wilkinson, *Henry Moore: Writings and Conversations* (Oakland, CA: University of California Press 2002), 122.

31 Susan Hiller, " 'Truth' and 'Truth to Material': Henry Moore," in *Susan Hiller: The Provisional Texture of Reality: Selected Talks and Texts, 1977–2007*, ed. Alexandra M. Kokoli (Zurich: JRP|Ringier, 2008), 67.

32 Susan Hiller, "Art and Anthropology/ Anthropology and Art," in *Thinking about Art: Conversations with Susan Hiller*, ed. Barbara Einzig, 18; Hiller, "I Don't Care What It's Called," in *Thinking about Art: Conversations with Susan Hiller*, ed. Barbara Einzig, 45–46.

33 Susuan Hiller, "Art and Anthropology," 24.

34 Guy Brett, "Analysis and Ecstasy," in *Susan Hiller: Recall*, 35–36; Sarah Kent, "Susan Hiller," in *Hayward Annual 78*, ed. Lucy Lippard (London: Arts Council of Great Britain, 1978), 46–47.

35 Susan Hiller, "An Audience for Art: A Conversation with Colin Gardner," in *Thinking about Art*, 152; Hiller, "Collaborative Meaning: Art as Experience," in *Thinking about Art*, 182; James Lingwood, "Introduction," in *Susan Hiller: Recall*, 9.

36 Susan Hiller, "The Provisional Texture of Reality (On Andrei Tarkovsky)," in *Susan Hiller: The Provisional Texture of Reality: Selected Talks and Texts, 1977–2007*, ed. Alexandra M. Kokoli (Zurich: JRP|Ringier, 2008), 27.

Index